A First Course in Ordinary Differential Equations

A First Course in Ordinary Differential Equations

Suman Kumar Tumuluri

CRC Press
Taylor & Francis Group
Boca Raton London New York

CRC Press is an imprint of the
Taylor & Francis Group, an **informa** business

First edition published 2021

by CRC Press

6000 Broken Sound Parkway NW, Suite 300, Boca Raton, FL 33487-2742

and by CRC Press

2 Park Square, Milton Park, Abingdon, Oxon, OX14 4RN

© 2021 Taylor & Francis Group, LLC

CRC Press is an imprint of Taylor & Francis Group, LLC

Library of Congress Cataloging-in-Publication Data

ISBN: [978-0-8153-5983-8] (hbk)
ISBN: [978-1-003-15375-7] (ebk)

Typeset in Computer Modern font
by KnowledgeWorks Global Ltd.

To my parents

Seshu Kumari

S. R. Sastry Tumuluri

Contents

Preface

Differential equations are indispensable tools for modeling most of the natural phenomena. At the same time this subject is also at the center of mathematical analysis. The aim of this book is to provide elementary methods for obtaining explicit, implicit, and approximate solutions to ordinary differential equations (ODEs) and then to introduce qualitative analysis. This book adopts a theoretical point of view to meet the standards required for students at the Masters level, while not compromising in terms of problem solving. Sufficient number of solved examples are given in every chapter so that the students understand and apply of the results efficiently. The readers may find that some of the methods covered in the first few chapters are already familiar. However, we address here question like: why do these methods work? After understanding the subtle details of the methods, one can use them with more confidence.

This book arose from the lecture notes prepared for the masters students in the School of Mathematics and Statistics, University of Hyderabad, where most of the material presented in Chapters 2 to 7 is covered in a one semester course. In this book, the ideas of calculus, real analysis, and linear algebra are encountered frequently. A systematic study of ODEs offers an excellent opportunity to interweave the threads of aforementioned disciplines of mathematics. There are a good number of well-established and successful books dealing with the subject of ODEs. However, the purpose of this book is to present material which is suitable for a one semester course, accessible to students of masters level, containing an abundant collection of problems with varying levels of difficulty with due diligence to accommodate enough rigor for students.

The book is organized as follows. Chapter 2 begins with the review of some elementary methods to find solutions to first order ODEs along with rigorous justification of those methods. Further, well-posedness of Cauchy problem, and initial value problem are discussed. Comparison theorems and their applications are presented. Chapter 3 deals with higher order linear ODEs. After presenting methods to find explicit solutions to ODEs with constant/variable coefficients, we study some theoretical aspects like oscillation behavior of solutions to ODEs. Chapter 4 treats boundary value problems for second order linear ODEs. Green's function is introduced here and its existence and uniqueness results are proved. Using Green's function, boundary value problems are solved. The last part of the section is dedicated to the study of the eigenvalue problems.

Chapter 5 introduces the concept of the exponential of a matrix. Solutions to the systems of first order ODEs with constant coefficients are given in terms of the exponential of matrices. The notion of a fundamental matrix is developed for the case of variable coefficients. Chapter 6 uses the results proved in Chapter 5. Moreover, it treats the stability of critical points of 2×2 systems of linear ODEs with constant coefficients. It also deals with Lyapunov's methods to establish the stability of nonlinear systems. At the end of this chapter, necessary conditions, and sufficient conditions for the existence of closed paths are discussed. Chapter 7 contains the power series method to find the solutions to regular, as well as some singular second order linear ODEs. Legendre's and Bessel's equations are also studied in detail. In Chapter 8, the Laplace transforms are computed for various functions. These are used to solve the linear ODEs with initial conditions and boundary conditions. Chapter 9 is concerned with the numerical study of first order (system of) ODEs.

Students are advised to workout all the given examples and exercises in the book. After reading and understanding the content of the book, in principle, the students are expected to:

- Solve ODEs in the standard forms described in the book with initial, boundary conditions.

- Analyze whether a given ODE or system of ODEs is well-posed or not.

- Analyze the asymptotic behavior of solutions to any first order autonomous ODE without solving it.

- Construct Green's functions for second order linear differential operators (whose solutions to the corresponding homogeneous problems can be found) with various boundary conditions.

- Solve homogeneous/non-homogeneous linear systems with constant coefficients by computing the matrix exponentials.

- Find the nature of the isolated critical points of any 2×2 linear autonomous system with constant coefficients.

- Guess a Lyapunov energy functional for a given simple 2×2 nonlinear autonomous system.

- Analyze the nature of the simple critical points of nonlinear systems.

- Study the existence of closed paths of a given 2×2 autonomous system.

- Compute the power series solutions to second order linear ODEs when the coefficients are analytic.

- Compute the Laplace transforms and the inverse Laplace transforms of various elementary functions.

- Find approximate solutions to initial value problems using the Euler method and the Runge–Kutta method.

Though enough care is taken to keep the book error-free, some errors might have crept in. For whatever imperfections remain, my apologies to the readers. Any errata that arises will be uploaded on my home page http://mathstat.uohyd.ac.in/people/profile/suman-kumar. Finally, I would welcome any comments or suggestions regarding this book by email to suman.hcu@gmail.com.

Acknowledgments. First of all, I am grateful for the facilities and the research environment provided by the School of Mathematics and Statistics during the preparation of this manuscript. Moreover, I would like to thank UGC for supporting our School under SAP-DSA 1. I would like to record my gratitude to Professor T. Amaranath who encouraged me to convert my lecture notes into the present book form. I was first introduced to much of the material here by Professors R. Radha and A. S. Vasudeva Murthy, to whom I convey my much belated thanks for their teaching. I would like to thank Professors T. Amaranath, B. Sri Padmavati, R. Radha and Dr. S. Ilangovan for reading the entire manuscript, and Dr. Archana S. Morye, Professor K. S. Mallikarjuna Rao, Dr. S. Sivaji Ganesh for reading some portions of it. In particular, I am indebted to them for their numerous helpful suggestions. Thanks are also due to my past and present doctoral students Dr. Bhargav Kumar Kakumani, Devendra, and Joydev Halder. I also thank Ms. Aastha Sharma, CRC Press, for her constant support. But for her, this book would not have been completed by the scheduled time. Special thanks are due to Ms. Shikha Garg and Mr. Shashi Kumar, CRC Press. Over the last two years I have been able to think little of except writing this book. My wife, Madhuri, has managed everything at home with a smile. My heartfelt thanks to her for this, and of course for everything. I also thank my daughter Sri Sarada and my son Sri Krishna for their enduring affection and occasional and charming naughtiness which energized me from time to time, while writing this book.

Suman Kumar Tumuluri
Hyderabad, 2021

Chapter 1

Introduction

1.1 Ordinary differential equations

The term 'equatio differentialis' (differential equations) was first used by Leibniz in 1676 to denote a relationship between the differentials of two variables. Very soon, this restricted usage was abandoned. Roughly speaking, differential equations are the equations involving one or more dependent variables (unknowns) and their derivatives/partial derivatives. If the unknown in the differential equation is a function of only one variable, then such differential equation is called an ordinary differential equation (ODE).

Notation: Unless specified otherwise, the unknown in the differential equation is denoted by y. Let $\mathbb{R}$ denote the set of real numbers, and J be an open interval in $\mathbb{R}$. Throughout the book we denote the derivative of the function $y : J \to \mathbb{R}$ with respect to x by either

$$\frac{d}{dx}y(x) \text{ or } \frac{dy}{dx}(x) \text{ or } y'(x).$$

When there is no ambiguity regarding the argument in the function y, we denote the derivative simply with $\frac{dy}{dx}$ or y'. Similarly, let y'' and y''' denote the second and the third derivative of y, respectively. In general, for $k \in \mathbb{N}$, $y^{(k)}$ or $\frac{d^k y}{dx^k}$ denotes the k-th order derivative of y.

With this notation, examples of ODEs are

$$\frac{d}{dx}y(x) = \left(\frac{d^2}{dx^2}y(x)\right)^5 + y^2(x), \ x \in (0, 1), \tag{1.1}$$

$$y' = 3y^2 + (\sin x)y + \log(\cos^2 y), \ x \in \mathbb{R}. \tag{1.2}$$

The order of an ODE is the largest number k such that the k-th order derivative of the unknown is present in the ODE. For example, the order of (1.1) is two.

At the beginning, it may look like tools from the integral calculus are sufficient to study ODEs. But very soon one realizes that to develop methods to solve or analyze them, one needs notions from subjects like analysis, linear algebra, etc. In fact, the study of differential equations motivated crucial development of many areas of mathematics: the theory of Fourier series and more general orthogonal expansions, integral transformations, Hilbert spaces, and Lebesgue integration to name a few.

1.2 Applications of ODEs

Many laws in physics, chemistry, biology etc., can be easily expressed using differential equations. One of the reasons for this is the following. The quantity $y'(x)$ can be interpreted as the rate of change of the quantity y with respect to the quantity x. In many natural phenomena, there is a relationship between the unknowns (which are relatively difficult to measure), the rate of change of the unknowns with respect to a known quantity, and the other known quantities (which are easy to measure) that govern the process. When this relationship is expressed in mathematics, it turns out to be a (system of) differential equation(s). Therefore the study of ODEs is crucial in understanding physical sciences. In fact, much of the theory developed in ODEs owes to the questions/situations raised in the study of subjects like mechanics, astronomy, electronics etc.

Listing all the available ODE models in any branch of science is an impossible task. Therefore in this chapter, we present a few ODE models which arise from physics and biology which can be solved or analyzed using the material in the book. We begin with models from physics.

Example 1.2.1 (Radioactivity and half-life). Let $N(t)$ denote the number of radioactive active atoms in a substance of a fixed quantity at time t. Then a model for the decay of the number of radioactive atoms is

$$\frac{d}{dt}N(t) = -kN(t), \ t > 0, \tag{1.3a}$$

$$N(t_0) = N_0, \tag{1.3b}$$

where $k > 0$. Equation (1.3b) is known as the initial condition. This kind of models are studied in detail in Chapter 2, Subsection 2.1.3. One can easily verify that the solution to (1.3a) is

$$N(t) = N_0 e^{-k(t-t_0)}, \ t > t_0. \tag{1.4}$$

The half-life of a specific radioactive isotope is defined as the time taken for half of its radioactive atoms to decay. In fact, the half-life is independent of the quantity of the radioactive material. We now calculate the half-life of an isotope using (1.3a) if k is known explicitly. For, it is enough to find T at which $N(T) = \frac{N_0}{2}$. From (1.4) we have

$$N(T) = N_0 e^{-k(T-t_0)} = \frac{N_0}{2}.$$

After rearranging the terms we get that the half-life T_{hl} is

$$T_{hl} = T - t_0 = \frac{\log 2}{k}.$$

Radiocarbon dating. Any living matter constantly takes carbon from the air and the ratio of the number of atoms of radioactive carbon ($^{14}C_6$) and stable carbon ($^{12}C_6$) in it is constant. Once the specimen (it could be a tree or animal etc.) is dead then the atoms of $^{14}C_6$ isotope in it begin to decay as given in (1.3a)–(1.3b). Suppose the specimen is dead at time t_0 at which it has N_0 atoms of $^{14}C_6$. If we know that at present (at time t) the number of atoms of $^{14}C_6$ in the sample is N_1, then from (1.4) we have

$$N_1 = N_0 e^{-k(t-t_0)}$$

which readily implies that the individual is alive up to time

$$t_0 = t + \frac{\log(\frac{N_1}{N_0})}{k}.$$

It is worth mentioning that W. Libby received the Nobel prize in 1960 for the discovery of the phenomenon of radiocarbon dating. □

We now move on to the situation where the second order linear ODEs are encountered.

Example 1.2.2 (Harmonic oscillator). Let a ball of mass M be attached to a wall by means of a spring with spring constant k as shown in Figure 1.1. If

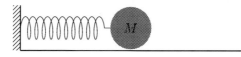

FIGURE 1.1: Harmonic oscillator

the ball is displaced by a distance $y(t)$ from the equilibrium at time t, then the spring exerts a restoring force $-ky$ on the ball. Moreover, the velocity and the acceleration of the ball are given by y' and y'', respectively. Suppose we neglect the damping forces (due to internal or external resistance), then by Newton's law of motion, we have

$$My''(t) = -ky(t), \ t > 0. \tag{1.5}$$

We now improve (1.5) by including the effect of damping forces due to the viscosity of the medium in which the ball moves. It is natural to assume that the damping force at any time is proportional to the velocity of the ball. Thus the damping force is $-cy'$ for some $c > 0$ and (1.5) becomes

$$My''(t) = -cy'(t) - ky(t), \ t > 0. \tag{1.6}$$

The explicit solutions to models in (1.5)–(1.6) can be computed using the methods developed in Chapter 3 (Section 3.1). □

A simple extension of the situation given in the previous example gives a system of linear equations. This is given in the next example.

Example 1.2.3 (Coupled harmonic oscillator). Let S_1 and S_2 be two springs with spring constants k_1 and k_2, respectively. Assume that one end of S_1 is attached to a wall and the other end is attached to a ball of mass m_1. There is another ball of mass m_2 which is attached to the ball of mass m_1 by the spring S_2 as shown in Figure 1.2. Let $y_1(t)$ and $y_2(t)$ be the displacements

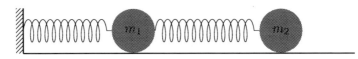

FIGURE 1.2: Coupled harmonic oscillator

of the balls with masses m_1 and m_2 from the equilibrium, respectively, at time t from the equilibrium. Since the ball with mass m_1 is attached to both springs, there are two restoring forces acting upon it: one from S_1 given by k_1y_1 and the other from S_2 given by $k_2(y_2 - y_1)$. The ball with mass m_2 has the restoring force from S_2 given by $k_2(y_2 - y_1)$. If we neglect the damping forces, then Newton's law gives the following equations of motion of the balls

$$\begin{cases} m_1y_1''(t) = -k_1y_1(t) + k_2(y_2(t) - y_1(t)), \\ m_2y_2''(t) = -k_2(y_2(t) - y_1(t)). \end{cases} \tag{1.7}$$

This system can be written as a 4×4 system of first order ODEs which can be solved using the methods described in Chapter 5.   □

We now provide a couple of ODE models from population dynamics. One of the first models in the population biology using differential equations is due to Malthus (see [22]). Here we present that model.

Example 1.2.4. Let $N(t)$ denote the population of a single species in a region (or environment) at time t. Suppose B denotes the fertility rate, i.e., the average number of off springs that an individual can produce in a unit time. Similarly, let μ denote the mortality rate. Then the Malthus model is given by the following population balance law: the rate of change of population at t is equal to the difference between the number of newborns at t and the number of individuals dead at t. Thus

$$\frac{d}{dt}N(t) = BN(t) - \mu N(t), \ t > 0. \tag{1.8}$$

Let $\alpha = B - \mu$ denote the Malthusian intrinsic growth parameter. Then notice that the Malthus model for the population is the same as (1.3a) with $k = -\alpha$. One can easily check that the solution N to the Malthus model goes to infinity (resp. to zero) if $\alpha > 0$ (resp. $\alpha < 0$) as t tends to infinity.   □

In fact, from the asymptotic behavior of the solution to the Malthus model, it is clear that we do not observe this phenomenon in reality. Later, Verhulst proposed a refinement to the Malthus model by introducing the notion of the carrying capacity K of environment (see [23, 29]). Roughly speaking, K is the maximum number of individuals who can survive in the environment.

Example 1.2.5. Verhulst model is a nonlinear first order equation given by

$$\frac{d}{dt}N(t) = \alpha N(t)(1 - \frac{N(t)}{K}), \ t > 0. \tag{1.9}$$

Computation of the solution to (1.9) can be done from the separation of variables method (see Chapter 2) and asymptotic behavior of the solution is also discussed in Example 2.5.13 of Chapter 2. □

We now present some important models which are studied by many engineers and mathematicians.

Example 1.2.6 (Pendulum equation). Let $\theta(t)$ be the angular displacement of a pendulum of length l from the mean position at time t. Then the equation of motion is

$$\frac{d^2}{dt^2}\theta(t) + \frac{g}{l}\sin(\theta(t)), \ t > 0, \tag{1.10}$$

where g is the acceleration due to gravity which is approximately equal to $9.8m/sec^2$. For small oscillations (θ small), we have $\sin\theta \approx \theta$ and (1.10) is approximated by the linear equation

$$\theta'' + \frac{g}{l}\theta = 0.$$

Example 1.2.7 (The Duffing equation). A model for the dynamics of a spring pendulum is given by the following nonlinear equation called Duffing equation:

$$My''(t) = -cy'(t) - ky(t) + \alpha y^3(t), \ t > 0. \tag{1.11}$$

It is easy to notice that it is a nonlinear (in particular, cubic) perturbation of the model of damped oscillator (1.6). □

In fact, equation (1.6) (of course, with different interpretation to the constants involved) arises in the modeling of electrical circuits. A perturbation of this model is introduced by Rayleigh and studied in detail by van der Pol. We present that model in the following example.

Example 1.2.8 (The van der Pol equation). A model which was used extensively in radio-engineering (radars) during the World War II is the following equation which is popularly known as the van der Pol equation:

$$y''(t) = -y'(t) - y(t) + \alpha y^2(t)y'(t), \ t > 0. \tag{1.12}$$

An interesting contribution to a non-homogeneous version of this equation is due to Cartwright et al. (see [5]). □

Example 1.2.9 (The Lorenz system). In [20], under some additional hypothesis, Lorenz obtained the following model as a simplified version of a more complicated model in the weather forecasting:

$$\begin{cases} x'(t) = -ax(t) + ay(t), \\ y'(t) = bx(t) - y(t) - x(t)z(t) \\ z'(t) = -bz(t) + x(t)y(t). \end{cases} \tag{1.13}$$

This is one of the early models which exhibited *chaotic* dynamics.   □

Example 1.2.10 (The Lotka–Volterra system). Suppose $x(t)$ and $y(t)$ denote the population of preys (say rats) and their predators (say cats) in a region. Assume that the fertility rate of preys is b and their mortality rate is proportional to the availability of predators. On the other hand, let μ be the mortality rate of predators and the fertility rate of them is proportional to the availability of the food resources (prey). In the absence of internal competition the model reads as

$$\begin{cases} x'(t) = bx(t) - c_1 x(t)y(t), \ t > 0, \\ y'(t) = -\mu y(t) + c_2 x(t)y(t), \end{cases} \tag{1.14}$$

where $c_1, c_2 > 0$. This is popularly known as the prey–predator model. In fact, if $x(0) > 0$ and $y(0) > 0$ then the solution to system (1.14) is periodic (see [29]). For analysis of behavior of solutions to linear/nonlinear systems see [2, 12, 13].   □

These examples elucidate the importance of ODEs in modeling different phenomena in various branches of science and engineering. For more models, one can refer to [3, 23]. For many interesting historical facts related to mathematicians worked in differential equations see [32]. We conclude this chapter with these examples and wish the readers a more joyful reading of the remaining text.

Chapter 2

First order ODEs

In this chapter, we briefly review some methods to solve the first order ordinary differential equations (ODEs). Then we discuss the well-posedness of the first order ODEs. Moreover, we study a few elementary differential inequalities and their applications. Toward the end of the chapter, we focus on the asymptotic behavior of scalar autonomous ODEs.

2.1 A review of some basic methods

We present some elementary methods to solve ODEs in some special forms. We begin with the simplest one among these methods called 'the method of separation of variables.' Later, we discuss a certain class of equations which will be reduced to the form of equations that can be solved using the method of separation of variables. Then we introduce the notion of exact equations and finally conclude this section with discussion on linear equations.

2.1.1 Separation of variables

Consider the ODE of the form

$$\frac{d}{dx}y(x) = \frac{f(x)}{g(y(x))}. \tag{2.1}$$

We assume that $f : (a_0, a_1) \to \mathbb{R}$ and $g : (b_0, b_1) \to (0, \infty)$ are continuous functions. We also assume that there exists y_0 in the interval (b_0, b_1) such that

$$g(y_0) \neq 0. \tag{2.2}$$

We define a function $F : (a_0, a_1) \times (b_0, b_1) \to \mathbb{R}$ by

$$F(x, y) = \int_{y_0}^{y} g(\xi)d\xi - \int_{x_0}^{x} f(s)ds, \ x \in (a_0, a_1), \ y \in (b_0, b_1).$$

Since f and g are continuous, F is a C^1-function. Moreover for every $x_0 \in (a_0, a_1)$ we have

$$\frac{\partial F}{\partial y}(x_0, y_0) = g(y_0) \neq 0.$$

Therefore by the implicit function theorem (see Appendix C) there exists $\delta > 0$ and a C^1-function $\phi : (x_0 - \delta, x_0 + \delta) \to \mathbb{R}$ such that

$$F(x, \phi(x)) = \int_{y_0}^{\phi(x)} g(\xi)d\xi - \int_{x_0}^{x} f(s)ds = F(x_0, y_0), \ x \in (x_0 - \delta, x_0 + \delta). \quad (2.3)$$

One can easily prove that ϕ is a solution to (2.1). For, on differentiating (2.3) with respect to x (using the Leibniz rule of differentiation[1]) we get

$$\phi'(x)g(\phi(x)) - f(x) = 0, \ x \in (x_0 - \delta, x_0 + \delta).$$

This proves that the function ϕ which is implicitly given by the relation $F(x, y) = F(x_0, y_0)$, is a solution to (2.1). In other words, the relation

$$\int^{y} g(y)dy = \int^{x} f(x)dx + c, \ c \in \mathbb{R}, \quad (2.4)$$

where the above integrals are indefinite integrals, defines a solution to (2.1). We now present some examples where this technique is demonstrated.

Example 2.1.1. Solve $\dfrac{d}{dx}y(x) = \dfrac{y(x)}{x}$, $x > 0$.

Solution. We notice that $\phi \equiv 0$ is a solution to the given equation. We now seek nonzero solutions to the given ODE. On comparing the given ODE with (2.1) we get $f(x) = \frac{1}{x}$ and $g(y) = \frac{1}{y}$, $y \neq 0$. Then a solution to the given problem is implicitly given by relation (2.4), i.e.,

$$\int^{y} g(y)dy = \int^{x} f(x)dx + c,$$

or

$$\int^{y} \frac{1}{y}dy = \int^{x} \frac{1}{x}dx + c.$$

On integrating we obtain

$$\log|y| = \log x + \log c, \ c > 0. \quad (2.5)$$

From this relation one can solve for y explicitly in terms of x to get $y = \tilde{c}x$ and hence

$$\phi(x) = \tilde{c}x, \ \tilde{c} \in \mathbb{R}\backslash\{0\}, \ x > 0,$$

is a solution to the given equation. $\qquad\square$

Example 2.1.2. Solve $\dfrac{d}{dx}y(x) = \dfrac{x}{y(x)}$, $x > 0$.

Solution. As before, on comparing the given ODE with (2.1) we get $f(x) = x$

---

[1] $\dfrac{d}{dx}\displaystyle\int_{\alpha(x)}^{\beta(x)} H(x, t)dt = \displaystyle\int_{\alpha(x)}^{\beta(x)} \dfrac{\partial H}{\partial x}(x, t)dt + H(x, \beta(x))\dfrac{d\beta}{dx}(x) - H(x, \alpha(x))\dfrac{d\alpha}{dx}(x).$

and $g(y) = y$. The solution is given by

$$\int^y y\,dy = \int^x x\,dx + c, \ c \in \mathbb{R},$$

or

$$\frac{y^2}{2} = \frac{x^2}{2} + c, \ c \in \mathbb{R}. \tag{2.6}$$

We now solve (2.6) for different values of c to get solutions to the given ODE. If $c = 0$, then from (2.6) we find that $y = \pm x$. Therefore

$$\phi_1(x) = x, \ \phi_2(x) = -x, \ x > 0,$$

are two solutions to the given ODE. Similarly for $c > 0$ from (2.6) we obtain that $y = \pm\sqrt{2c + x^2}$. Thus

$$\phi_3(x) = \sqrt{2c + x^2}, \ \phi_4(x) = -\sqrt{2c + x^2}, \ x > 0,$$

are solutions. Finally, if $c < 0$ then

$$\phi_5(x) = \sqrt{2c + x^2}, \ \phi_6(x) = -\sqrt{2c + x^2}, \ x > \sqrt{2|c|},$$

are solutions. □

It is worth mentioning that it is not always possible to write y as a function of x as we have done in the previous examples. In that case we simply say that the solution to the given ODE is implicitly given by (2.4). The following example demonstrates this feature.

Example 2.1.3. Solve $\dfrac{d}{dx}y(x) = \dfrac{1 + 3x^2}{2 + \sin(y(x))}$, $x \in \mathbb{R}$.

Solution. Here $f(x) = 1 + 3x^2$ and $g(y) = 2 + \sin y$. In order to find the solution we consider

$$\int^y (2 + \sin y)\,dy = \int^x (1 + 3x^2)\,dx + c.$$

or

$$2y - \cos y = x + x^3. \tag{2.7}$$

From (2.7), it is not possible to write y using elementary functions of x. Therefore we conclude that a solution to the given problem is implicitly given by (2.7). □

Many times it may not be possible to separate the variables in an ODE. However, in some cases by certain change of variables, ODEs can be reduced to the case in which one can use the method of separation of variables. We now present a situation where the dependent variable in the given ODE is changed to get a simpler ODE. In particular, we consider the ODE of the form

$$y'(x) = H(x, y(x)), \tag{2.8}$$

where H is such that

$$H(\lambda x, \lambda y) = H(x, y), \ (x, y) \neq (0, 0), \lambda \neq 0. \tag{2.9}$$

Functions which satisfy (2.9) are called homogeneous functions of degree[2] zero. From (2.9), we note that $H(x,y) = H(1, \frac{y}{x})$, $x \in \mathbb{R}\backslash\{0\}$, $y \in \mathbb{R}$. We assume that ϕ is a solution to (2.8) and set

$$v(x) = \frac{\phi(x)}{x}. \qquad (2.10)$$

On differentiating we get

$$v'(x) = \frac{\phi'(x)}{x} - \frac{v(x)}{x} = \frac{H(1, v(x)) - v(x)}{x}. \qquad (2.11)$$

We now solve (2.11) using the method of separation of variables. For, let

$$f(x) = \frac{1}{x}, \quad g(v) = \frac{1}{H(1,v) - v}.$$

Then a solution to (2.11) is implicitly given by

$$\int^v \frac{dv}{H(1,v) - v} = \int^x \frac{dx}{x} + c. \qquad (2.12)$$

Therefore a solution to (2.8) can be written using (2.10) and (2.12). This method was first introduced by Leibniz in 1693.

We now provide a couple of examples to elucidate this method.

Example 2.1.4. Solve $y'(x) = \dfrac{x^2 + y^2(x)}{x^2 + xy(x)}$, $x > 0$.

Solution. Let $H(x,y) = \dfrac{x^2 + y^2}{x^2 + xy}$ and observe that

$$H(\lambda x, \lambda y) = \frac{\lambda^2 x^2 + \lambda^2 y^2}{\lambda^2 x^2 + \lambda^2 xy} = H(x,y), \ \lambda \neq 0, \ (x,y) \neq (0,0).$$

Let ϕ be a solution to the given equation and we set $\phi(x) = xv(x)$. Then from (2.11) we have

$$v' = \frac{H(1,v) - v}{x} = \frac{\frac{1+v^2}{1+v} - v}{x} = \frac{1 - v}{x(1+v)}. \qquad (2.13)$$

Using the method of separation of variables, a solution to (2.13) is given by

$$\int^v \frac{1+v}{1-v} dv = \int^x \frac{dx}{x} + c,$$

or

$$-v - 2\log|1 - v| = \log(|x|) + c.$$

Therefore a solution to the given problem is implicitly given by

---

[2]A function H is said to be a homogeneous function of degree n, $n \in \mathbb{N} \cup \{0\}$ if $H(\lambda x, \lambda y) = \lambda^n H(x,y)$, $(x,y) \neq (0,0), \lambda \neq 0$.

$$\frac{y}{x} = -2\log(|1 - \frac{y}{x}|) - \log(|x|) - c,$$

where $c \in \mathbb{R}$. □

Example 2.1.5. Solve $y'(x) = \dfrac{x^2 + xy(x) + y^2(x)}{x^2}$, $x > 0$.

Solution. Let $H(x, y) = \dfrac{x^2 + xy + y^2}{x^2}$ and notice that

$$H(\lambda x, \lambda y) = \frac{\lambda^2 x^2 + \lambda^2 xy + \lambda^2 y^2}{\lambda^2 x^2} = H(x, y), \ \lambda \neq 0, \ (x, y) \neq (0, 0).$$

Let ϕ be a solution to the given equation and $\phi(x) = xv(x)$. Then from (2.11) we have

$$v' = \frac{H(1, v) - v}{x} = \frac{1 + v^2}{x}. \tag{2.14}$$

Using the method of separation of variables, a solution to (2.14) is given by

$$\int^v \frac{dv}{1 + v^2} = \int^x \frac{dx}{x} + c, \ c \in \mathbb{R},$$

or

$$\arctan v = \log(x) + c.$$

Therefore $v(x) = \tan(\log(x) + c)$ is a solution to (2.14). Hence

$$\phi(x) = x\tan(\log(x) + c), \ c \in \mathbb{R},$$

is a solution to the given problem. □

Example 2.1.6. Solve $y'(x) = \dfrac{4x + y(x) - 5}{x + y(x) - 2}$.

Solution. Suppose we set $H(x, y) = \dfrac{4x + y(x) - 5}{x + y(x) - 2}$, then H does not satisfy condition (2.9). So we cannot employ the method used in the previous examples.

In order to convert H to the form given in (2.9), we introduce new variables

$$X = x - h, \ Y(X) = y(X + h) - k,$$

where h and k will be chosen later. Using the chain rule we obtain

$$\frac{d}{dX}Y(X) = \frac{dy}{dx}(X + h)\frac{d}{dX}(X + h) = \frac{dy}{dx}(X + h).$$

Hence the given equation becomes

$$\frac{d}{dX}Y(X) = \frac{4X + Y(X) + 4h + k - 5}{X + Y(X) + h + k - 2}.$$

We choose (h, k) such that

$$4h + k - 5 = 0, \ h + k - 2 = 0.$$

After solving the above linear equations, we get $h = 1, \ k = 1$.
Hence with $x = X + 1, \ y = Y + 1$ the given equation reduces to

$$\frac{d}{dX}Y(X) = \frac{4X + Y(X)}{X + Y(X)}. \tag{2.15}$$

We now define

$$\tilde{H}(X, Y) = \frac{4X + Y}{X + Y},$$

and one can easily verify that $\tilde{H}$ is a homogeneous function of degree zero.
In order to solve (2.15), we use the standard procedure. In particular, we set
$\phi_1(X) = XV(X)$, where ϕ_1 is a solution to (2.15). Then we have

$$V'(X) = \frac{\tilde{H}(1, V) - V}{X} = \frac{4 - V^2}{X(1 + V)}. \tag{2.16}$$

Using the method of separation of variables, a solution to (2.16) is given by

$$\int^V \frac{(1 + V)dV}{4 - V^2} = \int^X \frac{dX}{X} + c, \ c \in \mathbb{R},$$

or

$$\int^V \frac{3dV}{4(2 - V)} - \int^V \frac{dV}{4(2 + V)} = \int^X \frac{dX}{X} + c.$$

After simplifying we get

$$X^4(2 - V)^3(2 + V) = C.$$

Thus a solution to (2.15) is given by

$$(2X - Y)^3(2X + Y) = C.$$

Hence we obtain that

$$(2x - y - 1)^3(2x + y - 3) = C, \ C \in \mathbb{R},$$

is a solution to the given ODE. □

In general, we can use the technique employed in the previous example to
solve an ODE of the form

$$y'(x) = \frac{a_1 x + b_1 y(x) + c_1}{a_2 x + b_2 y(x) + c_2}, \ a_1 b_2 - a_2 b_1 \neq 0. \tag{2.17}$$

That is, we first introduce new variables $X = x - h$ and $Y = y - k$ where h,
k satisfy

$$\begin{cases} a_1 h + b_1 k + c_1 = 0, \\ a_2 h + b_2 k + c_2 = 0. \end{cases}$$

Using this transformation, (2.17) reduces to

$$Y' = \frac{a_1 X + b_1 Y}{a_2 X + b_2 Y}$$

which can be solved using the standard procedure.
On the other hand, suppose $a_1 b_2 - a_2 b_1 = 0$. Then we set $Y = a_1 x + a_2 y$ and write an ODE for Y which is indeed easy to solve. This case is explained in the following example.

Example 2.1.7. Solve $y'(x) = \dfrac{10x + 2y(x) - 10}{5x + y(x) - 2}$.

Solution. It is easy to verify that the right hand side of the given ODE is not a function of degree zero. Moreover, we cannot use the technique employed in the previous example (why?). Let ϕ be a solution to the given equation. Then we introduce a new function $Y(x) = 5x + \phi(x)$. Thus we get

$$\frac{dY}{dx} = 5 + \frac{d\phi}{dx} = 5 + \frac{2Y - 10}{Y - 2} = \frac{7Y - 20}{Y - 2},$$

whose solution is given by

$$\int^Y \frac{Y - 2}{7Y - 20} dY = \int^x dx + c, \ c \in \mathbb{R},$$

or

$$\frac{Y}{7} + \frac{6}{49} \log |7Y - 20| = x + c.$$

Hence we obtain that

$$-14x + 7y + 6 \log |35x + 7y - 20| = c, \ c \in \mathbb{R},$$

is a solution to the given problem. ☐

2.1.2 Exact equations

In this subsection, we present another special form of differential equations called *exact equations* which can be solved easily. Let M, N be continuous functions in a rectangle

$$R = \{(x, y) : |x - x_0| \leq a, \ |y - y_0| \leq b\},$$

and N does not vanish in R. An ODE of the form

$$N(x, y(x))y'(x) + M(x, y(x)) = 0, \tag{2.18}$$

is said to be *exact* if there exists a C^1-function $F : R \to \mathbb{R}$ such that

$$\frac{\partial F}{\partial x}(x, y) = M(x, y), \quad \frac{\partial F}{\partial y}(x, y) = N(x, y), \ (x, y) \in R. \tag{2.19}$$

Example 2.1.8. Show that $y(x)y'(x) + x = 0$ is an exact equation.

Solution. In order to prove this, we first compare the given equation with (2.18) to get $M(x, y) = x$ and $N(x, y) = y$. It is easy to verify that

$$F(x,y) = \frac{x^2 + y^2}{2},$$

satisfies (2.19). Hence the given equation is exact.   □

We now establish the connection between F and the solutions to (2.18). To this end, we suppose (2.18) is exact and F is known to us. We observe that $\frac{\partial F}{\partial y} = N \neq 0$, in R. Let $(\tilde{x}, \tilde{y}) \in \mathbb{R}^2$ satisfy $|x_0 - \tilde{x}| < a$ and $|y_0 - \tilde{y}| < b$. Then by the implicit function theorem there exists an interval $(\tilde{x} - \delta, \tilde{x} + \delta)$, which is denoted by J, and a C^1-function $\phi : J \to \mathbb{R}$ such that

$$F(x, \phi(x)) = F(\tilde{x}, \tilde{y}), \ x \in J. \tag{2.20}$$

Claim. The function ϕ is a solution to (2.18).
For, on differentiating (2.20) with respect to x we get

$$\frac{\partial F}{\partial x}(x, \phi(x)) + \frac{\partial F}{\partial y}(x, \phi(x))\phi'(x) = 0, \ x \in J.$$

Thus we have

$$M(x, \phi(x)) + N(x, \phi(x))\phi'(x) = 0, \ x \in J,$$

which proves that ϕ is a solution to (2.18). Hence the claim is proved.

Now, we shall revisit Example 2.1.8 and solve the ODE therein.

Example 2.1.9. Solve $y(x)y'(x) + x = 0$.

Solution. We first recall that the given equation is exact and the corresponding F is given by $F(x,y) = \frac{x^2+y^2}{2}$ (see Example 2.1.8). From the discussion we had so far, the function which is implicitly given by $F(x,y) = c$ is a solution to the given equation. Therefore

$$x^2 + y^2 = 2c, \ c > 0,$$

gives us a solution. In particular, for $c > 0$, the function

$$\phi(x) = \sqrt{2c - x^2}, \ |x| < \sqrt{2c},$$

is a solution to the given equation.   □

From the discussion we had so far, for a given equation of the form (2.18), it is *a priori* not clear whether it is exact unless we can find a suitable F. So our next objective is to find an easily verifiable condition on M and N which guarantees that (2.18) is exact.
In order to do that, we assume that M and N are C^1-functions on

$$R_1 = \{(x,y) : |x - x_0| < a, \ |y - y_0| < b\}.$$

Moreover, we assume that (2.18) is exact. Then there exists F such that (2.19) holds. Since M and N are assumed to be C^1-functions, F is a C^2-function. As the partial derivatives of any C^2-function commute (see Appendix C), we have

$$\frac{\partial M}{\partial y}(x,y) = \frac{\partial^2 F}{\partial y \partial x}(x,y) = \frac{\partial^2 F}{\partial x \partial y}(x,y) = \frac{\partial N}{\partial x}(x,y), \ (x,y) \in R_1.$$

Therefore we obtain

$$\frac{\partial M}{\partial y}(x,y) = \frac{\partial N}{\partial x}(x,y), \ (x,y) \in R_1. \tag{2.21}$$

Conversely, we assume that M and N are C^1-functions such that (2.21) holds.
Claim. There exists F such that (2.19) holds.
In order to prove the claim we construct F explicitly. Without loss of generality we assume that $N(x_0, y_0) \neq 0$. If such an F exists, then

$$
\begin{aligned}
F(x,y) - F(x_0,y_0) &= F(x,y) - F(x,y_0) + F(x,y_0) - F(x_0,y_0) \\
&= \int_{y_0}^{y} \frac{\partial F}{\partial y}(x,\xi)d\xi + \int_{x_0}^{x} \frac{\partial F}{\partial x}(\eta,y_0)d\eta \\
&= \int_{y_0}^{y} N(x,\xi)d\xi + \int_{x_0}^{x} M(\eta,y_0)d\eta.
\end{aligned}
$$

This motivates us to choose

$$F(x,y) = \int_{y_0}^{y} N(x,\xi)d\xi + \int_{x_0}^{x} M(\eta,y_0)d\eta, \ (x,y) \in R_1. \tag{2.22}$$

For this choice of F, we have

$$\frac{\partial F}{\partial y}(x,y) = N(x,y), \ (x,y) \in R_1.$$

On the other hand, we observe that

$$N(x,\xi) = N(x_0,\xi) + \int_{x_0}^{x} \frac{\partial N}{\partial x}(\eta,\xi)d\eta, \tag{2.23}$$

$$M(\eta,y_0) = M(\eta,y) - \int_{y_0}^{y} \frac{\partial M}{\partial y}(\eta,\xi)d\xi. \tag{2.24}$$

Using (2.22)–(2.24) we have

$$
\begin{aligned}
F(x,y) &= \int_{y_0}^{y} N(x_0,\xi)d\xi + \int_{y_0}^{y}\int_{x_0}^{x} \frac{\partial N}{\partial x}(\eta,\xi)d\eta d\xi \\
&\quad + \int_{x_0}^{x} M(\eta,y)d\eta - \int_{x_0}^{x}\int_{y_0}^{y} \frac{\partial M}{\partial y}(\eta,\xi)d\xi d\eta \\
&= \int_{y_0}^{y} N(x_0,\xi)d\xi + \int_{x_0}^{x} M(\eta,y)d\eta, \tag{2.25}
\end{aligned}
$$

and the last equality follows from (2.21). Finally from (2.25) we obtain

$$\frac{\partial F}{\partial x}(x,y) = M(x,y), \ (x,y) \in R_1.$$

This completes the proof of the claim.

Henceforth, if M and N are C^1-functions, then we verify condition (2.21) in order to examine whether the given ODE (2.18) is exact. A corresponding F can be found using (2.22) or (2.25).

Example 2.1.10. Solve $(y^3(x) + x^2)y'(x) + (x^2 + 2xy(x)) = 0$.

Solution. By comparing with (2.18) we have
$$N = y^3 + x^2, \ M = x^2 + 2xy.$$
We take $(x_0, y_0) = (0, 1)$ so that $N(x_0, y_0) \neq 0$. We now verify that
$$\frac{\partial N}{\partial x} = 2x = \frac{\partial M}{\partial y}.$$
Hence the given ODE is exact. In order to find the solution, we first find F using formula (2.22). Hence we have
$$
\begin{aligned}
F(x, y) &= \int_1^y (x^2 + \xi^3)d\xi + \int_0^x (\eta^2 + 2\eta)d\eta \\
&= x^2 y - x^2 + \frac{y^4}{4} - \frac{1}{4} + \frac{x^3}{3} + x^2.
\end{aligned}
$$
Therefore a solution to the given ODE is implicitly given by
$$x^2 y + \frac{y^4}{4} + \frac{x^3}{3} = c, \ c \in \mathbb{R}.$$
This completes the solution. $\square$

Example 2.1.11. Solve $(y(x) + 3x^2 y^2(x))y'(x) + (x + 2xy^3(x)) = 0$.

Solution. As before, we first compare the given equation with (2.18) to get
$$N = y + 3x^2 y^2, \ M = x + 2xy^3.$$
We take $(x_0, y_0) = (0, 1)$ so that $N(x_0, y_0) \neq 0$. It is easy to verify that
$$\frac{\partial N}{\partial x} = 6xy^2 = \frac{\partial M}{\partial y},$$
which implies that the given equation is exact. From formula (2.22) we have
$$
\begin{aligned}
F(x, y) &= \int_1^y (\xi + 3x^2 \xi^2)d\xi + \int_0^x 3\eta d\eta \\
&= \frac{y^2}{2} - \frac{1}{2} + x^2 y^3 - x^2 + \frac{3x^2}{2}.
\end{aligned}
$$
Therefore a solution to the given ODE is implicitly given by
$$x^2 y^3 + \frac{y^2}{2} + \frac{x^2}{2} = c, \ c \in \mathbb{R}.$$
This completes the solution. $\square$

2.1.3　Linear ODEs

Another important type of ODE which can be solved easily is the linear equation (both homogeneous and non-homogeneous). Let J be a closed interval and $P : J \to \mathbb{R}$ be a continuous function. An equation of the form

$$y'(x) + P(x)y(x) = 0 \tag{2.26}$$

is called a *first order linear homogeneous ODE*. If Q is a nonzero continuous function on J, then

$$y'(x) + P(x)y(x) = Q(x) \tag{2.27}$$

is called a *first order linear non-homogeneous ODE*. Any first order ODE that we consider in this chapter which is not in any of the forms (2.26) or (2.27) is called a *nonlinear ODE*.

There are many ways to solve (2.26). One of them is to apply the method of separation of variables. On comparing (2.26) with (2.1), we get

$$f(x) = -P(x), \quad g(y) = \frac{1}{y}.$$

Therefore a solution to (2.26) is implicitly given by

$$\int^y \frac{dy}{y} = -\int^x P(x)dx + \tilde{c}, \ \tilde{c} \in \mathbb{R},$$

$$y = e^{\tilde{c}}e^{-\int^x P(x)dx}.$$

From the previous relation, we directly obtain that

$$\phi(x) = ce^{-\int^x P(x)dx}, \ c \in \mathbb{R}, \tag{2.28}$$

is a solution to (2.26). We now describe another way of obtaining the solution given in (2.28). Let ϕ be a solution to (2.26). On substituting ϕ in (2.26) and multiplying with $e^{\int^x P(x)dx}$ on both sides, we arrive at

$$e^{\int^x P(x)dx}\frac{d\phi(x)}{dx} + \frac{d}{dx}\left(e^{\int^x P(x)dx}\right)\phi(x) = 0,$$

or

$$\frac{d}{dx}\left(\phi(x)e^{\int^x P(x)dx}\right) = 0.$$

Therefore we get

$$\phi(x)e^{\int^x P(x)dx} = c, \ c \in \mathbb{R}.$$

This is the same as (2.28).

Using the same strategy we find a solution to (2.27). Let ψ be a solution to (2.27). As before, on substituting ψ in (2.27) and multiplying with $e^{\int^x P(x)dx}$, we obtain

$$\frac{d}{dx}\left(\psi(x)e^{\int^x P(x)dx}\right) = Q(x)e^{\int^x P(x)dx}.$$

On integrating we get

$$\psi(x)e^{\int^x P(x)dx} = c + \int^x Q(x)e^{\int^x P(x)dx}dx,$$

and thus we obtain the required solution

$$\psi(x) = ce^{-\int^x P(x)dx} + e^{-\int^x P(x)dx}\int^x Q(x)e^{\int^x P(x)dx}dx, \quad c \in \mathbb{R}. \quad (2.29)$$

From (2.28) we observe that the solution ϕ to (2.26) with $\phi(x_0) = y_0$ is

$$\phi(x) = y_0 e^{-\int_{x_0}^x P(s)ds}. \quad (2.30)$$

Similarly, from (2.29) the solution ψ to (2.27) with $\psi(x_0) = y_0$ is given by

$$\psi(x) = y_0 e^{-\int_{x_0}^x P(s)ds} + e^{-\int_{x_0}^x P(s)ds}\int_{x_0}^x Q(t)e^{\int_{x_0}^t P(s)ds}dt. \quad (2.31)$$

Example 2.1.12. Solve the following equations:

(i) $xy'(x) + y(x) = x^2$, $y(1) = 2$, (ii) $\cos(x)y'(x) + \sin(x)y(x) = 2\tan(x)$.

Solution. (i) We first write the given equation in the standard form of linear ODEs, i.e., in the form of (2.27) to get

$$y'(x) + \frac{y(x)}{x} = x.$$

Then we have $P(x) = \frac{1}{x}$, $Q(x) = x$, and $(x_0, y_0) = (1, 2)$. On substituting P, Q, and (x_0, y_0) in formula (2.31), we find that

$$\begin{aligned}
\psi(x) &= 2e^{-\int_1^x \frac{ds}{s}} + e^{-\int_1^x \frac{ds}{s}}\int_1^x te^{\int_1^t \frac{ds}{s}}dt \\
&= \frac{2}{x} + \frac{1}{x}\int_1^x t^2 dt \\
&= \frac{5}{3x} + \frac{x^2}{3},
\end{aligned}$$

is the solution to the given problem.

(ii) As before, we write the given equation in the form of (2.27) to obtain

$$y'(x) + \tan(x)y(x) = 2\tan(x)\sec(x).$$

We observe that $P(x) = \tan(x)$, $Q(x) = 2\tan(x)\sec(x)$. On substituting P and Q in formula (2.29) we find that

$$\begin{aligned}
\psi(x) &= ce^{-\int^x \tan(x)dx} + e^{-\int^x \tan(x)dx}\int^x 2\tan(t)\sec(t)e^{-\int^t \tan(x)dx}dt \\
&= c\cos(x) + \cos(x)\int^x 2\tan(t)\sec^2(t)dt \\
&= c\cos(x) + \cos(x)\tan^2(x),
\end{aligned}$$

where $c \in \mathbb{R}$ is a solution to the given problem.   □

Example 2.1.13. Assume that P, Q are continuous functions and $n \in \mathbb{Z}$. Then solve

$$y'(x) + P(x)y(x) = Q(x)y^n(x), \quad x \in J, \tag{2.32}$$

where J is an open interval.

Solution. For different values of n, we convert the given ODE to linear ODEs whose solution is already known to us.

If $n = 0$, then the given equation reduces to (2.27) whose solution is given in (2.29).

On the other hand let $n = 1$. Then the given equation becomes

$$y'(x) + (P(x) - Q(x))y(x) = 0, \quad x \in J,$$

which is a linear ODE. The solution in this case is given by

$$\phi(x) = ce^{\int^x \left(Q(x) - P(x) \right) dx}, \quad c \in \mathbb{R}.$$

Finally, we suppose $n \notin \{0, 1\}$. Let ψ be a solution to the given equation which never vanishes in J. On substituting ψ in the given ODE and dividing by ψ^n, we get

$$\frac{1}{\psi^n(x)} \frac{d\psi(x)}{dx} + P(x)\psi^{1-n}(x) = Q(x), \quad x \in J.$$

We now set $z(x) = \psi^{1-n}(x)$. Then $z'(x) = (1-n)\psi^{-n}(x)\psi'(x)$ and thus we have

$$\frac{1}{1-n} \frac{dz}{dx}(x) + P(x)z(x) = Q(x), \quad x \in J.$$

From (2.29) we obtain that

$$z(x) = ce^{\int^x (n-1)P(x)dx} + e^{\int^x (n-1)P(x)dx} \int^x (1-n)Q(x)e^{\int^x (1-n)P(x)dx} dx, \quad c \in \mathbb{R}.$$

Hence the solution to the given equation is given by

$$\psi(x) = \left[ce^{\int^x (n-1)P(x)dx} + e^{\int^x (n-1)P(x)dx} \int^x (1-n)Q(x)e^{\int^x (1-n)P(x)dx} dx \right]^{\frac{1}{1-n}}, \tag{2.33}$$

where $c \in \mathbb{R}$. $\qquad \square$

Often, equation (2.32) is called as the Bernoulli equation if $n \notin \{0, 1\}$.

2.2   Well-posedness

Throughout this chapter, we assume that every interval that we consider has a positive length[3]. We assume that J and Ω are open intervals in $\mathbb{R}$. Let

---

[3]If I is a bounded interval, then its length is defined as $(\sup I - \inf I)$, otherwise its length is defined as infinity.

$\bar{J}$ and $\bar{\Omega}$ denote the smallest closed intervals containing J and Ω, respectively. Let $f : \bar{J} \times \bar{\Omega} \to \mathbb{R}$ be a function. Consider the problem

$$\begin{cases} y'(x) = f(x, y(x)), \ x \in J, \\ y(x_0) = y_0. \end{cases} \tag{2.34}$$

Definition 2.2.1. Let $J_1 \subseteq \bar{J}$ be an interval containing x_0. We say that a function $\phi : J_1 \to \mathbb{R}$ is said to be a *solution* to (2.34) if

(i) $\phi \in C(J_1) \cap C^1(J_1^o)$, where J_1^o is the interval $(\inf J_1, \sup J_1)$,

(ii) $\phi(x) \in \Omega$, $x \in J_1$,

(iii) on substituting $y = \phi$ in (2.34) we get an identity in J_1.

Moreover, if $J_1 \backslash \{x_0\} \subset J \backslash \{x_0\}$, then we say that ϕ is a local solution. Otherwise it is called a global solution. If J_1 is of the form $[x_0, x_1]$ or $[x_0, x_1)$, then we say that ϕ is a right solution. If J_1 is of the form $[x_1, x_0]$ or $(x_1, x_0]$, then we say that ϕ is a left solution. If $x_0 \in J_1^o$ then we say that ϕ is a bilateral solution. If $J = (x_0, x_1)$ where $x_1 \in \mathbb{R} \cup \{\infty\}$, then (2.34) is said to be an initial value problem (IVP) and we deal with the right solutions in the study of IVPs. On the other hand, if $x_0 \in J$ then (2.34) is said to be a Cauchy problem. We usually seek bilateral solutions while studying Cauchy problems.

In fact, one of the main theorems of this chapter is to prove the existence of a bilateral (right) solutions to Cauchy problems (IVPs).

Example 2.2.2. Consider the Cauchy problem

$$\begin{cases} y'(x) = y^2(x), \ x \in \mathbb{R}, \\ y(0) = 1. \end{cases} \tag{2.35}$$

Using the method of separation of variables we obtain that a solution to the given problem is

$$\phi(x) = \frac{1}{1-x},$$

which is valid in $(-\infty, 1)$. Hence it is a bilateral solution which cannot be 'extended' beyond $x = 1$. Thus ϕ is a local but not a global solution to the given problem. $\qquad\square$

Unless specified otherwise, whenever we say that ϕ is a solution to a Cauchy problem (or IVP) (2.34), we mean that it is a *bilateral (or right) solution*.

Any proper mathematical model of any natural phenomena should admit a unique solution because the physical quantity, which is the unknown in the ODE, indeed exists and is unique. According to Hadamard, a problem is said to be *well-posed* if it admits a unique solution which continuously depends on the given data. Otherwise, the problem is said to be *ill-posed*. In the present context, in order to prove well-posedness of (2.34) we show that there exists a unique solution to it which depends continuously on the data (x_0, y_0).

Before that, we consider the following example which gives us an idea about the regularity of f to have existence of solutions.

Example 2.2.3. Let a and b be positive numbers. Show that there exists no right solution to the IVP:
$$\begin{cases} y'(x) = f(y(x)), \ x > 0, \\ y(0) = 0, \end{cases}$$
where
$$f(y) = \begin{cases} a, \ y < 0, \\ -b, \ y \geq 0. \end{cases}$$

Solution. On the contrary, we suppose that there exists $\phi : [0, x_1) \to \mathbb{R}$, which is a solution to the given problem. Since the possible number of values that ϕ' can assume is two (namely a or $-b$), owing to the Darboux theorem[4], we get that ϕ' is a constant.

Let $\phi'(x) = a$, for $x \in [0, x_1)$. Then ϕ is increasing and hence $\phi(x) > 0$ in $[0, x_1)$. If $\phi(x) > 0$, then $\phi'(x) = -b$, which is a contradiction to our assumption that $\phi'(x) = a$ in $[0, x_1)$. Using the same argument we can get a contradiction if we assume that $\phi'(x) = -b$, $x \in [0, x_1)$.

This shows that there is no right solution to the given problem. □

This example suggests that if f is discontinuous, then we may not have a solution to the IVP. Henceforth, unless specified otherwise, we assume that f is continuous. In the next lemma we establish the equivalence between Cauchy problem/IVP (2.34) and an integral equation which is crucial to the existence theory.

Lemma 2.2.4. *Assume that $x_0 \in J_1 \subseteq \bar{J}$ and $f : \bar{J} \times \mathbb{R} \to \mathbb{R}$ is a continuous function. Let $\phi : J_1 \to \mathbb{R}$ be a solution to (2.34) then ϕ satisfies*

$$\phi(x) = y_0 + \int_{x_0}^{x} f(t, \phi(t))dt, \ x \in J_1. \tag{2.36}$$

Conversely, if $\phi \in C(J_1)$ satisfies (2.36) then ϕ is a solution to (2.34).

Proof. We first assume that ϕ is a solution to (2.34). Then by integrating the identity obtained by substituting $y = \phi$ in (2.34) from x_0 to x, we arrive at (2.36).

On the other hand, since both f and ϕ are assumed to be continuous, the integrand on the right hand side of (2.36) is continuous on J_1. This implies that the right hand side of (2.36) is indeed in $C^1(J_1)$ and so is ϕ. Hence we can differentiate (2.36) on both sides to get $\phi'(x) = f(x, \phi(x))$, $x \in J_1$. Finally, on putting $x = x_0$ in (2.36), we get $\phi(x_0) = y_0$. This completes the proof. □

Remark 2.2.5. If f is k-times continuously differentiable, and $\phi \in C(J_1)$ is a solution to (2.36), then $\phi \in C^{k+1}(J_1)$. For, we first notice that when $k = 0$ the result is proved in Lemma 2.2.4. We next assume the result for some

---

[4]If $g : [a, b] \to \mathbb{R}$ is a differentiable function and $g'(x_1) < k < g'(x_2)$, then there exists c between x_1 and x_2 such that $g'(c) = k$, i.e., g' has the intermediate value property.

$k = m \in \mathbb{N}$. To prove the result for $k = m + 1$, we assume that f is $(m + 1)$-times continuously differentiable. From the induction hypothesis, it follows that $\phi \in C^{m+1}(J_1)$. Therefore the integrand in the right hand side of (2.36) is $(m + 1)$-times continuously differentiable which readily gives $\phi \in C^{m+2}(I)$.

In order to prove the existence of solution to (2.34), we define a sequence of functions which approximate a solution to (2.34). This sequence is called Picard's sequence (z_n) of successive approximations corresponding to (2.34) and is given by the iteration formula

$$z_0(x) = y_0, \ x \in J, \tag{2.37}$$

$$z_n(x) = y_0 + \int_{x_0}^{x} f(t, z_{n-1}(t))dt, \ n \in \mathbb{N}, \ x \in J. \tag{2.38}$$

Example 2.2.6. Compute Picard's sequence of successive approximations corresponding to

$$\begin{cases} y'(x) = ay(x), \ x > 0, \\ y(0) = 3, \end{cases}$$

where a is a constant.

Solution. We begin with $z_0 \equiv 3$. Then using (2.38) we get

$$z_1(x) = 3 + \int_0^x az_0 dt = 3 + 3\int_0^x a dt = 3(1 + ax),$$

and

$$z_2(x) = 3 + \int_0^x az_1 dt = 3 + \int_0^x 3a(1 + at)dt = 3 + 3ax + \frac{3a^2x^2}{2!}.$$

By observing the pattern, we guess that

$$z_n(x) = 3 + 3ax + \frac{3a^2x^2}{2!} + \cdots + \frac{3a^n x^n}{n!}, \ n \in \mathbb{N}. \tag{2.39}$$

In fact, we prove (2.39) by induction. To that end, let (2.39) hold for $n = k$. Then

$$\begin{aligned} z_{k+1}(x) &= 3 + \int_0^x az_k(t)dt \\ &= 3 + \int_0^x \left(3a + 3a^2t + \frac{3a^3t^2}{2!} + \cdots + \frac{3a^{n+1}t^n}{n!}\right)dt \\ &= 3 + 3ax + \frac{3a^2x^2}{2!} + \cdots + \frac{3a^{n+1}x^{n+1}}{(n+1)!}, \end{aligned}$$

which proves (2.39). Moreover, we readily see that for each $x > 0$, $z_n \to 3e^{ax}$ as $n \to \infty$. $\qquad \square$

Example 2.2.7. Compute the first four terms in Picard's sequence of successive approximations corresponding to the IVP

$$\begin{cases} y' = x^2 + y^2, \ x > 0, \\ y(0) = 0. \end{cases}$$

Solution. We begin with $z_0 \equiv 0$. From (2.38) we obtain

$$z_1(x) = 0 + \int_0^x (t^2 + z_0^2)dt = \frac{x^3}{3},$$

$$z_2(x) = 0 + \int_0^x (t^2 + z_1^2)dt = \frac{x^3}{3} + \frac{x^7}{63}$$

and

$$z_3(x) = 0 + \int_0^x (t^2 + z_2^2)dt = \frac{x^3}{3} + \frac{x^7}{63} + \frac{2x^{11}}{2079} + \frac{x^{15}}{59535}.$$

This completes the solution. □

Before stating the main theorem of this section, we give a couple of definitions and a standard result in real analysis. Let

$$d_\infty(f,g) = \sup_{x \in [a,b]} |f(x) - g(x)|, \ f, g \in C([a,b]).$$

Definition 2.2.8. A sequence (f_n) of continuous functions on $[a,b]$ is said to be a Cauchy sequence in $C([a,b], d_\infty)$ if for a given $\epsilon > 0$, there exists $n_0 \in \mathbb{N}$ such that $d_\infty(f_m, f_n) < \epsilon$ whenever $m, n \geq n_0$.

Definition 2.2.9. A sequence (f_n) in $C([a,b])$ is said to converge in $C([a,b], d_\infty)$ if there exists $f \in C([a,b])$ such that that the following holds: for given $\epsilon > 0$ there exists $n_0 \in \mathbb{N}$ such that $d_\infty(f_n, f) < \epsilon$ whenever $n \geq n_0$. This f is called the limit of (f_n) and is denoted by $f_n \to f$ as $n \to \infty$.

It is worth noting that convergence in $C([a,b], d_\infty)$ is the same as the uniform convergence on $[a,b]$.

Theorem 2.2.10. *A sequence (f_n) is a Cauchy sequence in $C([a,b], d_\infty)$ if and only if it is convergent in $C([a,b], d_\infty)$. In other words, $C([a,b], d_\infty)$ is complete.*

Proof. For a proof of this result see [17, 31]. □

We are ready to state the main theorem of this section which is due to Cauchy and Lipschitz. Some books name this result as Picard's theorem. This theorem deals with the existence and uniqueness of a solution to Cauchy problem. A small modification in the statement and its proof will give us an existence and uniqueness result for IVP (2.34).

Theorem 2.2.11 (Cauchy–Lipschitz). *We assume that f is continuous on*

$$R = \{(x,y) : |x - x_0| \leq a, \ |y - y_0| \leq b\}, \text{ for some } a, b > 0$$

and $M = \sup\limits_{(x,y) \in R} |f(x,y)|$. We next assume that f satisfies the Lipschitz condition, i.e., there exists $L > 0$ such that

$$|f(x, y_1) - f(x, y_2)| \leq L|y_1 - y_2|, \ (x, y_1), \ (x, y_2) \in R. \tag{2.40}$$

Then the sequence (z_n) of Picard's successive approximations converges uniformly on $[x_0 - \delta, x_0 + \delta]$ where $\delta = \min\{a, \frac{b}{M}\}$. Moreover the limit of z_n, say ϕ, is a unique solution to Cauchy problem (2.34).

Proof. The proof is divided into five main steps. We give a detailed proof of each step as follows. We first let $\tilde{J}$ denote the interval $[x_0 - \delta, x_0 + \delta]$.

Step 1. Here we prove that (2.38) is well defined.
In order to do that we show that for such a choice of δ, the sequence of successive approximations given by (2.37)–(2.38) satisfies

$$|z_n(x) - y_0| \leq b, \ x \in \tilde{J}.$$

For, we notice that when $n = 0$, the above statement is obviously true. We assume that $|z_n(x) - y_0| \leq b$, for some $n \in \mathbb{N}$. Then for $x \in [x_0, x_0 + \delta]$, we have

$$|z_{n+1}(x) - y_0| = \left| \int_{x_0}^{x} f(t, z_n(t))dt \right| \leq \int_{x_0}^{x} |f(t, z_n(t))|dt \leq M\delta \leq b.$$

Similarly, we can prove that $|z_{n+1}(x) - y_0| \leq b$ when $x \in [x_0 - \delta, x_0]$.

Step 2. In this step, we show that

$$|z_n(x) - z_{n-1}(x)| \leq \frac{ML^{n-1}}{n!}|x - x_0|^n, \ x \in \tilde{J}, \ n \in \mathbb{N}. \tag{2.41}$$

As in Step 1, we first prove (2.41) for $x \in [x_0, x_0 + \delta]$. For, we begin with the observation that for $n = 1$, we have

$$|z_1(x) - z_0(x)| = \left| \int_{x_0}^{x} f(t, y_0)dt \right| \leq \int_{x_0}^{x} |f(t, y_0)|dt \leq M(x - x_0).$$

We assume that (2.41) holds in $[x_0, x_0 + \delta]$ for $n = k$ and consider

$$
\begin{aligned}
|z_{k+1}(x) - z_k(x)| &\leq \int_{x_0}^{x} \left| f(t, z_k(t)) - f(t, z_{k-1}(t)) \right| dt \\
&\leq L \int_{x_0}^{x} |z_k(t) - z_{k-1}(t)| dt \\
&\leq \frac{ML^k}{k!} \int_{x_0}^{x} (t - x_0)^k dt = \frac{ML^k}{(k+1)!}(x - x_0)^{k+1}.
\end{aligned}
$$

This proves (2.41) when $x \in [x_0, x_0 + \delta]$. Using the same argument, one can easily extend the result to the entire interval $\tilde{J}$.

Step 3. We now show that z_n converges in $\left(C(\tilde{J}), d_\infty \right)$.

From Step 1 and continuity of f on R, we obtain that $z_n \in C(\tilde{J})$. Since $(C(\tilde{J}), d_\infty)$ is complete. it is enough to show that (z_n) is a Cauchy sequence. We set

$$S_N = \sum_{n=0}^{N} \frac{ML^{n-1}}{n!} \delta^n, \ N = 0, 1, 2, \ldots.$$

Clearly, $(S_N) \to \frac{M}{L} e^{L\delta}$ as $N \to \infty$. For $n > m \geq 0$, we have

$$
\begin{aligned}
|z_n(x) - z_m(x)| &\leq \sum_{k=m+1}^{n} |z_k(x) - z_{k-1}(x)| \\
&\leq \sum_{k=m+1}^{n} \frac{ML^{k-1}}{k!} |x - x_0|^k \\
&\leq |S_n - S_m|,
\end{aligned}
\tag{2.42}
$$

for $x \in \tilde{J}$. Since (S_N) is a Cauchy sequence in $\mathbb{R}$, (z_n) is a Cauchy sequence in $(C(\tilde{J}), d_\infty)$. Therefore (z_n) is a convergent sequence in $(C(\tilde{J}), d_\infty)$, i.e., there exists $\phi \in C(\tilde{J})$ such that

$$z_n \to \phi \text{ in } (C(\tilde{J}), d_\infty).$$

Step 4. Here we show that ϕ is a solution to Cauchy problem (2.34). In view of Lemma 2.2.4, it is enough to show that ϕ satisfies (2.36). For $x \in [x_0, x_0 + \delta]$, $n \in \mathbb{N}$, from the triangle inequality we get

$$\left| \phi(x) - y_0 - \int_{x_0}^{x} f(t, \phi(t))dt \right| \leq |\phi(x) - z_n(x)| + \int_{x_0}^{x} |f(t, z_n(t)) - f(t, \phi(t))|dt.$$

Let $\epsilon > 0$ be a given arbitrary constant. We choose $n_0 \in \mathbb{N}$ such that $d_\infty(z_n, \phi) < \epsilon$. for $n \geq n_0$. For this choice of n_0 we obtain

$$\left| \phi(x) - y_0 + \int_{x_0}^{x} f(t, \phi(t))dt \right| \leq \epsilon + \int_{x_0}^{x} L|z_{n_0}(t) - \phi(t)|dt \leq \epsilon + L\delta\epsilon.$$

Since ϵ is arbitrary, we readily have

$$\phi(x) = y_0 + \int_{x_0}^{x} f(t, \phi(t))dt, \ x \in [x_0, x_0 + \delta].$$

Using a similar argument we can extend this result to $\tilde{J}$. Hence ϕ satisfies (2.36) in $\tilde{J}$.

Step 5. Uniqueness of a continuous solution to (2.36) is an immediate consequence of (2.40) and the Gronwall lemma (see Lemma 2.3.2). Here we provide an outline of the proof of uniqueness. Let ϕ and $\tilde{\phi}$ be two solutions to (2.36).

Then for $x_0 \leq x \leq x_0 + \delta$, we get

$$
\begin{aligned}
|\phi(x) - \tilde{\phi}(x)| &= \left| \int_{x_0}^{x} \left(f(t, \phi(t)) - f(t, \tilde{\phi}(t)) \right) dt \right| \\
&\leq \int_{x_0}^{x} |f(t, \phi(t)) - f(t, \tilde{\phi}(t))| \, dt \\
&\leq \int_{x_0}^{x} L |\phi(t) - \tilde{\phi}(t)| \, dt.
\end{aligned}
$$

From Remark 2.3.3, it immediately follows that $\phi(x) = \tilde{\phi}(x)$, $x \in [x_0, x_0 + \delta]$. Using a similar argument, one can prove that $\phi(x) = \tilde{\phi}(x)$ in $[x_0 - \delta, x_0]$. Thus we have the uniqueness of the solution.
This completes the proof of the theorem.   □

By doing necessary modifications in the proof, we can obtain existence and uniqueness result when (2.34) is an IVP. We next prove a very useful result regarding the uniqueness of the solution.

Lemma 2.2.12. *Assume hypotheses of Theorem 2.2.11. Let I_1, I_2 be two subintervals of $[x_0 - a, x_0 + a]$. Suppose $\phi_1 : I_1 \to \mathbb{R}$ and $\phi_2 : I_2 \to \mathbb{R}$ are two solutions to (2.34). Then we have $\phi_1(x) = \phi_2(x)$, $x \in I_1 \cap I_2$.*

Proof. If (2.34) is an IVP, then the result holds true trivially. We now assume that (2.34) is a Cauchy problem. Let $I_1 \cap I_2 = [c, d]$. From Theorem 2.2.11 there exists a unique solution to (2.34) in a neighborhood of x_0, say $[x_0 - \delta, x_0 + \delta]$. Without loss of generality, we assume that $[x_0 - \delta, x_0 + \delta] \subset [c, d]$.
Claim. $\phi_1(x) = \phi_2(x)$, $x \in [c, x_0]$.

For, we suppose x_1 is the smallest number in (c, x_0) such that $\phi_1(x) = \phi_2(x)$, $x \in [x_1, x_0]$. We consider the problem

$$
\begin{cases}
z'(x) = f\big(x, z(x)\big), & x \in \mathbb{R}, \\
z(x_1) = \phi_1(x_1).
\end{cases}
\tag{2.43}
$$

Since f satisfies the hypotheses of Theorem 2.2.11, there exists $\delta_1 > 0$ such that there is a unique solution to (2.43) in $[x_1 - \delta_1, x_1 + \delta_1]$. Hence we get $\phi_1(x) = \phi_2(x)$, $x \in [x_1 - \delta_1, x_0] \subset (c, d)$ which is a contradiction to the fact that x_1 is the smallest such number. Therefore we have $\phi_1(x) = \phi_2(x)$, $x \in [c, x_0]$.
Using the same arguments one can prove that $\phi_1(x) = \phi_2(x)$, $x \in [x_0, d]$. The cases $I_1 \cap I_2 = [c, d)$ or $(c, d]$ or (c, d) can be dealt with in the same way.   □

In the next couple of examples, we consider initial value problems which do not satisfy the hypotheses of the Cauchy–Lipschitz theorem.

Example 2.2.13. Discuss the existence and uniqueness of solution to the following initial value problem

$$\begin{cases} \dfrac{d}{dx} y(x) = y^{\frac{2}{3}}(x), \ x > 0, \\[2mm] y(0) = 0. \end{cases}$$

Solution. Let $R = [0, a] \times [-b, b]$, where $a, b > 0$. On comparing the given IVP with (2.34), we observe that $f(x, y) = y^{\frac{2}{3}}$, $(x, y) \in R$. We first show that f does not satisfy the Lipschitz condition (2.40) in R. Suppose there exists $L > 0$ such that

$$|f(x, y_1) - f(x, y_2)| = |y_1^{\frac{2}{3}} - y_2^{\frac{2}{3}}| \le L|y_1 - y_2|, \ (x, y_1), (x, y_2) \in R. \quad (2.44)$$

On substituting $y_1 = \min\{b, (2L)^{-3}\}$ and $y_2 = 0$ in (2.44), we arrive at $|y_1|^{\frac{2}{3}} \le L|y_1|$ or

$$1 \le L y_1^{\frac{1}{3}} \le L(2L)^{-1},$$

which is a contradiction. Hence f does not satisfy (2.40).
Therefore Theorem 2.2.11 is not applicable. Using the method of separation of variables, we obtain that a solution is implicitly given by

$$\int^y y^{\frac{-2}{3}} dy - \int^x dx = c,$$

or

$$3y^{\frac{1}{3}} - x = c.$$

On writing y in terms of x we get $y = \frac{(x+c)^3}{27}$ and hence

$$\phi(x) = \frac{(x + c)^3}{27}, \ x \ge 0, \quad (2.45)$$

is a solution to the given equation for a suitable constant c. We substitute $\phi(0) = 0$ in (2.45) to find that $c = 0$.

On the other hand, it is easy to verify that $\phi \equiv 0$ is also a solution to the given equation. Moreover, the given equation admits infinitely many solutions given by

$$\phi_\tau(x) = \begin{cases} 0, \ 0 \le x \le \tau, \\[2mm] \dfrac{(x - \tau)^3}{27}, \ x > \tau, \end{cases}$$

where τ is an arbitrary positive constant. $\qquad \square$

We can replace the domain in Example (2.2.13) by $(-1, 1)$ to conclude that there are infinitely many solutions to the corresponding Cauchy problem. In fact, we can replace $f(x, y) = y^{\frac{2}{3}}$ in Example (2.2.13) by $f(x, y) = |y|^a$ for any $a \in (0, 1)$ and get infinitely many solutions. Therefore we conclude that in an IVP or Cauchy problem, if f is continuous but does not satisfy (2.40), then we may have more than one solution.

On the other hand, there are examples where f is a continuous function but does not satisfy the Lipschitz condition (2.40) and there exists a unique solution to the Cauchy problem/IVP. Hence f satisfying the Lipschitz condition is a sufficient condition but not a necessary condition for uniqueness of solutions to Cauchy problem/IVP. The following example supports this claim.

Example 2.2.14. Discuss the existence and uniqueness of solution to the Cauchy problem $y'(x) = f(y(x))$, $x \in \mathbb{R}$, $y(0) = 0$, where

$$f(y) = \begin{cases} y \sin\left(\frac{1}{y}\right), & y \neq 0, \\ 0, & y = 0. \end{cases}$$

Solution. Let $R = [-a, a] \times [-b, b]$, where $a, b > 0$. We first show that f does not satisfy the Lipschitz condition (2.40) in R. For, suppose that there exists $L > 0$ such that

$$|f(x, y) - f(x, z)| \leq L|y - z|, \quad (x, y), (x, z) \in R. \tag{2.46}$$

The strategy used in the previous example to get a contradiction will not work here. So we use another method. We choose $n_0 \in \mathbb{N}$ such that $4n_0\pi > \frac{1}{b}$. Moreover we take a sequence of ordered pairs $(y_n, z_n) = \left(\frac{1}{4n\pi}, \frac{1}{4n\pi + \frac{\pi}{2}}\right)$, $n \geq n_0$ so that $f(0, y_n) = 0$ and $f(0, z_n) = z_n$. On substituting $(x, y, z) = (0, y_n, z_n)$ in (2.46), we get

$$\left|0 - \frac{1}{4n\pi + \frac{\pi}{2}}\right| \leq L \left|\frac{1}{4n\pi} - \frac{1}{4n\pi + \frac{\pi}{2}}\right|, \quad n \geq n_0.$$

After simplification we arrive at the condition

$$8n \leq L, \quad \forall n \geq n_0,$$

which is a contradiction. Hence f does not satisfy (2.40).
We next show that $\psi \equiv 0$ is a unique solution to the given Cauchy problem. Suppose not, then there exists $h > 0$ such that $\phi : [-h, h] \to \mathbb{R}$ is a nonzero solution to the given Cauchy problem. Then there exists $\tilde{x} \in [-h, h]$ such that $\phi(\tilde{x}) \neq 0$. Without loss of generality, we assume that $\phi(\tilde{x}) > 0$. Let $n_1 \in \mathbb{N}$ be such that $\frac{1}{n_1\pi} < \phi(\tilde{x})$. Since $\phi(0) = 0$, by the intermediate value theorem, there exists $0 < x_2 < \tilde{x}$ such that $\phi(x_2) = \frac{1}{2n_1\pi}$. We assume that S denotes the interval $[\frac{1}{3n_1\pi}, \frac{1}{n_1\pi}]$ and observe that f is continuously differentiable on S. Thus by the mean value theorem we have

$$|f(u_1) - f(u_2)| = |f'(c)||u_1 - u_2| \leq M|u_1 - u_2|, \quad u_1, u_2 \in S,$$

where $M = \sup_{v \in S} |f'(v)|$. Therefore (2.40) holds on $\mathbb{R} \times S$. We now consider the problem

$$\begin{cases} z'(x) = f(z(x)), & x \in \mathbb{R}, \\ z(x_2) = \frac{1}{2n_1\pi}. \end{cases} \tag{2.47}$$

We notice that both the constant function $z(x) = \frac{1}{2n_1\pi}, x \in \mathbb{R}$ and ϕ are solutions to (2.47). Since f satisfies the hypotheses of Lemma 2.2.12, it immediately follows that $\phi(x) = \frac{1}{2n_1\pi}, x \in [-h, h]$. This is a contradiction to the fact that $\phi(0) = 0$. Hence $\phi \equiv 0$ is the only solution to the given problem. □

We now state a theorem (without proof) due to Peano which assures the existence of a solution to the Cauchy problem when f is continuous.

Theorem 2.2.15 (Peano). *Let f be a real valued continuous function on $R = \{(x, y) : |x - x_0| \leq a, |y - y_0| \leq b\}$. Then the Cauchy problem (2.34) has a solution on $[x_0 - \epsilon, x_0 + \epsilon]$, for some $\epsilon > 0$.*

If f is merely a continuous map, then the Picard sequence of successive approximations may not converge (see Exercise 2.17). Thus a a new type of approximate sequence is needed. For a proof see for instance [7, 35]. On the other hand, notice that the Peano existence theorem does not say anything about the uniqueness of the solution to (2.34). Examples 2.2.13–2.2.14 demonstrate this fact.

2.2.1 Continuable solutions

So far, we have seen the existence and uniqueness of a bilateral (resp. right) solution to Cauchy problem (resp. IVP) in a neighborhood of x_0, provided f is a continuous function satisfying (2.40). We now ask some natural questions like:
(1) Can we have existence of a solution on an interval which contains the neighborhood of x_0 given in Theorem 2.2.11?
(2) When can we have a global solution?
Before we answer these questions we prove the following useful result:

Theorem 2.2.16 (Concatenation). *Let J be an interval in $\mathbb{R}$. Assume that $[x_0, x_2] \subset J$, $\phi_1 : [x_0, x_1] \to \mathbb{R}$ and $\phi_2 : [x_1, x_2] \to \mathbb{R}$ are solutions to the IVPs*

$$\begin{cases} y'(x) = f(x, y(x)), \ x \in J, \\ y(x_0) = y_0 \end{cases} \quad (2.48)$$

and

$$\begin{cases} y'(x) = f(x, y(x)), \ x \in J, \\ y(x_1) = \phi_1(x_1), \end{cases} \quad (2.49)$$

respectively. Then

$$z(x) = \begin{cases} \phi_1(x), \ x_0 \leq x < x_1, \\ \phi_2(x), \ x_1 \leq x \leq x_2, \end{cases}$$

is also a solution to (2.48).

Proof. We use the equivalence of the IVP and the corresponding integral equation which is established in Lemma 2.2.4. Since ϕ_1 and ϕ_2 are solutions to (2.48) and (2.49), respectively, the corresponding integral equations that they satisfy are given by

$$\phi_1(x) = y_0 + \int_{x_0}^{x} f(t, \phi_1(t))dt, \ x \in [x_0, x_1], \tag{2.50}$$

$$\phi_2(x) = \phi_1(x_1) + \int_{x_1}^{x} f(t, \phi_2(t))dt, \ x \in [x_1, x_2]. \tag{2.51}$$

From the definition, it is clear that z is continuous. To complete the proof, it is enough to prove that z satisfies the corresponding integral equations in $[x_0, x_2]$. Since $z = \phi_1$ on $[x_0, x_1]$, we can rewrite (2.50) as

$$z(x) = y_0 + \int_{x_1}^{x} f(t, z(t))dt, \ x \in [x_0, x_1].$$

On the other hand, for $x \in [x_1, x_2]$ from (2.50)–(2.51) we have

$$\begin{aligned} z(x) &= \phi_2(x) = y_0 + \int_{x_0}^{x_1} f(t, \phi_1(t))dt + \int_{x_1}^{x} f(t, \phi_2(t))dt \\ &= y_0 + \int_{x_0}^{x} f(t, z(t))dt. \end{aligned}$$

Therefore z is a solution to (2.48). □

Let J be an interval, $\Omega \subseteq \mathbb{R}$ an open interval and $x_0 \in J$, $y_0 \in \Omega$. Assume that $f : J \times \Omega \to \mathbb{R}$ is a continuous function. Consider the problem

$$\begin{cases} y'(x) = f(x, y(x)), \ x \in J, \\ y(x_0) = y_0. \end{cases} \tag{2.52}$$

Definition 2.2.17. A right solution $\phi : [x_0, x_1) \to \mathbb{R}$ to (2.52) is said to be continuable at the right if there exists another right solution $\tilde{\phi} : [x_0, x_2) \to \mathbb{R}$ with $x_1 < x_2$ and $\phi(x) = \tilde{\phi}(x)$ for each $x \in [x_0, x_1)$. If a solution is not continuable at the right, then that solution is said to be saturated at the right.

Similarly, one can define continuable at the left and saturated at the left solutions to (2.52).

Example 2.2.18. Consider the IVP $\begin{cases} y' = y^2, \ x > 0, \\ y(0) = \frac{1}{2}. \end{cases}$

It is easy to verify that the function $\phi(x) = \frac{1}{2-x}$, $x \in [0, 2)$, is a solution to the IVP. Moreover ϕ is saturated (or not continuable) at the right. Similarly, $\psi(x) = \frac{1}{2-x}$, $x \in (2, 3]$, is saturated at the left on $(2, 3]$. □

This example provides us an idea that enables us to characterize continuable solutions. Details are given in the following theorem.

Theorem 2.2.19. *Let J be an interval, $\Omega \subseteq \mathbb{R}$ an open interval and $x_0 \in J$, $y_0 \in \Omega$. Assume that $f : J \times \Omega \to \mathbb{R}$ is a continuous function. Then a solution $\phi : [x_0, x_1) \to \mathbb{R}$ to (2.52) is continuable at the right if and only if*

(i) $x_1 < \sup J$,

(ii) $\lim\limits_{x \to x_1^-} \phi(x)$ *exists and belongs to Ω.*

Proof. Let ϕ be continuable at the right. Then from the definition of continuable solution at the right, $(i) - (ii)$ follow immediately.
Conversely, we assume that $\phi_1 : [x_0, x_1) \to \mathbb{R}$ is a solution to (2.52) such that $(i) - (ii)$ hold. Let l denote the limit in (ii), i.e.,

$$\lim_{x \to x_1^-} \phi_1(x) = l.$$

We now consider

$$\begin{cases} y'(x) = f(x, y(x)), x \in J, \\ y(x_1) = l. \end{cases} \qquad (2.53)$$

From Theorem 2.2.15, there exists a solution to (2.53) in a neighborhood of x_1. Let $\phi_2 : [x_1, x_2) \to \mathbb{R}$ be one such solution. In view of Theorem 2.2.16, it follows that the concatenate function

$$z(x) = \begin{cases} \phi_1(x), & x_0 \leq x < x_1, \\ \phi_2(x), & x_1 \leq x \leq x_2, \end{cases}$$

is also a solution to (2.52). Hence ϕ_1 is a solution which is continuable at the right. $\qquad \square$

After understanding Theorem 2.2.19, the reader is advised to come up with a criterion on solutions to (2.52), to be 'continuable at the left.'
We now turn our attention to the global existence of solutions to (2.52). From Section 2.1, we recall that the linear equation

$$y'(x) + P(x)y(x) = Q(x),$$

where $P, Q \in C([x_0 - a, x_0 + a])$, has the solution

$$\phi(x) = y_0 e^{-\int_{x_0}^{x} P(t)dt} + \int_{x_0}^{x} Q(t)e^{\int_{x_0}^{t} P(s)ds} dt, \quad x \in [x_0 - a, x_0 + a].$$

If we write the linear equation in the standard form, i.e., $y'(x) = f(x, y(x))$ then we have $f(x, y) = Q(x) - P(x)y$. We then observe that

$$|f(x, y_1) - f(x, y_2)| = |P(x)(y_1 - y_2)| \leq M|y_1 - y_2|, \quad y_1, y_2 \in \mathbb{R},$$

where $M = \sup\limits_{|x - x_0| \leq a} |P(x)|$. This is nothing but Lipschitz condition (2.40) without any restriction on the second variable. Motivated by this observation, we prove the following result.

Theorem 2.2.20 (Global existence)**.** *Let $a > 0$ and $f : [x_0 - a, x_0 + a] \times \mathbb{R} \to \mathbb{R}$ be continuous. Assume that there exists $L > 0$ such that*

$$|f(x, y_1) - f(x, y_2)| \leq L|y_1 - y_2|, \; x \in [x_0 - a, x_0 + a], \; y_1, y_2 \in \mathbb{R}. \quad (2.54)$$

Then there exists a unique solution $\phi : [x_0 - a, x_0 + a] \to \mathbb{R}$ to

$$y' = f(x, y), \; y(x_0) = y_0.$$

Proof. We show that the Picard sequence of successive approximations given by (2.37)–(2.38) converges in $[x_0 - a, x_0 + a]$. The proof closely follows that of Theorem 2.2.11.

Step 1 of the proof of Theorem 2.2.11 follows from the continuity of f on $[x_0 - a, x_0 + a] \times \mathbb{R}$. We then set $M = \max\limits_{|x - x_0| \leq a} |f(x, y_0)|$ and use the same arguments in Step 2 of the proof of Theorem 2.2.11 to show that (2.41) holds for $|x - x_0| \leq a$.

The remainder of the proof of is mere repetition of Steps 3–5 of the proof of Theorem 2.2.11, with $\tilde{J}$ replaced by $[x_0 - a, x_0 + a]$. This completes the proof. $\qquad\qquad \square$

For an extensive study of topics like well-posedness, saturated solutions, global existence, regularity with respect to the parameters etc., under weak hypotheses on f, readers can refer [10, 35].

2.3 Differential inequalities

In this section, we study linear differential inequalities which are used quite often in the analysis of ordinary and partial differential equations. In particular, we state and prove a very important inequality of this type due to Gronwall. First we prove the differential version of this inequality and next we prove the integral version. We then move on to the approximate results using these inequalities.

Lemma 2.3.1 (Gronwall's lemma–differential form)**.** *Assume that $\mu, \nu : [x_0, \infty) \to \mathbb{R}$ are continuous functions. If*

$$\eta'(x) \leq \mu(x)\eta(x) + \nu(x), \; x \geq x_0, \quad (2.55)$$

then

$$\eta(x) \leq \eta(x_0)e^{\int_{x_0}^{x} \mu(\tau)d\tau} + \int_{x_0}^{x} \nu(t)e^{\int_{t}^{x} \mu(\tau)d\tau}dt, \; x \geq x_0. \quad (2.56)$$

Proof. The proof that we present here uses the technique that is used to solve the first order non-homogeneous linear ODE. We multiply (2.55) with

$e^{-\int_{x_0}^t \mu(\tau)d\tau}$ to arrive at

$$e^{-\int_{x_0}^t \mu(\tau)d\tau}\left(\eta'(t)-\mu(t)\eta(t)\right) \le \nu(t)e^{-\int_{x_0}^t \mu(\tau)d\tau},$$

or

$$\frac{d}{dt}\left(e^{-\int_{x_0}^t \mu(\tau)d\tau}\eta(t)\right) \le \nu(t)e^{-\int_{x_0}^t \mu(\tau)d\tau}.$$

On integrating with respect to t from x_0 to x, we get

$$e^{-\int_{x_0}^x \mu(\tau)d\tau}\eta(x)-\eta(x_0) \le \int_{x_0}^x \nu(t)e^{-\int_{x_0}^t \mu(\tau)d\tau}dt.$$

After re-arrangement of the terms, we obtain

$$\eta(x) \le \eta(x_0)e^{\int_{x_0}^x \mu(\tau)d\tau} + \int_{x_0}^x \nu(t)e^{(\int_{x_0}^x - \int_{x_0}^t)\mu(\tau)d\tau}dt,$$

which is the same as (2.56). This completes the proof. $\square$

Using Gronwall's lemma in the differential form, we now prove the Gronwall lemma in the integral form.

Lemma 2.3.2 (Gronwall's lemma–integral form). *Assume that* $\alpha,\beta,y :$ $[x_0,\infty)\to\mathbb{R}$ *are continuous functions and* $\alpha(x)>0$, $x\ge x_0$. *If*

$$y(x) \le \int_{x_0}^x \alpha(t)y(t)dt + \beta(x), \quad x\ge x_0, \tag{2.57}$$

then

$$y(x) \le \beta(x) + \int_{x_0}^x \alpha(t)\beta(t)e^{\int_t^x \alpha(\tau)d\tau}dt, \quad x\ge x_0. \tag{2.58}$$

Moreover, if β *is a constant function, then*

$$y(x) \le \beta e^{\int_{x_0}^x \alpha(\tau)d\tau}, \quad x\ge x_0. \tag{2.59}$$

Proof. We begin with setting $\eta(x) = \int_{x_0}^x \alpha(t)y(t)dt$, $x\ge x_0$. Then we have $\eta'(x)=\alpha(x)y(x)$ and $\eta(x_0)=0$. Further on multiplying (2.57) with α, we arrive at

$$\eta'(x) \le \alpha(x)\eta(x)+\alpha(x)\beta(x), \quad x>x_0.$$

In view of the Gronwall lemma–differential form, it follows that

$$\eta(x) \le \int_{x_0}^x \alpha(t)\beta(t)e^{\int_t^x \alpha(\tau)d\tau}dt.$$

Therefore we obtain

$$y(x) \le \beta(x)+\eta(x) \le \beta(x) + \int_{x_0}^x \alpha(t)\beta(t)e^{\int_t^x \alpha(\tau)d\tau}dt.$$

This proves (2.58). Now if β is a constant, then (2.58) reduces to

$$
\begin{aligned}
y(x) &\leq \beta + \beta \int_{x_0}^{x} \alpha(t) e^{\int_{t}^{x} \alpha(\tau)d\tau} dt \\
&= \beta - \beta \int_{x_0}^{x} \frac{d}{dt} e^{\int_{t}^{x} \alpha(\tau)d\tau} dt \\
&= \beta - \beta \left(1 - e^{\int_{x_0}^{x} \alpha(\tau)d\tau}\right) = \beta e^{\int_{x_0}^{x} \alpha(\tau)d\tau}.
\end{aligned}
$$

This completes the proof of the lemma.   □

Before we discuss applications of the Gronwall lemma a couple of remarks are in order.

Remark 2.3.3. If y is a nonnegative continuous function satisfying

$$
y(x) \leq \int_{x_0}^{x} \alpha(t)y(t)dt, \ x \geq x_0,
$$

for some nonnegative continuous function α, then from (2.59) we get that y is a constant zero function, i.e., $y(x) = 0$, $x \geq x_0$.

Remark 2.3.4. The domain $[x_0, \infty)$ in which hypotheses (2.55), (2.57) hold can be replaced with $[x_0, x_0 + h]$, $h > 0$. In that case, (2.56), (2.58), and (2.59) hold in $[x_0, x_0 + h]$.

2.3.1   Applications of Gronwall's lemma

In the analysis of ordinary and partial differential equations there are a few standard techniques to prove approximation results. They are: (*i*) to find an estimate of the derivative of the error term in terms of the error and use the Gronwall lemma-differential form; (*ii*) to estimate the error term in terms of the indefinite integral and use the Gronwall lemma-integral form.

Using the latter technique, we next show that if the initial condition is 'slightly perturbed' then the solution to (2.52) 'cannot change quickly' if f satisfies the Lipschitz condition. Though all the results we prove in this subsection are valid for both Cauchy problems and IVPs, as before, we present the results and their proofs for Cauchy problems.

Theorem 2.3.5. *Assume that J, Ω are open intervals in $\mathbb{R}$ and $x_0 \in J$. Let $f : \bar{J} \times \bar{\Omega} \to \mathbb{R}$ be a continuous function satisfying*

$$
|f(x, y_1) - f(x, y_2)| \leq L|y_1 - y_2|, \ x \in \bar{J}, \ y_1, y_2 \in \bar{\Omega}. \tag{2.60}
$$

Assume that ϕ and $\tilde{\phi}$ are solutions to the Cauchy problems

$$
\begin{cases} y'(x) = f(x, y(x)), \ x \in J, \\ y(x_0) = y_0, \end{cases} \quad and \quad \begin{cases} \tilde{y}'(x) = f(x, \tilde{y}(x)), \ x \in J, \\ \tilde{y}(x_0) = \tilde{y}_0, \end{cases}
$$

respectively. Then

$$|\phi(x) - \tilde{\phi}(x)| \le |y_0 - \tilde{y}_0|e^{L|x-x_0|}, \tag{2.61}$$

in the intersection of domains of ϕ and $\tilde{\phi}$.

Proof. Let the intersection of the domains of ϕ and $\tilde{\phi}$ be $\tilde{J}$. We denote by $I_1 = \tilde{J} \cap \{x \ge x_0\}$ and $I_2 = \tilde{J} \cap \{x \le x_0\}$. Let $x \in I_1$. Using (2.36), the triangle inequality and (2.60), we readily obtain

$$\begin{aligned}
|\phi(x) - \tilde{\phi}(x)| &\le |y_0 - \tilde{y}_0| + \left| \int_{x_0}^{x} (f(t, \phi(t)) - f(t, \tilde{\phi}(t))) dt \right| \\
&\le |y_0 - \tilde{y}_0| + \int_{x_0}^{x} |f(t, \phi(t)) - f(t, \tilde{\phi}(t))| dt \\
&\le |y_0 - \tilde{y}_0| + L \int_{x_0}^{x} |\phi(t) - \tilde{\phi}(t)| dt.
\end{aligned}$$

In view of (2.59) we get

$$|\phi(x) - \tilde{\phi}(x)| \le |y_0 - \tilde{y}_0|e^{L(x-x_0)}, \ x \in I_1.$$

Using the same argument we can show that (2.61) holds for $x \in I_2$. This completes the proof of the theorem. $\qquad\square$

Assume that $J, \Omega \subseteq \mathbb{R}$ are open intervals in $\mathbb{R}$ and $f : \bar{J} \times \Omega \to \mathbb{R}$ is a continuous function. We now introduce the notion of ϵ-approximate solution to

$$y'(x) = f(x, y(x)), \ x \in J. \tag{2.62}$$

Definition 2.3.6. Assume that $\epsilon > 0$. We say that a continuous and piecewise C^1 function[5] ϕ_ϵ is an ϵ-appropriate solution to (2.62) in J if

(i) $(x, \phi_\epsilon(x)) \in J \times \Omega$,

(ii) $|\phi_\epsilon'(x) - f(x, \phi_\epsilon(x))| \le \epsilon$, at every x where ϕ_ϵ is C^1.

An ϵ-approximate solution to (2.62) when $\epsilon = 0$ is taken to be the solution ϕ to it. We now show that ϕ_ϵ is close to the solution ϕ whenever f satisfies the Lipschitz condition (2.40).

Theorem 2.3.7. *Assume the hypotheses of Theorem 2.2.11. Assume that ϕ and ϕ_ϵ are solution and ϵ-approximate solution to (2.62) with $\phi(x_0) = y_0$ and $\phi_\epsilon(x_0) = y_1$, respectively. Then*

$$|\phi_\epsilon(x) - \phi(x)| \le |y_1 - y_0|e^{L|x-x_0|} + \frac{\epsilon}{L}(e^{L|x-x_0|} - 1), \ x \in \tilde{J}, \tag{2.63}$$

where $\tilde{J}$ is an interval in which ϕ exists.

---

[5]A function $g : J \to \mathbb{R}$ is said to be piecewise C^1 if for every closed and bounded subinterval $\tilde{J}$ of J, there exists a finite set (possibly the empty set), say S, such that (i) g is differentiable in $\tilde{J} \backslash S$ and g' is also continuous in $\tilde{J} \backslash S$, (ii) at each point in S both the left and the right hand derivatives exist but are not equal.

Proof. As before, let $I_1 = \tilde{J} \cap \{x \geq x_0\}$ and $I_2 = \tilde{J} \cap \{x \leq x_0\}$. We prove the result only for $x \in I_1$. Using the fundamental theorem of integral calculus, we have

$$\phi_\epsilon(x) = y_1 + \int_{x_0}^x \phi'_\epsilon(t)dt.$$

On the other hand, using the integral equation formulation, we notice that ϕ satisfies (2.36), i.e.,

$$\phi(x) = y_0 + \int_{x_0}^x f(t, \phi(t))dt.$$

Using the triangle inequality, Lipschitz condition and the definition of ϵ-approximate solution, we obtain

$$
\begin{aligned}
|\phi_\epsilon(x) - \phi(x)| &\leq |y_1 - y_0| + \left|\int_{x_0}^x (\phi'_\epsilon(t) - f(t, \phi(t)))dt\right| \\
&\leq |y_1 - y_0| + \left|\int_{x_0}^x [\phi'_\epsilon(t) - f(t, \phi_\epsilon(t))\right. \\
&\qquad \left. + f(t, \phi_\epsilon(t)) - f(t, \phi(t))]dt\right| \\
&\leq |y_1 - y_0| + \int_{x_0}^x |\phi'_\epsilon(t) - f(t, \phi_\epsilon(t))|dt \\
&\qquad + \int_{x_0}^x |f(t, \phi_\epsilon(t)) - f(t, \phi(t))|dt \\
&\leq |y_1 - y_0| + \int_{x_0}^x \epsilon dt + \int_{x_0}^x L|\phi_\epsilon(t) - \phi(t)|dt \\
&= |y_1 - y_0| + \epsilon(x - x_0) + L\int_{x_0}^x |\phi_\epsilon(t) - \phi(t)|dt.
\end{aligned}
$$

From the Gronwall lemma (integral form) we get

$$
\begin{aligned}
|\phi_\epsilon(x) - \phi(x)| &\leq |y_1 - y_0| + \epsilon(x - x_0) + L\int_{x_0}^x (|y_1 - y_0| + \epsilon(t - x_0))e^{L(x-t)}dt \\
&= |y_1 - y_0|e^{L(x-x_0)} + \frac{\epsilon}{L}(e^{L(x-x_0)} - 1), \quad x \in I_1.
\end{aligned}
$$

Using similar arguments presented for $x \in I_1$, one can easily prove (2.63) for $x \in I_2$. □

In the next result we estimate the difference between two ϵ-approximate solutions.

Theorem 2.3.8. *Assume the hypotheses of Theorem 2.2.11. Let ϕ_{ϵ_i} be an ϵ_i-approximate solution to (2.62) with $\phi_{\epsilon_i}(x_0) = y_i$, for $i = 1, 2$. Then*

$$|\phi_{\epsilon_1}(x) - \phi_{\epsilon_2}(x)| \leq |y_1 - y_2|e^{L|x-x_0|} + \frac{\epsilon_1 + \epsilon_2}{L}(e^{L|x-x_0|} - 1), \quad x \in J. \quad (2.64)$$

Proof. We assume that $J_1 = J \cap \{x > x_0\}$ and use the fundamental theorem of integral calculus to have

$$\phi_{\epsilon_i}(x) = y_i + \int_{x_0}^x \phi'_{\epsilon_i}(t)dt, \ x \in J_1, \ i = 1, 2.$$

Using the triangle inequality and the definition of ϵ-approximate solutions we obtain

$$|\phi_{\epsilon_1}(x) - \phi_{\epsilon_2}(x)| \leq |y_1 - y_2| + \left|\int_{x_0}^x \left(\phi'_{\epsilon_1}(t) - \phi'_{\epsilon_2}(t)\right)dt\right|$$

$$\leq |y_1 - y_2| + \left|\int_{x_0}^x \left[\phi'_{\epsilon_1}(t) - f(t, \phi_{\epsilon_1}(t))\right.\right.$$

$$\left.\left. + f(t, \phi_{\epsilon_1}(t)) - f(t, \phi_{\epsilon_2}(t)) + f(t, \phi_{\epsilon_1}(t)) - \phi'_{\epsilon_2}(t)\right]dt\right|$$

$$\leq |y_1 - y_2| + \int_{x_0}^x |\phi'_{\epsilon_1}(t) - f(t, \phi_{\epsilon_1}(t))|dt$$

$$+ \int_{x_0}^x |f(t, \phi_{\epsilon_1}(t)) - f(t, \phi_{\epsilon_2}(t))|dt$$

$$+ \int_{x_0}^x |f(t, \phi_{\epsilon_2}(t)) - \phi'_{\epsilon_2}(t)|dt$$

$$\leq |y_1 - y_2| + \int_{x_0}^x \epsilon_1 dt + \int_{x_0}^x L|\phi_{\epsilon_1}(t) - \phi_{\epsilon_2}(t)|dt + \int_{x_0}^x \epsilon_2 dt$$

$$= |y_1 - y_2| + (\epsilon_1 + \epsilon_2)(x - x_0) + L\int_{x_0}^x |\phi_{\epsilon_1}(t) - \phi_{\epsilon_2}(t)|dt.$$

In view of the Gronwall lemma (integral form) we get

$$|\phi_{\epsilon_1}(x) - \phi_{\epsilon_2}(x)| \leq |y_1 - y_2| + (\epsilon_1 + \epsilon_2)(x - x_0)$$

$$+ L\int_{x_0}^x \left(|y_1 - y_2| + (\epsilon_1 + \epsilon_2)(t - x_0)\right)e^{L(x-t)}dt$$

$$= |y_1 - y_2| + (\epsilon_1 + \epsilon_2)(x - x_0)$$

$$- \int_{x_0}^x \left(|y_1 - y_2| + (\epsilon_1 + \epsilon_2)(t - x_0)\right)\frac{d}{dt}(e^{L(x-t)})dt.$$

Now integration by parts immediately gives us (2.64) when $x \in J_1$. The case $x \in J \cap \{x < x_0\}$ can be dealt with in a similar manner. □

2.4 Comparison results

In this section, we prove a nonlinear version of the Gronwall lemma. The results which will be discussed here are quite useful in subsequent chapters (for instance Chapter 6).

Theorem 2.4.1. *Let* $f : \mathbb{R}^2 \to \mathbb{R}$ *be a continuous function satisfying the global Lipschitz condition, i.e., there exists* $L > 0$ *such that*

$$|f(x, y_1) - f(x, y_2)| \leq L|y_1 - y_2|, \ x, y_1, y_2 \in \mathbb{R}. \tag{2.65}$$

Assume that $\alpha, \phi \in C^1([x_0, \infty))$. *Moreover* α *and* ϕ *satisfy*

$$\begin{cases} \alpha'(x) < f(x, \alpha(x)), \ x \geq x_0, \\ \alpha(x_0) = \alpha_0, \end{cases} \qquad \begin{cases} \phi'(x) = f(x, \phi(x)), \ x \geq x_0, \\ \phi(x_0) = y_0, \end{cases}$$

respectively. If $\alpha_0 \leq y_0$, *then* $\alpha(x) < \phi(x), \ x > x_0$.

Proof. The proof is divided into two steps.

Step 1. In this step we prove the theorem when $\alpha_0 < y_0$.
On the contrary, we suppose that there exists $\tilde{x} > x_0$ such that $\alpha(\tilde{x}) \geq \phi(\tilde{x})$. Let $S = \{x \in [x_0, \infty) : \alpha(x) = \phi(x)\}$. Then clearly S is a nonempty set. Let $x_1 = \inf S$. We define

$$\eta(x) = \alpha(x) - \phi(x), \ x \geq x_0.$$

We now show that

$$\eta(x_1) = 0. \tag{2.66}$$

For, let $\eta(x_1) > 0$. Since $\eta(x_0) < 0$, there exists $x_0 < x_2 < x_1$ such that $\eta(x_2) = 0$. Thus we get $x_2 \in S$ and $x_2 < x_1$, which is a contradiction. We now suppose $\eta(x_1) < 0$. Then there exists $\delta > 0$ such that $\eta(x) < 0$ in $[x_1, x_1 + \delta)$. Since $x_1 = \inf S$, there exists $x_1 < \tilde{x}_1 < x_1 + \delta$ such that $\tilde{x}_1 \in S$, which is a contradiction. This proves (2.66).

We next prove

$$\eta(x) < 0, \ x \in [x_0, x_1). \tag{2.67}$$

For, if $\eta(x) \geq 0$ for some $x \in (x_0, x_1)$, then there exists $\tilde{x}_2 \in (x_0, x_1)$ such that $\eta(\tilde{x}_2) = 0$. Thus we have $\tilde{x}_2 < \inf S$ and $\tilde{x}_2 \in S$, which is a contradiction. This proves (2.66).

From (2.66)–(2.67), we obtain

$$\eta'(x_1) = \lim_{x \to x_1^-} \frac{\eta(x_1) - \eta(x)}{x_1 - x} = \lim_{x \to x_1^-} \frac{-\eta(x)}{x_1 - x} \geq 0. \tag{2.68}$$

On the other hand, we find

$$\eta'(x_1) = \alpha'(x_1) - \phi'(x_1) < f(x_1, \alpha(x_1)) - f(x_1, \phi(x_1)) = 0,$$

which is a contradiction to (2.68). Therefore if $\alpha_0 < y_0$, then $\alpha(x) < \phi(x)$, for every $x > x_0$.

Step. 2 In this step we establish that if $\alpha_0 = y_0$ then $\alpha(x) < \phi(x), \ x > x_0$. To this end, we notice that $\eta(x_0) = 0$ and consider

$$\begin{aligned} \eta'(x_0) &= \alpha'(x_0) - \phi'(x_0) \\ &< f(x_0, \alpha(x_0)) - f(x_0, \phi(x_0)) \\ &= 0. \end{aligned}$$

Therefore, η is decreasing in a neighborhood of x_0, i.e., there exists $\delta_3 > 0$ such that $\alpha(x) < \phi(x)$, $x \in (x_0, x_0 + \delta_3)$. Now take any $\hat{x} \in (x_0, x_0 + \delta_3)$ and use Step 1, to obtain $\alpha(x) < \phi(x)$, for $x > \hat{x}$. Since $\hat{x} \in (x_0, x_0 + \delta_3)$ is arbitrary, we arrive at $\alpha(x) < \phi(x)$, $x > x_0$. This completes the proof. $\qquad\square$

Remark 2.4.2. From the hypotheses in Theorem 2.4.1 one should not jump to a quick conclusion that $\alpha'(x) < \phi'(x)$, $x \geq x_0$. This is a common mistake many students (sometimes even the brighter students also!) make. We give a counter example here. Set $f(x, y) = -y + \sin^2 y$, $x_0 = y_0 = 0$, and $\alpha(x) = -e^{-x}$, $x \geq 0$. Clearly f, α satisfy the hypotheses of Theorem 2.4.1. Moreover, we observe that $\phi \equiv 0$ is the solution to $\phi'(x) = f(x, \phi(x))$, $\phi(0) = 0$. Finally, note that $\alpha'(x) > \phi'(x)$, $x > 0$.

Using similar arguments employed in Theorem 2.4.1, it is straightforward to prove the corresponding result when $\alpha' > f$ and the details are given in the following result.

Theorem 2.4.3. *Let $f : \mathbb{R}^2 \to \mathbb{R}$ be as in Theorem 2.4.1. Assume that α and ϕ satisfy*

$$\begin{cases} \alpha'(x) > f(x, \alpha(x)), & x \geq x_0, \\ \alpha(x_0) = \alpha_0, \end{cases} \qquad \begin{cases} \phi'(x) = f(x, \phi(x)), & x \geq x_0, \\ \phi(x_0) = y_0, \end{cases}$$

respectively. If $\alpha_0 \geq y_0$, then $\alpha(x) > \phi(x)$, $x > x_0$.

Proof. The proof is left to the reader. $\qquad\square$

In the next theorem we prove a stronger version of Theorem 2.4.1.

Theorem 2.4.4. *Let $f : \mathbb{R}^2 \to \mathbb{R}$ be as in Theorem 2.4.1. Assume that α satisfies the differential inequality*

$$\begin{cases} \alpha'(x) \leq f(x, \alpha(x)), & x \geq x_0, \\ \alpha(x_0) = \alpha_0, \end{cases} \tag{2.69}$$

and ϕ is the solution to

$$\begin{cases} y'(x) = f(x, y(x)), & x \geq x_0, \\ y(x_0) = y_0. \end{cases} \tag{2.70}$$

If $\alpha_0 \leq y_0$, then $\alpha(x) \leq \phi(x)$, $x > x_0$.

Proof. Let $\epsilon > 0$ and $f_\epsilon = f + \epsilon$. Then we have

$$\alpha'(x) \leq f(x, \alpha(x)) < f_\epsilon(x, \alpha(x)), \quad x \geq x_0.$$

We assume that ϕ_ϵ satisfies

$$\begin{cases} \phi'_\epsilon(x) = f_\epsilon(x, \phi_\epsilon(x)), & x \geq x_0, \\ \phi_\epsilon(x_0) = y_0. \end{cases}$$

In view of Theorem 2.4.1, we readily obtain

$$\alpha(x) < \phi_\epsilon(x), \ x > x_0. \tag{2.71}$$

On the other hand, we notice that ϕ_ϵ is an ϵ-approximate solution to (2.70). From Theorem 2.3.7, it follows that

$$|\phi(x) - \phi_\epsilon(x)| \leq \frac{\epsilon}{L} e^{L(x-x_0)}, \ x \geq x_0. \tag{2.72}$$

Using (2.71)–(2.72) we get

$$\alpha(x) < \phi(x) + \tfrac{\epsilon}{L} e^{L(x-x_0)}, \ x \geq x_0.$$

Since $\epsilon > 0$ is arbitrary, it is easy to prove that

$$\alpha(x) \leq \phi(x), \ x > x_0,$$

(details are left to the reader). This completes the proof. □

Using the same arguments which are used to prove Theorem 2.4.4 one can easily prove the following result. We omit the proof.

Theorem 2.4.5. *Let $f : \mathbb{R}^2 \to \mathbb{R}$ be as in Theorem 2.4.1. Assume that α satisfies the differential inequality*

$$\begin{cases} \alpha'(x) \geq f\big(x, \alpha(x)\big), \ x \geq x_0, \\ \alpha(x_0) = \alpha_0, \end{cases} \tag{2.73}$$

and ϕ is the solution to

$$\begin{cases} y'(x) = f\big(x, y(x)\big), \ x \geq x_0, \\ y(x_0) = y_0. \end{cases} \tag{2.74}$$

If $\alpha_0 \geq y_0$, then $\alpha(x) \geq \phi(x)$, $x > x_0$.

Remark 2.4.6. If f satisfies (2.40) in a rectangle R defined in Theorem 2.2.11 instead of (2.65), then the results that are proved in Theorems 2.4.1–2.4.5 hold in a neighborhood of x_0 where the solutions to the ODEs in the theorems exist.

We now apply Theorems 2.4.1–2.4.5 to analyze the behavior of solutions to ODEs.

Example 2.4.7. Analyze the behavior of the solution to

$$\begin{cases} \dfrac{dy}{dx} = y^2 + x^2, \ x > 0, \\ y(0) = 1, \end{cases}$$

using the comparison theorems (Theorems 2.4.1–2.4.5).

Solution. Let ϕ be the saturated at the right solution to the given equation. Let J be the domain of ϕ. We begin with that observation that $\phi'(x) \geq 0$, $x \in J$ and hence ϕ is increasing. Since $\phi(0) = 1$, we obtain that $\phi(x) \geq 1$, $x \in J$. Consider the initial value problem

$$\begin{cases} \dfrac{dz}{dx} = z^2, \ x > 0, \\ z(0) = 1. \end{cases} \tag{2.75}$$

One can easily solve (2.75) to get $z(x) = \dfrac{1}{1-x}$, $x \in (0,1)$, is the solution to (2.75). Since ϕ satisfies

$$\begin{cases} \dfrac{d\phi}{dx} \geq \phi^2, \ x \in J, \\ \phi(0) = 1, \end{cases}$$

in view of Theorem 2.4.5 and Remark 2.4.6 (with $f(u,v) = v^2$), it follows that

$$\phi(x) \geq \frac{1}{1-x}, \ x \in (0,1) \cap J.$$

Therefore the solution to the given problem becomes unbounded in $(0,1) \cap J$. On the other hand, we observe that for the initial value problem

$$\begin{cases} \dfrac{dw}{dx} = 1 + w^2, \ x \in (0,1), \\ w(0) = 1, \end{cases} \tag{2.76}$$

the function $\psi(x) = \tan(x + \frac{\pi}{4})$, $x \in [0, \frac{\pi}{4})$, is the solution. Moreover ϕ satisfies

$$\begin{cases} \dfrac{d\phi}{dx} \leq 1 + \phi^2, \ x \in (0,1) \cap J, \\ \phi(0) = 1. \end{cases}$$

Using Theorem 2.4.4 and Remark 2.4.6, we get

$$\phi(x) \leq \tan(x + \tfrac{\pi}{4}), \ x \in [0, \tfrac{\pi}{4}) \cap J.$$

Finally, from Theorem 2.2.19 it follows that $[0, \frac{\pi}{4}) \subseteq J \subseteq [0,1)$. $\quad\square$

Example 2.4.8. Analyze the behavior of the solution to

$$\begin{cases} \dfrac{dy}{dx} = y^2 + x^2, \ x > 0, \\ y(0) = 0, \end{cases}$$

using the comparison theorems.

Solution. Let ϕ be the saturated at the right solution to the given IVP and J be its domain. We begin with that observation that $\phi'(x) \geq 0$, $x \in J$ and hence ϕ is increasing. Since $\phi(0) = 0$, we obtain that $\phi(x) \geq 0$, $x \in J$. We next consider the initial value problem

$$\begin{cases} \dfrac{dz}{dx} = x^2, \ x > 0, \\ z(0) = 0. \end{cases} \tag{2.77}$$

Then $z(x) = \dfrac{x^3}{3}$, $x \geq 0$, is a solution to (2.77). Moreover, ϕ satisfies

$$\begin{cases} \dfrac{d\phi}{dx} \geq x^2, \ x \in (0, \infty) \cap J, \\ \phi(0) = 0, \end{cases}$$

and from Theorem 2.4.5 (with $f(u, v) = u^2 + v^2$), it follows that

$$\phi(x) \geq \frac{x^3}{3}, \ x \in (0, \infty) \cap J. \tag{2.78}$$

On the other hand, we consider the initial value problem

$$\begin{cases} \dfrac{dw}{dx} = (x + w)^2, \ x > 0, \\ w(0) = 0. \end{cases} \tag{2.79}$$

To solve (2.79), we put $\tilde{w} = x + w$, to get

$$\frac{d\tilde{w}}{dx} = 1 + \frac{dw}{dx} = 1 + \tilde{w}^2.$$

Thus we obtain that $\tilde{w} = \tan x$, which in turn gives $w = -x + \tan x$, is the solution to (2.79). Since $\phi \geq 0$, $x > 0$, we find that

$$\begin{cases} \dfrac{d\phi}{dx} \leq (x + \phi)^2, \ x \in (0, \frac{\pi}{2}), \\ \phi(0) = 0. \end{cases}$$

Using Theorem 2.4.4 and Remark 2.4.6, we obtain

$$\phi(x) \leq -x + \tan x, \ x \in (0, \frac{\pi}{2}). \tag{2.80}$$

From estimates (2.78) and (2.80), we conclude that $J \subseteq [0, \frac{\pi}{2})$. □

Example 2.4.9. Analyze the behavior of the solution to

$$\begin{cases} \dfrac{dy}{dx} = y^2 - x, \ x > 0, \\ y(0) = 1, \end{cases}$$

using the comparison theorems.

Solution. As before, let ϕ be the right saturated solution to the given problem and J be its domain. Since we have the trivial estimate

$$-x \le \phi^2(x) - x \le \phi^2(x), \ x > 0,$$

it follows that

$$-x \le \phi'(x) \le \phi^2, \ x > 0.$$

Let z and w satisfy

$$\begin{cases} \dfrac{dz}{dx} = z^2, \ x \in (0,1), \\ z(0) = 1, \end{cases}$$

and

$$\begin{cases} \dfrac{dw}{dx} = -x, \ x > 0, \\ w(0) = 1, \end{cases}$$

respectively. An easy computation gives us that

$$z(x) = \frac{1}{1-x}, \ x \in (0,1); \quad w(x) = 1 - \frac{x^2}{2}, \ x \ge 0.$$

In view of the comparison theorems we get

$$\phi(x) \le \frac{1}{1-x}, \ x \in [0,1) \cap J,$$

and

$$\phi(x) \ge 1 - \frac{x^2}{2}, \ x \in [0, \infty) \cap J.$$

This completes the solution. □

Example 2.4.10. Show that the IVP

$$\begin{cases} \dfrac{dy}{dx} = y^2 + x, \ x > 0, \\ y(0) = 0 \end{cases}$$

has no solution defined on the entire interval $[0, 5)$.

Solution. Let $\phi : J \to \mathbb{R}$ be the right saturated solution to the given IVP and $z(x) = \tan^{-1} \phi(x)$. On the contrary, we suppose $[0, 5) \subseteq J$. In order to arrive at a contradiction, we show that there exists $x_0 \in (0, 5)$ such that $z(x_0) = \frac{\pi}{2}$. For, we notice that z satisfies

$$\begin{cases} \dfrac{dz}{dx} = \sin^2(z) + x \cos^2(z), \ x \in [0, 5), \\ z(0) = 0. \end{cases} \tag{2.81}$$

Since $z'(x) > 0$ and $z(0) = 0$, we have $z(x) > 0$, $x \in (0,5)$. Moreover, it is easy to verify that $z'(x) \geq 1$, $x \geq 1$. Using the comparison theorems we get

$$z(x) \geq x - 1 + z(1) > x - 1, \; \forall x \geq 1.$$

Since $z(3) \geq 2$, there exists $x_0 \in (1,3)$ such that $z(x_0) = \frac{\pi}{2}$. This is a contradiction. In fact, we proved that $J \subseteq [0,3)$. $\qquad\qquad\qquad\qquad\qquad\square$

2.5   The first order scalar autonomous equations

This section is dedicated to the study of the behavior of the solutions to differential equations of special forms. In order to do that, we first introduce Lipschitz functions of one variable.

Definition 2.5.1. Let J be an interval in $\mathbb{R}$. A function $g : J \to \mathbb{R}$ is said to be a Lipschitz function if there exists $L > 0$, called a Lipschitz constant of g, such that

$$|g(y_1) - g(y_2)| \leq L|y_1 - y_2|, \; y_1, y_2 \in J.$$

For example, $g : \mathbb{R} \to \mathbb{R}$, defined by $g(y) = |y|$, $y \in \mathbb{R}$, is a Lipschitz function. For, using the triangle inequality we have

$$|g(y_1) - g(y_2)| = \big||y_1| - |y_2|\big| \leq |y_1 - y_2|, \; y_1, y_2 \in \mathbb{R}.$$

Definition 2.5.2. Let J be an interval in $\mathbb{R}$. A function $g : J \to \mathbb{R}$ is said to be a locally Lipschitz function if for every closed[6] and bounded subset K of J, there exists $L > 0$ (which may depend on K) such that

$$|g(y_1) - g(y_2)| \leq L|y_1 - y_2|, \; y_1, y_2 \in K.$$

Remark 2.5.3. Every Lipschitz function is locally Lipschitz function and every locally Lipschitz function is continuous.

In the following lemma we provide sufficient conditions for functions to be Lipschitz and locally Lipschitz.

Lemma 2.5.4. *Suppose* $g : J \to \mathbb{R}$ *is a differentiable function. Then the following hold:*
(i) If g' *is bounded, then* g *is a Lipschitz function.*

(ii) If g' *is continuous, then* g *is a locally Lipschitz function.*

Proof. (i) Let $\sup\limits_{c \in J} |g'(c)| = M$. Then by the mean value theorem, we get

---

[6] A set $A \subseteq \mathbb{R}^n$ is said to be closed in $\mathbb{R}^n$ if (x_n) is a sequence in A such that $x_n \to x$ in $\mathbb{R}^n$ then $x \in A$.

$$|g(y_1) - g(y_2)| = |g'(c)||y_1 - y_2| \le M|y_1 - y_2|, \; y_1, y_2 \in J.$$

Therefore g is a Lipschitz function with Lipschitz constant M. This proves (i).

(ii) Let $K \subseteq J$ be a closed and bounded set. Then g' is bounded on K. Let $L_K = \max\limits_{x \in K} |g'(x)|$. In view of the mean value theorem, we get

$$|g(y_1) - g(y_2)| = |g'(c)||y_1 - y_2| \le L_K|y_1 - y_2|, \; y_1, y_2 \in K.$$

Hence g is a locally Lipschitz function. This proves (ii). $\square$

Example 2.5.5. In view of Lemma 2.5.4, functions like $\cos y$, $\sin y$, e^{-y^2}, $\frac{1}{1+y^2}$ etc, are Lipschitz functions on $\mathbb{R}$ because their derivatives are bounded. Similarly, every polynomial is a locally Lipschitz function.

Definition 2.5.6. Let $f : \mathbb{R} \to \mathbb{R}$ be continuous function. The equation of the form

$$y'(x) = f\big(y(x)\big), \; x \in \mathbb{R}, \tag{2.82}$$

is called a first order autonomous equation. A number $\tilde{y} \in \mathbb{R}$ is said to be a critical point of (2.82) if $f(\tilde{y}) = 0$.

Henceforth, unless specified otherwise, we assume that f is a locally Lipschitz function. This assumption guarantees the existence and uniqueness of a solution to (2.82). The rest of this section is dedicated to the study the behavior of the solutions to the first order autonomous equations, whereas in Chapters 5 and 6, we study the first order 2×2 autonomous system of equations.

Remark 2.5.7. Let $\tilde{y}$ be a critical point of (2.82). If ϕ is the solution to (2.82) with $\phi(x_0) = \tilde{y}$, then $\phi(x) = \tilde{y}$, $x \in \mathbb{R}$. For, we first observe that the constant function $\psi \equiv \tilde{y}$ is also a solution to (2.82) with $\psi(x_0) = \tilde{y}$. From the uniqueness of solutions to the Cauchy problem/IVP (see Cauchy–Lipschitz theorem), we have $\phi(x) \equiv \tilde{y}$, $x \in \mathbb{R}$. Conversely, if a constant function $\phi \equiv \tilde{y}$ is a solution to (2.82) then $\tilde{y}$ is a critical point of (2.82).

Henceforth, we analyze the dynamics of the solution to the initial value problem of the form

$$\begin{cases} y'(x) = f\big(y(x)\big), \; x > x_0, \\ y(x_0) = y_0, \end{cases} \tag{2.83}$$

as $x \to \infty$. In fact, we show that the asymptotic behavior of the solution ϕ to (2.83) depends on the location of y_0 with respect to the critical points of (2.83). The main advantage of doing this analysis is that we need not solve the ODE to find the behavior of the solution as $x \to \infty$.

To this end, we first prove a couple of useful results. We first prove that the critical points of (2.82) cannot be attained by any non-constant solution to (2.82).

Lemma 2.5.8. *Let $\tilde{y}$ be a critical point of (2.82). Suppose that ϕ is a solution to (2.83) and there exists a real number $x_1 \geq x_0$ such that $\phi(x_1) = \tilde{y}$. Then $\phi(x) = \tilde{y}$, $x \geq x_0$.*

Proof. On the contrary, we assume that ϕ is a non-constant function. Without loss of generality we assume that $x_1 \geq x_0$ is the smallest number such that $\phi(x_1) = \tilde{y}$. For the existence of a such number see Step 1 in the proof of Theorem 2.4.1. If $x_1 = x_0$, then from Remark 2.5.7, ϕ is a constant function, which is a contradiction.

Therefore we get $x_1 > x_0$. We now consider the auxiliary problem

$$\begin{cases} z'(x) = f\big(z(x)\big), \ x > x_0, \\ z(x_1) = \tilde{y}. \end{cases} \tag{2.84}$$

Then both ϕ and $z(x) = \tilde{y}$, $x > x_0$ are solutions to (2.84) in the interval $[x_0, x_1]$. By Lemma 2.2.12, we have $\phi(x) = \tilde{y}$, $x \in [x_0, x_1]$. This is a contradiction to the assumption that x_1 is the smallest number such that $\phi(x_1) = \tilde{y}$. Hence ϕ is a constant function.   □

We next show that the limit at infinity of any solution to (2.83), if it is exists in $\mathbb{R}$, must be a critical point.

Lemma 2.5.9. *Assume that ϕ is a solution to (2.83). If there exists $l \in \mathbb{R}$ such that $\lim_{x \to \infty} \phi(x) = l$, then $f(l) = 0$.*

Proof. Since $\phi(x) \to l$ as $x \to \infty$, we readily obtain that $f(\phi(x)) \to f(l)$ as $x \to \infty$.

We assume that $f(l) > 0$. Then we can choose $\epsilon > 0$ such that $f(u) > 0$ whenever $u \in [l - \epsilon, l + \epsilon]$. Let

$$m = \min_{u \in [l-\epsilon, l+\epsilon]} f(u) > 0.$$

Moreover, we choose $x_1 > x_0$ such that

$$l - \epsilon \leq \phi(x) \leq l + \epsilon, \text{ for } x \geq x_1.$$

Thus we have

$$\phi'(x) = f(\phi(x)) \geq m, \ x \geq x_1.$$

By the comparison theorem, we get

$$\phi(x) \geq m(x - x_1) + \phi(x_1), \ x \geq x_1,$$

which in turn implies that ϕ is unbounded. This is a contradiction. Therefore we obtain $f(l) \leq 0$.

Using the similar arguments we can get a contradiction if we assume $f(l) < 0$. Hence we get $f(l) = 0$.   □

We now present the main theorem of this section which describes the long time behavior of the solution to (2.83).

Theorem 2.5.10. *Let $f : \mathbb{R} \to \mathbb{R}$ be a Lipschitz function. Assume that ϕ is a solution to (2.83). Moreover, assume that the critical points of (2.83) are given by $\tilde{y}_1 < \tilde{y}_2 < \cdots < \tilde{y}_n$. Then the following hold true:*

(i) For $1 \le i \le n-1$, if $\tilde{y}_i < y_0 < \tilde{y}_{i+1}$ and $f(u) > 0$, $u \in (\tilde{y}_i, \tilde{y}_{i+1})$ then $\lim\limits_{x \to \infty} \phi(x) = \tilde{y}_{i+1}$.

(ii) For $1 \le i \le n-1$, if $\tilde{y}_i < y_0 < \tilde{y}_{i+1}$ and $f(u) < 0$, $u \in (\tilde{y}_i, \tilde{y}_{i+1})$ then $\lim\limits_{x \to \infty} \phi(x) = \tilde{y}_i$.

(iii) If $y_0 < \tilde{y}_1$, and $f(u) > 0$, $u < \tilde{y}_1$, then $\lim\limits_{x \to \infty} \phi(x) = \tilde{y}_1$.

(iv) If $y_0 < \tilde{y}_1$, and $f(u) < 0$, $u < \tilde{y}_1$, then $\lim\limits_{x \to \infty} \phi(x) = -\infty$.

(v) If $y_0 > \tilde{y}_n$, and $f(u) > 0$, $u > \tilde{y}_n$, then $\lim\limits_{x \to \infty} \phi(x) = \infty$.

(vi) If $y_0 > \tilde{y}_n$, and $f(u) < 0$, $u < \tilde{y}_n$, then $\lim\limits_{x \to \infty} \phi(x) = \tilde{y}_n$.

Proof. We first notice that ϕ cannot be a constant function. For, if ϕ is a constant function then that constant must be y_0. By substituting this constant solution in (2.83) we obtain that $f(y_0) = 0$, which is a contradiction.

(i) We first fix $1 \le i \le n-1$. Since $\phi(x_0) = y_0 \in (\tilde{y}_i, \tilde{y}_{i+1})$, in view of Lemma 2.5.8, ϕ attains neither $\tilde{y}_i$ nor $\tilde{y}_{i+1}$. Thus it follows that

$$\tilde{y}_i < \phi(x) < \tilde{y}_{i+1}, \ x \ge x_0.$$

Since $f(u) > 0$ whenever $u \in (\tilde{y}_i, \tilde{y}_{i+1})$, we find

$$\phi'(x) = f\big(\phi(x)\big) > 0, \ x \ge x_0.$$

Hence ϕ is increasing in (x_0, ∞). Therefore there exists $y_0 < l \le \tilde{y}_{i+1}$, such that

$$\lim_{x \to \infty} \phi(x) = l.$$

In view of Lemma 2.5.9, we must have $l = \tilde{y}_{i+1}$. This proves (i).

(ii) One can easily prove (ii) using the same arguments presented in the proof of (i). Therefore we omit the details.

(iii) Since $\phi(x_0) = y_0 < \tilde{y}_1$ and using Lemma 2.5.8, the solution ϕ cannot attain $\tilde{y}_1$. Thus it follows that

$$\phi(x) < \tilde{y}_1, \ x \ge x_0.$$

Therefore we have

$$\phi'(x) = f\big(\phi(x)\big) > 0, \ x > x_0,$$

or ϕ is increasing. Since ϕ is bounded above by y_1, there exists l with $y_0 < l \le y_1$ such that

$$\lim_{x \to \infty} \phi(x) = l.$$

Again, from Lemma 2.5.9, we should have $f(l) = 0$. This readily implies that $l = \tilde{y}_1$. This completes the proof of (iii).

(iv) By proceeding as in the proof of (iii), one can show that ϕ satisfies

$$\phi(x) < \tilde{y}_1, \ x \geq x_0.$$

Thus we have

$$\phi'(x) = f\big(\phi(x)\big) < 0, \ x > x_0.$$

In other words, ϕ is decreasing. Suppose ϕ is bounded below. Then there exists a real number $l < \tilde{y}_1$ such that $l = \inf_{x \geq x_0} \phi(x)$. Moreover, this gives us

$$\lim_{x \to \infty} \phi(x) = l.$$

From Lemma 2.5.9, we should have $f(l) = 0$. This is a contradiction. Therefore ϕ is not bounded below and $\phi(x) \to -\infty$ as $x \to \infty$. Thus (iv) is proved.

Proofs of (v) and (vi) are similar to those of (iv) and (iii), respectively. They are left to the reader as easy exercises.

This completes the proof of the theorem. $\qquad\qquad\qquad\qquad\qquad\square$

There is a simpler way to present all these results using geometry. We first draw a horizontal line called *phase line* representing the y−coordinate. We then mark all the critical points of $y'(t) = f(y(t))$ on the phase line with symbol •. Finally, we draw arrows on the phase line indicating the direction in which $y(x)$ is changing using the following rule. If f is negative in a segment (or ray), then draw left arrows in that segment (or ray). If f is positive in a segment (or ray), then draw right arrows in that segment (or ray). If we depict the graph of f against y (the same phase line), then it is easy to locate the regions where f does not change its sign. *Phase diagram* is a picture of the phase line on which the critical points and directions of motion of the solution are indicated.

We analyze the asymptotic behavior of (2.83) using the phase diagram as follows. Let y be the solution to (2.83). First identify y_0 on the phase diagram. If $f(y_0) < 0$, then $y(x)$ moves to the left of y_0 as x increases. In other words, $y(x)$ decreases as x increases. Moreover, if y_0 is between two critical points (see Figure 2.1), then $y(x)$ decreases to the smaller critical point. Similarly, if y_0 is between two critical points and $f(y_0) > 0$, then $y(x)$ increases to the larger critical point. If there are no critical points to the left of y_0 and $f(y_0) > 0$, then y increases to the smallest critical point. If there are no critical points to the right of y_0 and $f(y_0) > 0$, then y is increasing and unbounded (see Figure 2.1).

Similar analysis can be performed for the problems of the form

$$\begin{cases} y'(x) = f\big(y(x)\big), \ x < x_0, \\ y(x_0) = y_0. \end{cases} \tag{2.85}$$

The main difference is that we would like to analyze the behavior of the

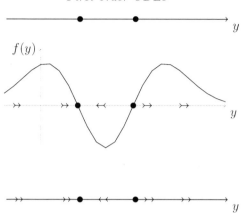

FIGURE 2.1: Top: Phase line with critical points denoted by •; Middle: A sketch of a typical f against y and the phase diagram for $y' = f(y)$. The arrows indicate the directions of y depending on the initial data y_0; Bottom: The corresponding phase diagram.

solution y as $x \to -\infty$. After understanding the proof of Theorem 2.5.10, the reader is advised to state and prove the corresponding theorem for (2.85).

A few remarks are in order before we see some applications of these results.

Remark 2.5.11. If $f : \mathbb{R} \to \mathbb{R}$ is a locally Lipschitz function with a finite number of zeros, then it follows that $(i) - (iii)$ and (vi) of Theorem 2.5.10 hold true. In fact, in this case we cannot say that the solution to (2.83) exists for each $x > x_0$. For example, the solution to $y' = y^2$, $y(0) = 1$ does not exist for $x > 1$. Thus (v) does not make sense in this case. Similarly, the solution to $y' = -y^2$, $y(0) = -1$ does not satisfy (iv).

Remark 2.5.12. If $f : \mathbb{R} \to \mathbb{R}$ is a locally Lipschitz function such that the set $\{\tilde{y} : f(\tilde{y}) = 0\}$ is infinite and discrete[7] in $\mathbb{R}$, then it follows that $(i) - (iii)$ and (vi) of Theorem 2.5.10 hold true.

For example, $f(u) = u \sin u$ is a function whose set of zeros is infinite and discrete in $\mathbb{R}$.

We now see some examples where we can find the asymptotic behavior of the solution to the ODEs without solving them.

Example 2.5.13. Discuss the asymptotic behavior of the solution to the Verhulst (or logistic) equation which models the population of a single species

$$y'(t) = \alpha y(t)(1 - \frac{y(t)}{K}), \ t > 0,$$

---

[7]An infinite set $S \subset \mathbb{R}$ is said to be discrete if there is no convergent sequence in S consisting of distinct members of it. For example $\mathbb{N}, \mathbb{Z}$ are discrete where as $\{\frac{1}{n} : n \in \mathbb{N}\}$ is not discrete.

where $\alpha > 0$ (Malthus parameter), $K > 0$ (the carrying capacity), and $y(t)$ denotes the population at time t.

Solution. As the given equation is a population model, we take the initial data to be nonnegative, i.e., $y(0) \geq 0$. Moreover, we note that the Verhulst equation is a scalar autonomous equation with the locally Lipschitz function

$$f(u) = \alpha u (1 - \frac{u}{K}), \ u \in \mathbb{R}.$$

It is easy to observe that $\tilde{y}_1 = 0$ and $\tilde{y}_2 = K$ are the critical points. Therefore if $y(0) = 0$ $(y(0) = K)$ then, due to Remark 2.5.7, the constant function $y \equiv 0$ $(y \equiv K)$ is a solution to the logistic equation.

We turn our attention to the non-constant solutions. We first notice that if $0 < u < K$, then $f(u) > 0$. Thus, in view of Remark 2.5.11 and Theorem 2.5.10(*i*), it follows that if $0 < y(0) < K$, then $y(t)$ increases with t and $y(t) \to K$ as $t \to \infty$. We next observe that $f(u) < 0$ if $u > K$. Therefore using Theorem 2.5.10(*vi*), we obtain that if $y(0) > K$, then $y(t)$ decreases with time t and $y(t) \to K$ as $t \to \infty$.

To summarize, we get that if $y(0) > 0$, then $y(t) \to K$ as $t \to \infty$. □

Example 2.5.14. Discuss the asymptotic behavior of the solution to the Allee population model

$$y'(t) = \alpha y(t) \left(1 - \frac{y(t)}{K_-}\right) \left(\frac{y(t)}{K_+} - 1\right), \ t > 0,$$

where $\alpha > 0$, $0 < K_- < K_+$.

Solution. As before, we take $y(0) \geq 0$. The Allee equation is an autonomous equation with the locally Lipschitz function

$$f(u) = \alpha u \left(1 - \frac{u}{K_-}\right) \left(\frac{u}{K_+} - 1\right).$$

It is evident that $\tilde{y}_1 = 0$, $\tilde{y}_2 = K_-$, and $\tilde{y}_3 = K_+$ are the critical points. If $y(0)$ is any of these critical points, then the solution is the constant function given by that critical point.

We now focus our attention on the non-constant solutions. As in the previous example, we use Remark (2.5.11). It is straightforward to observe that

$$f(u) \begin{cases} < 0, \ 0 < u < K_-, \\ > 0, \ K_- < u < K_+, \\ < 0, u > K_+. \end{cases}$$

From Theorem 2.5.10(*i*),(*ii*),(*vi*) the following conclusions can be drawn:

(*i*) If $0 < y(0) < K_-$, then $y(t)$ decreases to 0 as $t \to \infty$.

(*ii*) If $K_- < y(0) < K_+$, then $y(t)$ increases to K_+ as $t \to \infty$.

(*iii*) If $y(0) > K_+$, then $y(t)$ decreases to K_+ as $t \to \infty$.

This completes the solution. □

Example 2.5.15. Discuss the asymptotic behavior of the solution to

$$y'(x) = \sin\left(y(x)\right), \ x > 0.$$

Solution. The given ODE is an autonomous equation and $f(u) = \sin u$, is a Lipschitz function. Let ϕ be a solution to the given ODE. It is straightforward to get that $\tilde{y}_k = k\pi$, $k \in \mathbb{Z}$ are the critical points of the given ODE. If $\phi(0)$ is any of these critical points, then the solution is the constant function given by that critical point.

Regarding the non-constant solutions, from Theorem 2.5.10$(i),(ii)$ the following conclusions can be drawn:

(i) For $n \in \mathbb{Z}$, if $2n\pi < \phi(0) < (2n+1)\pi$, then $\phi(x) \to (2n+1)\pi$, as $x \to \infty$.

(ii) For $n \in \mathbb{Z}$, if $(2n+1)\pi < \phi(0) < (2n+2)\pi$, then $\phi(x) \to (2n+1)\pi$, as $x \to \infty$.

Thus we get $\phi(x) \to (2n+1)\pi$, as $x \to \infty$, whenever $2n\pi < \phi(0) < (2n+2)\pi$, $n \in \mathbb{Z}$. This completes the solution. $\quad\square$

We now present a couple of examples in which the comparison theorems are also needed apart from Theorem 2.5.10 to study the asymptotic behavior.

Example 2.5.16. Let f be a C^1-function satisfying

$$1 - u^2 < f(t, u) < 2(1 - u^2), \ t \geq 0, \ u \in \mathbb{R}. \tag{2.86}$$

Consider the IVP

$$\begin{cases} y'(x) = f(x, y(x)), \ x > x_0, \\ y(x_0) = y_0. \end{cases} \tag{2.87}$$

Find y_0 such that (2.87) has a non-constant, bounded solution in (x_0, ∞).

Solution. Let ϕ denote the solution to (2.87) in $[0, \infty)$. Since f satisfies (2.87), it is natural to consider the following initial value problems

$$\begin{cases} \alpha'(x) = 1 - \alpha^2(x), \\ \alpha(x_0) = y_0, \end{cases} \quad \begin{cases} \beta'(x) = 2\left(1 - \beta^2(x)\right), \\ \beta(x_0) = y_0, \end{cases}$$

and use the comparison theorems. We first, observe that $\tilde{y}_1 = -1$ and $\tilde{y}_2 = 1$ are the critical points of both the IVPs. If $y_0 \in (-1, 1)$, then in view of Theorem 2.5.10(ii), we obtain that both α and β are increasing. Moreover we have

$$\alpha(x), \beta(x) \in (-1, 1), \ x > x_0.$$

From the comparison theorems discussed in Section 2.4, it follows that

$$\alpha(x) < \phi(x) < \beta(x), \ x > x_0.$$

Therefore if $y_0 \in (-1, 1)$, then ϕ is non-constant and

$$\phi(x) \in (-1, 1), \ x > x_0.$$

This completes the proof. $\quad\square$

The reader is advised to find the set of all values that y_0 can take in the previous example so that the corresponding IVP is has a non-constant bounded solution.

Example 2.5.17. Let f be a C^1-function satisfying

$$2(1 - u^2) < f(t, u) < 2 - u^2.$$

Consider

$$\begin{cases} y'(x) = f(x, y(x)), & x > x_0, \\ y(x_0) = y_0. \end{cases} \tag{2.88}$$

Find y_0 such that (2.88) has a non-constant, bounded solution in (x_0, ∞).

Solution. Let ϕ denote the solution to (2.88) in $[0, \infty)$. As before, we consider two initial value problems

$$\begin{cases} \alpha'(x) = 2(1 - \alpha^2(x)), \\ \alpha(x_0) = y_0, \end{cases} \tag{2.89}$$

and

$$\begin{cases} \beta'(x) = 2 - \beta^2(x), \\ \beta(x_0) = y_0. \end{cases} \tag{2.90}$$

We observe that $\tilde{y}_1 = -1$ and $\tilde{y}_2 = 1$, are the critical points of (2.89) whereas $\tilde{y}_3 = -\sqrt{2}$ and $\tilde{y}_4 = \sqrt{2}$, are the critical points of (2.90).
If $y_0 \in (1, \sqrt{2})$, then in view of Theorem 2.5.10(vi) we get that α is decreasing and

$$1 < \alpha(x) < y_0 < \sqrt{2}, \ x > x_0.$$

Similarly, from Theorem 2.5.10(i), we obtain that β is increasing and

$$1 < y_0 < \beta(x) < \sqrt{2}, \ x > x_0.$$

Therefore we have

$$\alpha(x), \beta(x) \in (1, \sqrt{2}), \ x > x_0.$$

Due to the comparison theorems it follows that

$$\alpha(x) < \phi(x) < \beta(x), \ x > x_0.$$

Hence if $y_0 \in (1, \sqrt{2})$, then ϕ is non-constant and $1 < \phi(x) < \sqrt{2}, \ x \geq x_0$. $\square$

Exercise 2.1. Find the solutions to the following differential equations:

(i) $y'(x) = x^n y^m(x), \ n, m \in \mathbb{Z}.$

(ii) $y'(x) = \dfrac{e^{x+y(x)}}{1 + e^{y(x)}}.$

(iii) $y'(x) = \dfrac{y(x) + xe^{\frac{ky(x)}{x}}}{x}$, $k \in \mathbb{R}$.

(iv) $y'(x) = \dfrac{x^2 + 2xy(x) + 3y^2(x)}{x^2}$.

(v) $y'(x) = \dfrac{7x + 5y(x) - 17}{x + y(x) - 3}$.

(vi) $y'(x) = \dfrac{x + y(x) + 7}{x - y(x) + 8}$.

(vii) $y'(x) = \dfrac{3x + 10y(x) + 16}{6x + 20y(x) + 8}$.

$(viii)$ $y'(x) = \dfrac{x + y(x) + 7}{x + y(x) + 8}$.

(ix) $\left(x + x^2 + 2xy(x)\right)y'(x) + \left(y(x) + y^2(x) + 2xy(x)\right) = 0$.

(x) $\left(x\sec^2\left(xy(x)\right) + \sec(x) + \cos\left(y(x)\right)\right)y'(x) + y(x)\left(\sec^2\left(xy(x)\right) + \tan(x)\sec(x)\right) = 0$.

(xi) $y'(x) - \dfrac{xy(x)}{1 + x^2} = 1 + x^2$.

(xii) $y'(x) - \dfrac{2y(x)}{x} = e^{-3x}$.

$(xiii)$ $y'(x) + \dfrac{y(x)}{x} = xy^3(x)$.

(xiv) $y'(x) + xy(x) = 3xy^2(x)$.

Exercise 2.2. Find k such that $\dfrac{xy'(x)}{(x^2 + y^2(x))^k} - \dfrac{y}{(x^2 + y^2(x))^k} = 0$, $x > 1$, is exact.

Exercise 2.3. Solve $xy'(x) = y(x) + \dfrac{y^2(x) - x^2}{x^2 + 1}$.

Hint: Put $z(x) = \frac{y(x) - x}{y(x) + x}$.

Exercise 2.4. Let $g_1, g_2 : \mathbb{R} \to \mathbb{R}$ be continuous functions. By using appropriate substitution solve the following equations.

(i) $y^n(x)g_1(x) + g_2\left(\dfrac{y(x)}{x}\right)\left(y(x) - xy'(x)\right) = 0$.

(ii) $y'(x) = x^{n-1}g_1\left(\dfrac{y(x)}{x^n}\right)$.

Hint: For (i) put $z(x) = \frac{y(x)}{x}$ and for (ii) put $z(x) = \frac{y(x)}{x^n}$.

Exercise 2.5. A nonzero continuous function μ is said to be an *integrating factor* of

$$N\big(x, y(x)\big)y'(x) + M(x, y(x)) = 0, \tag{2.91}$$

if

$$\mu\big(x, y(x)\big)N\big(x, y(x)\big)y'(x) + \mu\big(x, y(x)\big)M(x, y(x)) = 0,$$

is exact. Assume that M, N are C^1-functions and answer the following questions.

(*i*) A C^1-function μ is an integrating factor of (2.91) if and only if μ is a solution to

$$N\frac{\partial \mu}{\partial x} - M\frac{\partial \mu}{\partial y} = \left(\frac{\partial M}{\partial y} - \frac{\partial N}{\partial x}\right)\mu.$$

(*ii*) An integrating factor of (2.91) is a function of x alone if and only if the term $\dfrac{1}{N}\left(\dfrac{\partial M}{\partial y} - \dfrac{\partial N}{\partial x}\right)$ is a function of x alone. Find the integrating factor in this case.

(*iii*) An integrating factor of (2.91) is a function of y alone if and only if the term $\dfrac{1}{M}\left(\dfrac{\partial M}{\partial y} - \dfrac{\partial N}{\partial x}\right)$ is a function of y alone. Find the integrating factor in this case.

Exercise 2.6. Solve $y'(x) = \max\{x, y(x)\}$, $x > 0$, subject to $y(0) = 0$.

Exercise 2.7. Let $f_1, f_2, f_3 : \mathbb{R} \to \mathbb{R}$ be continuous functions. An equation of the form

$$y'(x) + f_1(x)y^2(x) + f_2(x)y(x) + f_3(x) = 0 \tag{2.92}$$

is said to be a Riccati equation. If y_1 and y_2 are solutions to the above Riccati equation, then show that $z = y_1 - y_2$ satisfies a Bernoulli equation. Solve that Bernoulli equation.

Exercise 2.8. Suppose there are four distinct solutions y, y_1, y_2 and y_3 to (2.92), then show that their cross ratio $\dfrac{(y - y_2)(y_1 - y_3)}{(y_1 - y_2)(y - y_3)}$ is a constant.

Exercise 2.9. Suppose there exist some nonzero constants a, b, c such that $f_1(x) : f_2(x) : f_3(x) = a : b : c$, $x \in \mathbb{R}$, then solve (2.92).

Exercise 2.10. Show that three solutions with $y_1 > y_2 > y_3$ uniquely determine the coefficients in the Riccati equation (2.92).

Exercise 2.11. Let g be a C^1-function. In order to solve equations of the form

$$y(x) = g(x, y'(x)),$$

we first convert the given ODE to the standard form by differentiating it with respect to x. Thus we have

$$y'(x) = \frac{\partial g}{\partial x}(x, y'(x)) + \frac{\partial g}{\partial y'}(x, y'(x))y''(x).$$

We now put $p(x) = y'(x)$ to get an ODE in the form

$$p(x) = \frac{\partial g}{\partial x}(x, p(x)) + \frac{\partial g}{\partial y'}(x, p(x))p'(x).$$

Apply this method to show that Clairaut's equation

$$y(x) = xy'(x) + h(y'(x)),$$

where h is a C^1-function, has a solution of the form $\phi(x) = cx + h(c)$, $c \in \mathbb{R}$.

Exercise 2.12. Let $R = \{(x, y) : |x - x_0| \le a, \ |y - y_0| \le b\}$ be a rectangle in $\mathbb{R}^2$. Assume that $f : R \to \mathbb{R}$ is a differentiable function such that $\left|\frac{\partial f}{\partial y}\right|$ is bounded. Then show that f satisfies (2.40). Can we extend this result to the case of an infinite rectangle?

Exercise 2.13. Verify whether the following functions satisfy (2.40) in the prescribed regions.

(i) $f_1(x, y) = \alpha(x)q_1(\cos y) + \beta(x)q_2(\sin y)$, in $|x| < 1$, $y \in \mathbb{R}$, where α, β are continuous and q_1, q_2 are polynomials.

(ii) $f_2(x, y) = \alpha(x) + \beta(x)y$, in $|x| \le 1$, $y \in \mathbb{R}$, where α, β are continuous functions.

(iii) $f_3(x, y) = p(x)e^{-x^2 y}$, in $|x| \le 1$, $y > 0$, where α, β are continuous functions and p is a polynomial.

(iv) $f_4(x, y) = \alpha(x)q_1(\cos y) + \beta(x)q_2(\sin y)$, in $(x, y) \in \mathbb{R}^2$, where α, β are bounded functions and q_1, q_2 are polynomials.

(v) $f_5(x, y) = \alpha(x) + \beta(x)y$, in $(x, y) \in \mathbb{R}^2$, where α, β are bounded functions.

(vi) $f_6(x, y) = p(y)e^{-x^2 y^2}$, in $|x| \le 1$, $y \in \mathbb{R}$, where α, β are continuous functions and p is a polynomial.

Exercise 2.14. Give a pair of continuous functions $f, g : [0, \infty) \times \mathbb{R} \to \mathbb{R}$ such that

(i) $f(x, y_1) - f(x, y_2) = g(x, y_1) - g(x, y_2)$, $x \ge 0$, $y_1, y_2 \in \mathbb{R}$,

(ii) the initial value problem $y'(x) = f(x, y(x))$, $y(0) = 0$ has a unique solution and

(iii) the initial value problem $y'(x) = g(x, y(x))$, $y(0) = 0$ has more than one solution.

Hint: Take $f(x, y) = (1 - x)(1 + \sqrt{|y|})$, $g(x, y) = (1 - x)\sqrt{|y|}$.

Exercise 2.15. Find the first four terms of the sequence of Picard's successive approximation for the following problems:

(i) $y'(x) = -2xy(x)$, $y(0) = 1$.

(ii) $y'(x) = 1 - xy(x)$, $y(0) = 1$.

(iii) $y'(x) = y^2(x)$, $y(0) = 5$.

(iv) $y'(x) = 1 + y^2(x)$, $y(0) = 2$.

Exercise 2.16. Consider the sequence (z_n) defined by $z_0(x) = 0$ and

$$z_n(x) = 1 + \int_0^x z_{n-1}^2(t)dt, \ n \geq 1.$$

Show that z_n is a polynomial of degree $2^{n-1} - 1$. Does (z_n) converge? If so, what is the limit of it?

Exercise 2.17. Consider the function $f : [-1, 1] \times [-1, 1] \to \mathbb{R}$ given by

$$f(x,y) = \begin{cases} 2x, & \text{for } x \in [-1,1], \ y \in [-1,0], \\ 2x - \dfrac{4y}{x}, & \text{for } x \in [-1,1], \ 0 < y < x^2, \\ -2x, & \text{for } x \in [-1,1], \ x^2 \leq y \leq 1, \end{cases}$$

and the initial value problem (IVP) $y'(x) = f(x, y(x))$, $y(0) = 0$. Does this f satisfy the Lipschitz condition? Find the Picard sequence of successive approximation (z_n) corresponding to this f. Does it converge to a solution to the given IVP? If not, does there exist a subsequence of (z_n) which converges to a solution to the given IVP?

Exercise 2.18. Find the intervals containing $x = 0$ which is determined by Theorem 2.2.11 for the initial value problem $y'(x) = f(x, y(x))$, $y(0) = 0$, where f is given by

(i) $f : [-1, 1] \times [-1, 1] \to \mathbb{R}$ is given by $f(x, y) = x^2 + y^2$.

(ii) $f : [-\frac{1}{2}, \frac{1}{2}] \times [-1, 1] \to \mathbb{R}$ is given by $f(x, y) = 1 + y^2$.

Exercise 2.19. Let $f : \mathbb{R}^3 \to \mathbb{R}$ be a C^2-function with $f(x_0, y_0, z_0) = 0$ and $\frac{\partial f}{\partial z}(x_0, y_0, z_0) \neq 0$. Show that there exist an interval $[x_0 - \delta, x_0 + \delta]$ and a unique solution $y : [x_0 - \delta, x_0 + \delta] \to \mathbb{R}$ to the problem $f(x, y(x), y'(x)) = 0$, $y(x_0) = y_0$, $y'(x_0) = z_0$.

Hint: Use the implicit function theorem.

Exercise 2.20. Discuss the existence and uniqueness of solutions to the following problems for different values of y_0:

(i) $\begin{cases} xy'(x) = 10y(x), \ x \in \mathbb{R}, \\ y(0) = y_0. \end{cases}$
(ii) $\begin{cases} y'(x) = |y(x)|^\alpha, \ x > 0, \ \alpha \in (0, 1), \\ y(0) = y_0. \end{cases}$

Exercise 2.21. Let $(x_0, y_0) \in \mathbb{R}^2$, $a, b > 0$, and $R = \{(x, y) \in \mathbb{R}^2 : |x - x_0| \leq a, |y - y_0| \leq b\}$. Assume that $f : R \to \mathbb{R}$ is a continuous function. Suppose there exist two solutions to the Cauchy problem

$$\begin{cases} y' = f(x, y), \ |x - x_0| < a, \\ y(x_0) = y_0. \end{cases}$$

Then show that the Cauchy problem has infinitely many solutions.

Hint: Suppose ϕ_1, ϕ_2 are two solutions to the given Cauchy problem and there exists $c \in (x_0 - a, x_0 + a)$ such that $\phi_1(c) < \phi_2(c)$. If $\phi_1(c) < m < \phi_2(c)$, then show that there exists a solution ϕ to the given Cauchy problem such that $\phi(c) = m$.

Exercise 2.22 (Global existence). Suppose $f : \mathbb{R}^2 \to \mathbb{R}$ is a continuous function. Assume that for every $a > 0$, there exists $L > 0$ (possibly depending on a) such that

$$|f(x, y_1) - f(x, y_2)| \leq L|y_1 - y_2|, \ |x| \leq a, \ y \in \mathbb{R}.$$

Then show that for every (x_0, y_0), there exists a unique solution $\phi : \mathbb{R} \to \mathbb{R}$ to

$$\begin{cases} y'(x) = f(x, y(x)), \ x \in \mathbb{R}, \\ y(x_0) = y_0. \end{cases}$$

Exercise 2.23. Discuss the global existence of solution to the following problems with $y(0) = y_0$.

(i) $y'(x) = \dfrac{\sin^2 y(x)}{1 - x^2}$, $x \in (-1, 1)$.

(ii) $y'(x) = e^{-x^2 y^2(x)} + \dfrac{\sin(y(x))}{1 + x^2}$. $x \in (-1, 1)$.

(iii) $y'(x) = \alpha(x)p(\sin(y(x)))$, $x \in \mathbb{R}$, where α is a bounded continuous function and p is a polynomial.

Hint: Use Exercise 2.22.

Exercise 2.24. Let $F_1, F_2 : [a, b] \times \mathbb{R} \to \mathbb{R}$ be continuous functions. Assume that F_i satisfies

$$|F_i(x, y_1) - F_i(x, y_2)| \leq 10|y_1 - y_2|, \ x \in [a, b], \ y_1, y_2 \in \mathbb{R},$$

for some $i = 1, 2$. Moreover assume that $F_1(x, y) \leq F_2(x, y)$, $x \in [a, b]$, $y \in \mathbb{R}$. Let α and β be solutions to $\alpha'(x) = F_1(x, \alpha(x))$ and $\beta'(x) = F_2(x, \beta(x))$ in $x \in (a, b)$, respectively. If $\alpha(a) = \beta(a)$, then show that $\alpha(x) \leq \beta(x)$, $x \in [a, b]$.

Exercise 2.25. Find the asymptotic behavior of the solutions to the following equations for various values of $y(0)$:

(i) $y'(x) = (y^2(x) - 1)(y^2(x) - 4)$, $y(0) \in [-2, 2]$.

(ii) $y'(x) = 1 + \cos(y(x))$, $y(0) \in \mathbb{R}$.

(iii) $y'(x) = \sin(2y(x))$, $y(0) \in \mathbb{R}$.

Exercise 2.26. Suppose $f : \mathbb{R} \to \mathbb{R}$ is a C^1-function with a bounded derivative. If ϕ is a solution to $y'(x) = f(y(x))$ such that $\phi(x_1) = \phi(x_2)$ for some $x_1 \neq x_2$, then show that ϕ is a constant function.

Exercise 2.27. Find an unbounded, locally Lipschitz function $f : \mathbb{R} \to \mathbb{R}$ such that every solution to $y'(x) = f(y(x))$, $x \in \mathbb{R}$, is a global solution which is bounded.

Chapter 3

Higher order linear ODEs

In this chapter, we provide some methods to find the solutions to both homogeneous and non-homogeneous linear ODEs in a closed form[1]. This is possible only when the given ODE has a particular structure. For instance, one can find a solution in a closed form for any linear homogeneous ODE with constant coefficients whose order is less than or equal to four. After discussing the methods to solve linear ODEs with constant coefficients in detail, we turn our attention toward the linear ODEs with variable coefficients. In particular, we investigate the dimension of the solution space of linear ODEs and study the properties of the Wronskian of the solutions. Moreover, the oscillatory behavior of the solutions to the second order linear ODEs is discussed. Finally, the non-homogeneous linear ODEs are solved explicitly in terms of the the solutions to the corresponding homogeneous ODEs using the method of separation of variables.

3.1   ODEs with constant coefficients

For $k \in \mathbb{N}$, we denote by $y^{(k)}$ the k-th order derivative of y. In this section, we briefly review some methods to solve the homogeneous and non-homogeneous linear equations with constant coefficients of the following form

$$a_n y^{(n)}(x) + \cdots + a_1 y'(x) + a_0 y(x) = 0, \ a_n \neq 0, \ x \in \mathbb{R}, \qquad (3.1)$$

and

$$a_n y^{(n)}(x) + \cdots + a_1 y'(x) + a_0 y(x) = f(x), \ x \in \mathbb{R}, \qquad (3.2)$$

respectively, where $a_i \in \mathbb{R}$, $0 \leq i \leq n$ and f has some special structure which will be described later. We use the word linear because if ϕ_1 and ϕ_2 are solutions to (3.1) then $c_1 \phi_1 + c_2 \phi_2$ is also a solution to (3.1) for all $c_1, c_2 \in \mathbb{R}$.

---

[1] Exponential functions, trigonometric functions, polynomials, their finite products, sums, reciprocals, inverses, and compositions are called functions in a closed form. For example $\frac{e^x + 2 \log x}{(x^2+1)(2+\sin(1+x^3))}$ is a function in a closed form. For more details about this concept, the reader can refer to [30]

We denote by $D^k = \frac{d^k}{dx^k}$, $k \in \mathbb{N}$. Then (3.1) can be written as

$$a_n D^n y + \cdots + a_1 D y + a_0 y = 0, \ a_n \neq 0, \ x \in \mathbb{R}.$$

Then the polynomial $P(X)$ associated with the differential operator given in (3.1) is given by $P(X) = a_n X^n + \cdots + a_1 X + a_0$. Moreover we write (3.1) as $P(D)y = 0$.

3.1.1 Factorization of differential operators: homogeneous case

In this subsection, we find the solution to the ODE $P(D)y = 0$ in terms of the roots of the polynomial P. In order to do that, we first recall the following basic results.

Result 1. If all the coefficients of the non-constant polynomial P are real and $z \in \mathbb{C}$ is a root of P, then $\bar{z}$ is also a root of P. Hence there exist polynomials $Q_1, \ldots, Q_k$ such that degree of each Q_i, $1 \leq i \leq k$ is either one or two and

$$P(X) = Q_1(X)Q_2(X)\cdots Q_k(X).$$

Result 2. Let $T_i : V \to V$, $1 \leq i \leq k$ be linear operators defined on a real vector space V and $T_i T_j = T_j T_i$, $1 \leq i, j \leq k$. Let $K_i = \{v \in V : T_i(v) = 0\}$, $1 \leq i \leq k$. Then

$$\left\{ \sum_{i=1}^{k} c_i v_i : c_i \in \mathbb{R}, v_i \in K_i \right\} \subseteq \{v \in V : T_1 T_2 \cdots T_k(v) = 0\}.$$

We use these two results to find solutions to the differential equations with constant coefficients. We begin with the following useful result which is motivated by Result 2.

Lemma 3.1.1. *Assume that $I : C(\mathbb{R}) \to C(\mathbb{R})$ denotes the identity function. Consider the ODE*

$$(D - \alpha_1 I)(D - \alpha_2 I)\cdots(D - \alpha_n I)y = 0, \ x \in \mathbb{R}, \tag{3.3}$$

where $\alpha_i \in \mathbb{R}$ are distinct real numbers for $1 \leq i \leq n$. Then

$$\phi(x) = c_1 e^{\alpha_1 x} + \cdots + c_n e^{\alpha_n x}, \ x \in \mathbb{R}, \tag{3.4}$$

where $c_j \in \mathbb{R}$, $1 \leq j \leq n$ is a solution to (3.3).

Proof. We first set

$$P(D)y := (D - \alpha_1 I)(D - \alpha_2 I)\cdots(D - \alpha_n I)y, \ y \in C^n(\mathbb{R}).$$

Let $T_i = D - \alpha_i I$, $1 \leq i, j \leq n$. Then notice that each T_j is a linear operator and $T_i T_j = T_j T_i$, $1 \leq i, j \leq n$. Furthermore, we find that $\phi_j = e^{\alpha_j x}$, is a solution to $(D - \alpha_j I)\phi_j = 0$, $1 \leq j \leq n$ or $T_j(e^{\alpha_j x}) = 0$. In view of Result 2, it follows that ϕ given in (3.4) is a solution to (3.3). $\qquad\square$

Remark 3.1.2. A restatement of Lemma 3.1.1 is the following: Suppose $P(X)$ is an n-th degree polynomial with distinct real roots $\alpha_1,\ldots,\alpha_n$. Then ϕ given in (3.4) is a solution to the ODE $P(D)y = 0$.

A generalization of Lemma 3.1.1 is given in the following lemma.

Lemma 3.1.3. *Assume that $Q_1,\ldots,Q_n$ are polynomials and P is the product of $Q_1,\ldots,Q_n$. Suppose ϕ_j is a solution to the ODE $Q_j(D)y = 0$, $1 \leq j \leq n$. Then*

$$\phi(x) = c_1\phi_1(x) + \cdots + c_n\phi_n(x), \quad x \in \mathbb{R}, \tag{3.5}$$

where $c_j \in \mathbb{R}$, $1 \leq j \leq n$ is a solution to $P(D)y = 0$.

Proof. The proof is along the same lines of the proof of Lemma 3.1.1. On setting $T_j = Q_j(D)$, observing that each T_j is linear, T_i and T_j commute, and using Result 2, we obtain that ϕ given in (3.5) is a solution to $P(D)y = 0$. □

We now present some worked out examples where we use the results that we have proved so far.

Example 3.1.4. Solve $(D^2 - \alpha^2 I)y = 0$, $\alpha \in \mathbb{R}$, $x \in \mathbb{R}$.

Solution: The polynomial associated with the given ODE is $P(X) = X^2 - \alpha^2$. The roots of P are given by $\pm\alpha$. Hence from Remark 3.1.2, the function

$$\phi(x) = c_1 e^{\alpha x} + c_2 e^{-\alpha x}, \quad x \in \mathbb{R},$$

where $c_1, c_2 \in \mathbb{R}$ is a solution to the given ODE. □

Example 3.1.5. Solve $(D^2 - 9D + 20I)y = 0$, $x \in \mathbb{R}$.

Solution: The polynomial associated with the given ODE is $P(X) = X^2 - 9X + 20$. The roots of P are $\alpha_1 = 4$ and $\alpha_2 = 5$. Thus

$$\phi(x) = c_1 e^{4x} + c_2 e^{5x}, \quad x \in \mathbb{R}.$$

where $c_1, c_2 \in \mathbb{R}$ is a solution to the given ODE. □

Example 3.1.6. Solve $(D^2 + \alpha^2 I)y = 0$, $\alpha \in \mathbb{R}$, $x \in \mathbb{R}$.

Solution: The method which is used in the previous examples cannot be applied to find a solution to this equation because there are no real roots of the polynomial $P(X) = X^2 + \alpha^2$. By inspection we readily observe that both $\sin(\alpha x)$ and $\cos(\alpha x)$ are solutions to $(D^2 + \alpha^2 I)y = 0$. Hence we find that

$$\phi(x) = c_1 \cos(\alpha x) + c_2 \sin(\alpha x), \quad x \in \mathbb{R}, \ c_1, c_2 \in \mathbb{R}$$

is a solution to $(D^2 + \alpha^2 I)y = 0$. □

Example 3.1.7. If $a^2 < 4b$, then solve $(D^2 + aD + bI)y = 0$, $x \in \mathbb{R}$.

Solution: The polynomial corresponding to the given equation is $P(X) = X^2 + aX + b$, whose roots are not real numbers. We now use the following ansatz to find the solution by converting it to an equation which does not contain the first order derivative of the unknown. To this end, we set

$$\phi(x) = e^{-ax/2}v(x).$$

Then a straightforward computation gives us

$$\phi'(x) = e^{-ax/2}\left(v'(x) - \tfrac{a}{2}v(x)\right),$$

and

$$\phi''(x) = e^{-ax/2}\left(v''(x) - av'(x) + \tfrac{a^2}{4}v(x)\right).$$

Hence we have an identity

$$(D^2 + aD + bI)e^{-ax/2}v(x) = e^{-ax/2}\left(D^2 + b - \frac{a^2}{4}\right)v(x), \quad x \in \mathbb{R}. \qquad (3.6)$$

Since $e^{-ax/2} \neq 0$, from (3.6) it follows that $\phi = e^{-ax/2}v(x)$ is a solution to the given equation if and only if

$$\left(D^2 + b - \tfrac{a^2}{4}\right)v(x) = 0.$$

From Example 3.1.6, we get

$$v(x) = c_1 \cos\left(\sqrt{4b - a^2}\ \frac{x}{2}\right) + c_2 \sin\left(\sqrt{4b - a^2}\ \frac{x}{2}\right).$$

Hence a solution to the given equation is

$$\phi(x) = e^{-ax/2}\left(c_1 \cos\left(\sqrt{4b - a^2}\ \frac{x}{2}\right) + c_2 \sin\left(\sqrt{4b - a^2}\ \frac{x}{2}\right)\right), \quad x \in \mathbb{R}, \quad (3.7)$$

where $c_1, c_2 \in \mathbb{R}$. □

Remark 3.1.8. Consider the polynomial $P(X) = X^2 + aX + b$ corresponding to the ODE given in Example 3.1.7. Observe that the roots of P are given by $\alpha_\pm = -\dfrac{a}{2} \pm i\dfrac{\sqrt{4b - a^2}}{2}$. Let $\mathrm{Re}(\alpha_+)$, and $\mathrm{Im}(\alpha_+)$ denote the real and imaginary parts of α_+, respectively. Then the formula given in (3.7) can be written as

$$\phi(x) = e^{\mathrm{Re}(\alpha_+)x}\left(c_1 \cos\left(\mathrm{Im}(\alpha_+)x\right) + c_2 \sin\left(\mathrm{Im}(\alpha_+)x\right)\right), \quad x \in \mathbb{R}, \qquad (3.8)$$

where $c_1, c_2 \in \mathbb{R}$.

Example 3.1.9. Solve $(D^4 - 16I)y = 0$, $x \in \mathbb{R}$.

Solution: We first factorize the given differential operator (or the associated polynomial) to get

$$(D - 2I)(D + 2I)(D^2 + 4I)y = 0.$$

Using Lemma 3.1.3, a solution to the given equation is

$$\phi(x) = c_1\phi_1(x) + c_2\phi_2(x) + c_3\phi_3(x),$$

where
$$(D - 2I)\phi_1 = 0, \ (D + 2I)\phi_2 = 0, \ (D^2 + 4I)\phi_3 = 0$$

In view of Lemma 3.1.1 and Example 3.1.6, we obtain
$$\phi_1(x) = e^{2x}, \ \phi_2(x) = e^{-2x}, \ \phi_3(x) = A\cos(2x) + B\sin(2x), \ x \in \mathbb{R},$$

where A, B are arbitrary real constants. Therefore a solution to the given equation can be written as
$$\phi(x) = C_1 e^{2x} + C_2 e^{-2x} + C_3 \cos(2x) + C_4 \sin(2x), \ x \in \mathbb{R},$$

where C_i's, $1 \leq i \leq 4$, are arbitrary real numbers. $\qquad\square$

Example 3.1.10. Solve $(D^4 - 3D^3 - D^2 + 13D - 10I)y = 0, \ x \in \mathbb{R}$.

Solution: We first factorize the polynomial associated to the given equation as
$$P(X) = (X - 1)(X + 2)(X^2 - 4X + 5).$$

The given ODE can be written as
$$(D - I)(D + 2I)(D^2 - 4D + 5I)y = 0.$$

Since roots of $X^2 - 4X + 5$ are $2\pm i$, from Remark 3.1.8 a solution to $(D^2 - 4D + 5I)y = 0$ is
$$\psi(x) = e^{2x}(A\cos x + B\sin x), \ x \in \mathbb{R}, \ A, B \in \mathbb{R}.$$

Owing to Lemmas 3.1.1 and 3.1.3, a solution to the given equation is
$$\phi(x) = c_1 e^x + c_2 e^{-2x} + e^{2x}(c_3 \cos x + c_4 \sin x), \ x \in \mathbb{R},$$

where c_1, c_2, c_3, and c_4 are arbitrary real constants. $\qquad\square$

Suppose the polynomial associated with the given n-th order homogeneous ODE with constant coefficients has a repeated root. Then we show that there exists another solution to the ODE which is not of the form $e^{\alpha x}\cos(\beta x)$ or $e^{\alpha x}\sin(\beta x)$ for any $\alpha, \beta \in \mathbb{R}$. For let P be a polynomial with real coefficients and $\alpha \in \mathbb{R}$. Then it is straightforward to observe that the following identity holds:
$$P(D)e^{\alpha x} = P(\alpha)e^{\alpha x}, \ \alpha \in \mathbb{R}. \tag{3.9}$$

We now focus on the case where we have repeated roots for the polynomial P.

Lemma 3.1.11. *Assume that $\alpha_k \in \mathbb{R}$ is a root of multiplicity m for a polynomial P with real coefficients. Then the functions*
$$e^{\alpha_k x}, \ xe^{\alpha_k x}, \ x^2 e^{\alpha_k x}, \ldots, x^{m-1}e^{\alpha_k x},$$

are solutions to $P(D)y = 0$.

Proof. We first assume that α_k is a root of P with multiplicity two, i.e., $(X - \alpha_k)^2$ divides $P(X)$. Then we readily have that $P(\alpha_k) = 0$ and $P'(\alpha_k) = 0$. On differentiating the identity in (3.9) with respect to α, we get another identity

$$\frac{d}{d\alpha}P(D)e^{\alpha x} = (P'(\alpha) + xP(\alpha))\,e^{\alpha x}.$$

Since $e^{\alpha x}$ is an infinite differentiable function of two variables, viz., α and x, we have

$$P(D)\frac{d}{d\alpha}e^{\alpha x} = (P'(\alpha) + xP(\alpha))\,e^{\alpha x}, \quad \alpha \in \mathbb{R}. \tag{3.10}$$

We put $\alpha = \alpha_k$ in (3.10) and use the fact that $P(\alpha_k) = 0$ and $P'(\alpha_k) = 0$ to obtain

$$P(D)\,(xe^{\alpha_k x}) = 0.$$

Thus we have proved that if α_k is a root of $P(X)$ with multiplicity two then $\phi(x) = xe^{\alpha_k x}$ is also a solution to $P(D)y = 0$. One can easily extend this argument to the case when $P(X)$ has a root α_k whose multiplicity is m. The details are left to the reader as an easy exercise. □

We now turn our attention toward the case when we have repeated complex roots for $P(X)$.

Lemma 3.1.12. *Let $(X^2 + aX + b)^k$, $k \in \mathbb{N}$ with $a^2 < 4b$, be a factor of an n-th degree polynomial $P(X)$. Then*

$$\phi(x) = e^{-ax/2}\left((c_1 + c_2 x + \cdots + c_k x^{k-1})\cos\left(\frac{\sqrt{4b - a^2}}{2}x\right)\right.$$

$$\left. + (c_{k+1} + c_{k+2}x + \cdots + c_{2k}x^{k-1})\sin\left(\frac{\sqrt{4b - a^2}}{2}x\right)\right), \quad x \in \mathbb{R}, \tag{3.11}$$

where $c_m \in \mathbb{R}$, $1 \le m \le 2k$, is a solution to $P(D)y = 0$, $x \in \mathbb{R}$.

Proof. Since $X^2 + aX + b$ is a factor of $P(X)$, from Example 3.1.7 a class of solutions to $P(D)y = 0$ is given by (3.7). To prove that there are solutions to $P(D)y = 0$ other than the ones given in (3.7), we set $v(x) = \sin(\beta x)$, $\beta \in \mathbb{R}$, in identity (3.6). Then we have

$$(D^2 + aD + bI)e^{-ax/2}\sin(\beta x) = e^{-ax/2}\left(-\beta^2 + b - \frac{a^2}{4}\right)\sin(\beta x).$$

Applying the operator $(D^2 + aD + bI)$ on both sides of the previous identity, we get

$$(D^2 + aD + bI)^2 e^{-ax/2}\sin(\beta x) = e^{-ax/2}\left(-\beta^2 + b - \frac{a^2}{4}\right)^2\sin(\beta x). \tag{3.12}$$

On differentiating (3.12) with respect to β and putting $\beta = \dfrac{\sqrt{4b - a^2}}{2}$, we obtain

$$(D^2 + aD + bI)^2\left(e^{-ax/2}x\sin\left(\frac{\sqrt{4b - a^2}}{2}x\right)\right) = 0, \quad x \in \mathbb{R}. \tag{3.13}$$

This proves that

$$\phi_1(x) = xe^{-ax/2} \sin\left(\frac{\sqrt{4b-a^2}}{2}x\right), \ x \in \mathbb{R}$$

is a solution to $P(D)y = 0$. We can replace $\sin(\beta x)$ by $\cos(\beta x)$ in the above calculations to obtain

$$\phi_2(x) = xe^{-ax/2} \cos\left(\frac{\sqrt{4b-a^2}}{2}x\right), \ x \in \mathbb{R}$$

is also a solution to $P(D)y = 0$. Therefore if the quadratic polynomial $(X^2 + aX+b)^2$, with $a^2 < 4b$, is a factor of $P(X)$, then for every $c_1, c_2, c_3,$ and $c_4 \in \mathbb{R}$ the function

$$\phi(x) = e^{-ax/2}\left((c_1 + c_2 x)\cos\left(\frac{\sqrt{4b-a^2}}{2}x\right) + (c_3 + c_4 x)\sin\left(\frac{\sqrt{4b-a^2}}{2}x\right)\right)$$

$$(3.14)$$

is a solution to $P(D)y = 0$.

By repeating the same procedure k times one can easily prove that if $(X^2 + aX+b)^k$, with $a^2 < 4b$, is a factor of $P(X)$, then ϕ given in (3.11) is a solution to $P(D)y = 0$. $\qquad\square$

Remark 3.1.13. Let $P(X) = (X^2 + aX + b)^k$, $a^2 < 4b$. Then we notice that the roots of P are $\alpha_\pm = -\dfrac{a}{2} \pm i\dfrac{\sqrt{4b-a^2}}{2}$. Let $\mathrm{Re}(\alpha_+)$ and $\mathrm{Im}(\alpha_+)$ denote the real and imaginary parts of α_+, respectively. In view of Lemma 3.1.12 the solution to $(D^2 + aD + bI)^k y = 0$, $x \in \mathbb{R}$ is given by (see formula (3.11))

$$\phi(x) = e^{\mathrm{Re}(\alpha_+)x}\left[(c_1 + c_2 x + \cdots + c_k x^{k-1})\cos\left(\mathrm{Im}(\alpha_+)x\right)\right.$$
$$\left. + (c_{k+1} + c_{k+2}x + \cdots + c_{2k}x^{k-1})\sin\left(\mathrm{Im}(\alpha_+)x\right)\right], \ x \in \mathbb{R}. \ (3.15)$$

where c_i's are real numbers.

Example 3.1.14. Solve $(D^4 - 1)^2 y = 0$.

Solution: The polynomial associated to the given equation is $P(X) = (X^4 - 1)^2$ and its factorization is $P(X) = (X-1)^2(X+1)^2(X^2+1)^2$. Hence the given ODE can be written as

$$(D-I)^2(D+I)^2(D^2+I)^2 y = 0.$$

Let $Q_1(X) = (X-1)^2$, $Q_2(X) = (X+1)^2$ and $Q_3(X) = (X^2+1)^2$. In view of Lemma 3.1.3 we first need to find solutions to $Q_i(D)y = 0$, $i = 1, 2, 3$. We now notice that $X = 1$ is a repeated root of $Q_1(X)$. From Lemma 3.1.11, we get that solution to $Q_1(D)y = 0$ is

$$\phi_1(x) = (c_1 + c_2 x)e^x, \ x \in \mathbb{R}, \ c_1, c_2 \in \mathbb{R}.$$

Similarly, a solution to $Q_2(D)y = 0$ is given by

$$\phi_2(x) = (c_3 + c_4 x)e^{-x}, \ x \in \mathbb{R}, \ c_3, c_4 \in \mathbb{R}.$$

Finally, a solution to $Q_3(D)y = 0$ is given by

$$\phi_3(x) = (c_5 + c_6x)\sin x + (c_7 + c_8x)\cos x, \quad x \in \mathbb{R},$$

where $c_5, \ldots, c_8 \in \mathbb{R}$. Hence the function

$$\phi(x) = (c_1 + c_2x)e^x + (c_3 + c_4x)e^{-x} + (c_5 + c_6x)\sin x + (c_7 + c_8x)\cos x, \quad x \in \mathbb{R},$$

where $c_1, \ldots, c_8 \in \mathbb{R}$, is a solution to the given equation. □

Example 3.1.15. Solve $(D^4 + 1)^2 y = 0$.

Solution: The polynomial $P(X) = (X^4 + 1)^2$ which is associated to the given equation has the factorization

$$P(X) = (X^2 + \sqrt{2}X + 1)^2(X^2 - \sqrt{2}X + 1)^2.$$

Let $Q_1(X) = (X^2 + \sqrt{2}X + 1)^2$ and $Q_2(X) = (X^2 - \sqrt{2}X + 1)^2$. Then the roots of Q_1 and Q_2 are equal to $\alpha_\pm = \frac{-1 \pm i}{\sqrt{2}}$ and $\beta_\pm = \frac{1 \pm i}{\sqrt{2}}$, respectively. Moreover the roots of Q_1 and Q_2 are repeated. Therefore solutions to $Q_1(D)y = 0$ and $Q_2(D)y = 0$ are

$$\phi_1 = e^{\frac{x}{\sqrt{2}}}\left((c_1 + c_2x)\sin(\tfrac{x}{\sqrt{2}}) + (c_3 + c_4x)\cos(\tfrac{x}{\sqrt{2}})\right), \quad x \in \mathbb{R},$$

and

$$\phi_1 = e^{-\frac{x}{\sqrt{2}}}\left((c_5 + c_6x)\sin(\tfrac{x}{\sqrt{2}}) + (c_7 + c_8x)\cos(\tfrac{x}{\sqrt{2}})\right), \quad x \in \mathbb{R},$$

where c_i's are real numbers. Hence a solution to the given equation is given by

$$\begin{aligned}
\phi(x) &= e^{\frac{x}{\sqrt{2}}}\left((c_1 + c_2x)\sin(\frac{x}{\sqrt{2}}) + (c_3 + c_4x)\cos(\frac{x}{\sqrt{2}})\right) \\
&\quad + e^{-\frac{x}{\sqrt{2}}}\left((c_5 + c_6x)\sin(\frac{x}{\sqrt{2}}) + (c_7 + c_8x)\cos(\frac{x}{\sqrt{2}})\right)
\end{aligned}$$

where $c_1, \ldots, c_8 \in \mathbb{R}$. □

3.1.2 Factorization of differential operators: non-homogeneous case

In this subsection, we study some methods to find a solution to the following non-homogeneous ODE with constant coefficients

$$\mathcal{L}[y] := (a_n D^n + a_{n-1}D^{n-1} + \cdots + a_0 I)y = f, \tag{3.16}$$

where $a_i \in \mathbb{R}$, $0 \le i \le n$ and f is a continuous function. Our objective is to find a function in the set $\mathcal{L}^{-1}(\{f\})$. Any member in the set $\mathcal{L}^{-1}(\{f\})$ is called a *particular solution* to $\mathcal{L}[y] = f$.

Remark 3.1.16. Suppose $f = f_1 + f_2$, $\mathscr{L}[y_1] = f_1$, and $\mathscr{L}[y_2] = f_2$. Then it immediately follows that $\mathscr{L}[y_1 + y_2] = \mathscr{L}[y_1] + \mathscr{L}[y_2] = f_1 + f_2 = f$. Therefore to find a solution to $\mathscr{L}[y] = f_1 + f_2$, we find solutions to $\mathscr{L}[y_i] = f_i$, $i = 1, 2$ and put $y = y_1 + y_2$. This is called the principle of superposition.

We now consider the case where the operator $\mathscr{L}$ is given by

$$\mathscr{L} = (D - \alpha_1 I)(D - \alpha_2 I) \cdots (D - \alpha_n I), \ \alpha_i \in \mathbb{R}.$$

Since the operator $\mathscr{L}$ is the composition of the operators $D - \alpha_i I$, $1 \leq i \leq n$, we readily get that

$$\mathscr{L}^{-1} = (D - \alpha_n I)^{-1} \cdots (D - \alpha_1 I)^{-1}.$$

We now find $(D - \alpha_1 I)^{-1} f$. For, we set $v = (D - \alpha_1 I)^{-1} f$, then v satisfies

$$(D - \alpha_1 I)v = f. \tag{3.17}$$

Equation (3.17) is a first order linear equation and from (2.29) its solution is given by

$$v(x) = e^{\alpha_1 x} \int^x f(x) e^{-\alpha_1 x} dx, \tag{3.18}$$

where the integral is the indefinite integral. One can apply formula (3.18) recursively to obtain $(D - \alpha_n I)^{-1} \cdots (D - \alpha_1 I)^{-1} f$.

Example 3.1.17. Solve $(D^2 - 5D + 6I)y = x$, $x \in \mathbb{R}$.

Solution: We factorize the given equation as $(D - 2I)(D - 3I)y = x$. By setting $f(x) = x$ and using formula (3.18), we find

$$(D - 3I)^{-1} f = e^{3x} \int^x x e^{-3x} dx = -e^{3x} \left(\frac{x e^{-3x}}{3} + \frac{e^{-3x}}{9} \right) = -\frac{x}{3} - \frac{1}{9}.$$

Again using formula (3.18), we compute

$$(D - 2I)^{-1}(D - 3I)^{-1} f = (D - 2I)^{-1} \left(-\frac{x}{3} - \frac{1}{9} \right) = \frac{x}{6} + \frac{5}{36}.$$

Hence

$$\phi(x) = \frac{x}{6} + \frac{5}{36}, \ x \in \mathbb{R}.$$

is a particular solution to the given ODE. □

3.1.2.1 Method of partial fractions

The method of finding $\mathscr{L}^{-1} f$ using successive application of formula (3.18) has a drawback. That is we can find $(D - \alpha_k)^{-1} \cdots (D - \alpha_1)^{-1} f$ only after computation of $(D - \alpha_1)^{-1} f$, $(D - \alpha_2)^{-1}(D - \alpha_1)^{-1} f, \ldots, (D - \alpha_{k-1})^{-1} \cdots (D - \alpha_1)^{-1} f$. This avoids 'parallel' computation of $(D - \alpha_k)^{-1} \cdots (D - \alpha_1)^{-1} f$. We illustrate what we mean by parallel computation by finding a solution to the problem given in Example 3.1.17. Before that, we prove the following elementary result.

Lemma 3.1.18. *Let J be an interval, $\alpha_1, \alpha_2 \in \mathbb{R}$, $f \in C^1(J)$, $(D-\alpha_1 I)y_1 = f$ and $(D - \alpha_2 I)y_2 = f$. Then we have*

$$(D - \alpha_1 I)(D - \alpha_2 I)\left(\frac{y_1 - y_2}{\alpha_1 - \alpha_2}\right) = f$$

or equivalently

$$(D - \alpha_2 I)^{-1}(D - \alpha_1 I)^{-1}f = \frac{1}{\alpha_1 - \alpha_2}\left((D - \alpha_1 I)^{-1} - (D - \alpha_2 I)^{-1}\right)f.$$

$$(3.19)$$

Proof. A straightforward calculation gives us that

$$(D - \alpha_1 I)(D - \alpha_2 I)\left(\frac{y_1 - y_2}{\alpha_1 - \alpha_2}\right) = \frac{(D - \alpha_2 I)f}{\alpha_1 - \alpha_2} - \frac{(D - \alpha_1 I)f}{\alpha_1 - \alpha_2} = f.$$

This proves the lemma. $\qquad\square$

Formally, we can write (3.19) as

$$\frac{1}{(D - \alpha_1 I)(D - \alpha_2 I)} = \frac{1}{\alpha_1 - \alpha_2}\left(\frac{1}{D - \alpha_1 I} - \frac{1}{D - \alpha_2 I}\right).$$

This is analogous to the formula of partial fractions when D is a real variable. Lemma 3.1.18 can be extended to various other cases by simply considering the corresponding identities in the partial fractions.

We now use this method to find a solution to the problem given in Example 3.1.17.

Example 3.1.17 (Revisited) Solve $(D - 2I)(D - 3I)y = x$, $x \in \mathbb{R}$, using the method of partial fractions.

Solution: Let $f(x) = x$. In order to compute $(D - 3I)^{-1}(D - 2I)^{-1}f$, we use formula (3.19) to get

$$\phi(x) = (D - 3I)^{-1}(D - 2I)^{-1}f = (D - 3I)^{-1}f - (D - 2I)^{-1}f.$$

An easy computation shows that

$$(D - 3I)^{-1}f = \frac{-x}{3} - \frac{1}{9}, \quad (D - 2I)^{-1}f = \frac{-x}{2} - \frac{1}{4}.$$

Therefore a particular solution to the given equation is

$$\phi(x) = \left(\frac{-x}{3} - \frac{1}{9}\right) - \left(\frac{-x}{2} - \frac{1}{4}\right) = \frac{x}{6} + \frac{5}{36}. \qquad\square$$

3.1.2.2   Power series method

If the non-homogeneous term f in equation (3.16) is a polynomial, then we use a special method to find a solution to (3.16). An advantage of this method is that we need not factorize the given differential operator. Before we begin explaining the method, consider the identity

$$(1 + a)\left(1 - a + \cdots + (-1)^n a^n\right) = 1 + (-1)^n a^{n+1}, \ a \in \mathbb{R}, \ n \in \mathbb{N}.$$

Suppose f and P are polynomials of degrees n_1 and n_2, respectively. Moreover assume that $P(0) \neq 0$ and $P(X) = P(0)\big(1+Q(X)\big)$. Then we have the identity

$$\big(I + Q(D)\big)\big(I - Q(D) + \cdots + (-1)^{n_1} Q^{n_1}(D)\big) f = \big(I + (-1)^{n_1} Q^{n_1+1}(D)\big) f = f,$$

because $Q^{n_1+1}(D)f = 0$. Therefore it follows that

$$\big(I + Q(D)\big)^{-1} f = \big(1 - Q(D) + \cdots + (-1)^{n_1} Q^{n_1}(D)\big) f,$$

and thus

$$\big(P(D)\big)^{-1} f = \frac{1}{P(0)}\big(1 - Q(D) + \cdots + (-1)^{n_1} Q^{n_1}(D)\big) f. \tag{3.20}$$

We illustrate this technique by giving some examples.

Example 3.1.19. Solve $(D^2 + D + 3I)y = x^2 + 4x,\ x \in \mathbb{R}$.

Solution: A solution to the given problem is

$$
\begin{aligned}
\phi(x) &= (D^2 + D + 3I)^{-1}(x^2 + 4x) \\
&= \frac{1}{3}\left(I + \frac{D^2 + D}{3}\right)^{-1} x^2 + \frac{4}{3}\left(I + \frac{D^2 + D}{3}\right)^{-1} x \\
&= \frac{1}{3}\left(I - \frac{D^2 + D}{3} + \frac{(D^2 + D)^2}{9}\right) x^2 + \frac{4}{3}\left(I - \frac{D^2 + D}{3}\right) x
\end{aligned}
$$

because $(D^2 + D)^k x^2 = 0$ for $k \geq 3$, and $(D^2 + D)^m x = 0$ for $m \geq 2$. Therefore

$$\phi(x) = \frac{1}{3}\left(x^2 - \frac{2 + 2x}{3} + \frac{2}{9}\right) + \frac{4}{3}\left(x - \frac{1}{3}\right) = \frac{x^2}{3} + \frac{10x}{9} - \frac{16}{27}$$

is a particular solution to the given ODE. $\square$

Example 3.1.20. Solve $(D^3 - 2D^2 + 5D)y = x^3,\ x \in \mathbb{R}$.

Solution: We need to compute

$$\phi(x) = (D^3 - 2D^2 + 5D)^{-1}x^3 = D^{-1}(D^2 - 2D + 5I)^{-1}x^3.$$

For, we consider

$$
\begin{aligned}
(D^2 - 2D + 5I)^{-1}x^3 &= \frac{1}{5}\left(I + \frac{D^2 - 2D}{5}\right)^{-1} x^3 \\
&= \frac{1}{5}\left(I - \frac{D^2 - 2D}{5} + \frac{(D^2 - 2D)^2}{25} - \frac{(D^2 - 2D)^3}{125}\right) x^3 \\
&= \frac{1}{5}\left(x^3 - \frac{6x - 6x^2}{5} + \frac{24x - 24}{25} + \frac{48}{125}\right) \\
&= \frac{x^3}{5} + \frac{6x^2}{25} - \frac{6x}{125} - \frac{72}{625}.
\end{aligned}
$$

Hence the function

$$\phi(x) = D^{-1}\left(\frac{x^3}{5} + \frac{6x^2}{25} - \frac{6x}{125} - \frac{72}{625}\right) = \frac{x^4}{20} + \frac{6x^3}{75} - \frac{6x^2}{250} - \frac{72x}{625}$$

is a particular solution to the given ODE.   □

3.1.2.3   Method of undetermined coefficients

We now present a method to find a solution to the non-homogeneous ODEs (3.16) when the right hand side $f(x)$ is one of the trigonometric functions $\sin(\alpha x)$ or $\cos(\alpha x)$ where $\alpha \in \mathbb{R}$. In this case, we assume that there is a solution is of the form

$$\phi(x) = A\sin(\alpha x) + B\cos(\alpha x),$$

where A and B are unknowns. We substitute this ϕ in the given differential equation and obtain a linear system of equations for A and B. We solve this system to find A, B, which in turn gives us the solution to the given differential equation. This is demonstrated in the following example.

Example 3.1.21. Solve $(D^2 + D + I)y = \sin(3x)$, $x \in \mathbb{R}$, using the method of undetermined coefficients.

Solution: In the method of undetermined coefficients, we begin with the ansatz

$$\phi(x) = A\cos(3x) + B\sin(3x), \quad x \in \mathbb{R}.$$

We need to determine A and B such that ϕ is a solution to the given equation. To this end, we compute

$$\phi'(x) = -3A\sin(3x) + 3B\cos(3x),$$
$$\phi''(x) = -9A\cos(3x) - 9B\sin(3x).$$

By substituting the functions ϕ, ϕ', and ϕ'' in the given equation, we get

$$(-8A + 3B)\cos(3x) - (3A + 8B)\sin(3x) = \sin(3x). \qquad (3.21)$$

By comparing the coefficients of $\sin(3x)$ and $\cos(3x)$ on both sides of (3.21), we obtain the system

$$-8A + 3B = 0, \quad -3A - 8B = 1.$$

On solving this system we get that the solution to this system of equations is $A = -\dfrac{3}{73}$ and $B = -\dfrac{8}{73}$. Hence the function

$$\phi(x) = -\frac{3}{73}\cos(3x) - \frac{8}{73}\sin(3x), \quad x \in \mathbb{R}$$

is a particular solution to the given equation.   □

If the non-homogeneous term f (or a part of it) in (3.16) is a solution to the corresponding homogeneous ODE, then the standard guess $\phi = A\cos(\alpha x) + B\sin(\alpha x)$ will not work. Instead, we need to begin with

$$\phi = Ax\cos(\alpha x) + Bx\sin(\alpha x).$$

If one of $x\cos(\alpha x)$ or $x\sin(\alpha x)$ is a solution to the corresponding homogeneous ODE, then we need to make our guess as

$$\phi = Ax^2\cos(\alpha x) + Bx^2\sin(\alpha x)$$

and so on. This is explained in the following example.

Example 3.1.22. Solve $(D^2 + I)y = \cos x$, $x \in \mathbb{R}$, using the method of undetermined coefficients.

Solution: We first begin with the ansatz $\phi_1(x) = A\cos(x) + B\sin(x)$, $x \in \mathbb{R}$ as in Example 3.1.21. We now compute ϕ_1'' and substitute in the given equation to get

$$-A\cos(x) - B\sin(x) + A\cos(x) + B\sin(x) = \cos(x)$$

which is not possible. The ansatz $\phi_1(x) = A\cos(x) + B\sin(x)$ did not work because it is a solution to the homogeneous problem $(D^2 + I)y = 0$. We now provide a modified ansatz, namely,

$$\phi(x) = Ax\cos(x) + Bx\sin(x), \quad x \in \mathbb{R}. \tag{3.22}$$

We need to find A and B such that (3.22) is a solution to the given equation. As before, we compute ϕ'' and substitute in the given equation to obtain

$$(D^2 + I)\phi(x) = -2A\sin(x) + 2B\cos(x) = \cos(x).$$

Hence we have $A = 0$ and $B = \dfrac{1}{2}$, which implies that

$$\phi(x) = \frac{x}{2}\sin x, \quad x \in \mathbb{R}$$

is a required particular solution. $\qquad\square$

Example 3.1.23. Solve $(D^2 + 4I)^2 y = \sin^2 x$, $x \in \mathbb{R}$, using the method of undetermined coefficients.

Solution: We first write the given ODE as

$$(D^2 + 4I)^2 y(x) = \frac{1 - \cos(2x)}{2}.$$

Then the solution to the given equation is $\phi = \phi_1 - \phi_2$ where

$$(D^2 + 4I)^2 \phi_1 = \frac{1}{2}, \quad x \in \mathbb{R},$$

and

$$(D^2 + 4I)^2 \phi_2 = \frac{\cos(2x)}{2}. \tag{3.23}$$

We use the method of power series to find ϕ_1, i.e.,

$$\phi_1(x) = (D^2 + 4I)^{-2}\frac{1}{2} = (D^2 + 4I)^{-1}\frac{1}{4}\left(I + \frac{D^2}{4}\right)^{-1}\frac{1}{2} = (D^2 + 4I)^{-1}\frac{1}{8} = \frac{1}{32}.$$

Since a solution to the $(D^2 + 4I)^2 u = 0$ is

$$u = (a_1 + a_2 x)\cos(2x) + (b_1 + b_2 x)\sin(2x),$$

we take a solution to (3.23) as

$$\phi_2(x) = Ax^2\cos(2x) + Bx^2\sin(2x), \quad x \in \mathbb{R}. \tag{3.24}$$

Then a straightforward computation yields

$$(D^2 + 4I)^2\phi_2 = -32A\cos(2x) - 32B\sin(2x).$$

Hence (3.24) is a solution to (3.23) if and only if $A = -\frac{1}{64}$, $B = 0$. Therefore

$$\phi(x) = \phi_1(x) - \phi_2(x) = \frac{1}{32} + \frac{x^2}{64}\cos(2x), \quad x \in \mathbb{R}$$

is a particular solution to the given equation. $\qquad\qquad\qquad\square$

In some special cases, one can avoid the method of undetermined coefficients and write the solution to the non-homogeneous equation without guessing. This is described in the following remark.

Remark 3.1.24. Since $D^2\sin(\alpha x) = -\alpha^2\sin(\alpha x)$, we obtain

$$P(D^2)\sin(\alpha x) = P(-\alpha^2)\sin(\alpha x)$$

for every polynomial P. From this we conclude that

$$\left(P(D^2)\right)^{-1}\sin(\alpha x) = \frac{1}{P(-\alpha^2)}\sin(\alpha x). \tag{3.25}$$

whenever $P(-\alpha^2) \neq 0$. Furthermore, one can replace sin with cos in (3.25).

Example 3.1.25. Solve $(D^6 + 2D^4 - 7D^2 + I)y = 2\cos^2(x)$, $x \in \mathbb{R}$.
Solution: We proceed as in Example 3.1.23 and write the solution to the given equation as $\phi = \phi_1 + \phi_2$, where ϕ_1 and ϕ_2 are solutions to

$$(D^6 + 2D^4 - 7D^2 + I)\phi_1 = 1,$$

and

$$(D^6 + 2D^4 - 7D^2 + I)\phi_2 = \cos(2x),$$

respectively. One can use the power series method to find $\phi_1 \equiv 1$. Let

$$P(X) = X^3 + 2X^2 - 7X + 1.$$

Then ϕ_2 is a solution to $P(D^2)\phi_2 = \cos(2x)$. Since $P(-4) = -3 \neq 0$, from Remark 3.1.24 we get

$$\phi_2(x) = \frac{1}{P(-2^2)}\cos(2x) = -\frac{1}{3}\cos(2x).$$

Therefore

$$\phi(x) = 1 - \frac{1}{3}\cos(2x), \ x \in \mathbb{R}$$

is a required particular solution. $\qquad\square$

3.1.2.4 Exponential shift rule

We begin with stating a very useful identity involving the exponential function and a differential operator with constant coefficients. To this end, using the principle of mathematical induction, one can easily prove that for $\alpha \in \mathbb{R}$, $n \in \mathbb{N}$, $g \in C^\infty(J)$ we have

$$D^n e^{\alpha x} g(x) = e^{\alpha x}(D + \alpha I)^n g(x), \ x \in J.$$

Therefore if $P(X)$ is any polynomial with real coefficients, then

$$P(D)e^{\alpha x} g(x) = e^{\alpha x} P(D + \alpha I)g(x). \tag{3.26}$$

From (3.26) we get the following identity

$$\left(P(D)\right)^{-1}\left(e^{\alpha x} f(x)\right) = e^{\alpha x}\left(P(D + \alpha I)\right)^{-1} f(x). \tag{3.27}$$

For, put $g = P(D + \alpha I)^{-1} f$ in (3.26) to obtain

$$P(D)\left(e^{\alpha x}\left(P(D + \alpha I)\right)^{-1} f(x)\right) = e^{\alpha x} f(x).$$

This readily implies identity (3.27). Formally, (3.27) is written as

$$\frac{1}{P(D)} e^{\alpha x} f(x) = e^{\alpha x} \frac{1}{P(D + \alpha I)} f(x).$$

We now consider a special case where $f \equiv 1$. We first recall identity (3.9)

$$P(D)e^{\alpha x} = P(\alpha)e^{\alpha x}, \ \alpha \in \mathbb{R}.$$

Suppose $\alpha_0 \in \mathbb{R}$ be such that $P(\alpha_0) \neq 0$, then from (3.9), we obtain

$$\left(P(D)\right)^{-1} e^{\alpha_0 x} = \frac{e^{\alpha_0 x}}{P(\alpha_0)}. \tag{3.28}$$

On the other hand, suppose $P(\alpha_0) = 0$ and $P'(\alpha_0) \neq 0$. Then we use the technique introduced in Lemma 3.1.11 to find a solution to $P(D)y = e^{\alpha_0 x}$. On differentiating (3.9) with respect to α and substituting $\alpha = \alpha_0$, we get

$$P(D)x e^{\alpha_0 x} = P'(\alpha_0)e^{\alpha_0 x}.$$

By rearranging, one gets

$$\left(P(D)\right)^{-1} e^{\alpha_0 x} = \frac{x e^{\alpha_0 x}}{P'(\alpha_0)}. \tag{3.29}$$

Similarly, if $P(\alpha_0) = P'(\alpha_0) = \cdots = P^{(k-1)}(\alpha_0) = 0$ and $P^{(k)}(\alpha_0) \neq 0$ then, we have

$$\left(P(D)\right)^{-1} e^{\alpha_0 x} = \frac{x^k e^{\alpha_0 x}}{P^{(k)}(\alpha_0)}, \quad k \in \mathbb{N}. \tag{3.30}$$

Therefore we use (3.27)–(3.30) to solve the linear differential equation with constant coefficients of the form $P(D)y = e^{\alpha x} f$. This is discussed in the following examples.

Example 3.1.26. Solve $(D^3 - 2D^2 + 7D + 2I)y = e^x$, $x \in \mathbb{R}$.

Solution: We notice that

$$P(X) = X^3 - 2X^2 + 7X + 2$$

is the polynomial associated with the given differential operator, i.e., the given ODE is $P(D)y = e^x$. Since $P(1) = 8 \neq 0$, in view of (3.28) we get that the function

$$\phi(x) = \left(P(D)\right)^{-1} e^x = \frac{e^x}{P(1)} = \frac{e^x}{8}, \quad x \in \mathbb{R}$$

is a particular solution to the given equation. □

Example 3.1.27. Solve $(D^3 - 6D^2 + 12D - 8I)y = e^{2x}$, $x \in \mathbb{R}$.

Solution: We notice that

$$P(X) = X^3 - 6X^2 + 12X - 8$$

is the polynomial associated with the given differential operator, i.e., the given ODE is $P(D)y = e^{2x}$. Since $P(2) = P'(2) = P''(2) = 0$ and $P'''(2) = 6 \neq 0$, in view of (3.30) we get that the function

$$\phi(x) = \left(P(D)\right)^{-1} e^{2x} = \frac{x^3 e^x}{P'''(2)} = \frac{x^3 e^{2x}}{6}, \quad x \in \mathbb{R}$$

is a particular solution to the given equation. □

Example 3.1.28. Solve $(D^2 - 7D + 12I)y = xe^{3x}$, $x \in \mathbb{R}$.

Solution: We first observe that

$$
\begin{aligned}
\phi(x) &= (D^2 - 7D + 12I)^{-1} \left(xe^{3x}\right) \\
&= e^{3x} \left((D + 3I)^2 - 7(D + 3I) + 12I\right)^{-1} x \\
&= e^{3x} (D^2 - D)^{-1} x.
\end{aligned}
$$

Moreover, using the power series method we get

$$(D^2 - D)^{-1}x = -D^{-1}(I - D)^{-1}x = -D^{-1}(I + D)x = -\frac{x^2}{2} - x.$$

Hence a particular solution to the given ODE is

$$\phi(x) = -\frac{1}{2}\left(x^2 + 2x\right)e^{3x}, \quad x \in \mathbb{R}.$$

This completes the solution. □

Example 3.1.29. Solve $(D^4 + 4D^3 + 6D^2 + 4D - 2I)y = e^{-x}\sin(3x)$.

Solution: We begin with the observation

$$
\begin{aligned}
\phi(x) &= (D^4 + 4D^3 + 6D^2 + 4D - 2I)^{-1}\left(e^{-x}\sin(3x)\right) \\
&= e^{-x}\left((D-I)^4 + 4(D-I)^3 + 6(D-I)^2 + 4(D-I) - 2I\right)^{-1}\sin(3x) \\
&= e^{-x}\left(D^4 - 3I\right)^{-1}\sin(3x).
\end{aligned}
$$

By setting $P(X) = X^2 - 3$ and using Remark 3.1.24, we readily get that

$$
\left(D^4 - 3I\right)^{-1}\sin(3x) = P(D^2)^{-1}\sin(3x) = \frac{1}{P(-9)}\sin(3x) = \frac{1}{78}\sin(3x).
$$

Therefore the function

$$
\phi(x) = \frac{e^{-x}}{78}\sin(3x), \ x \in \mathbb{R},
$$

is a particular solution to the given ODE. □

3.1.3 Euler's equation

Let $a, b \in \mathbb{R}$. The following linear ordinary differential equation

$$
x^2 y'' + axy' + by = 0, \ x > 0, \tag{3.31}
$$

is called the second order Euler equation. We solve the second order Euler equation by converting it into an ODE with constant coefficients. To this end, consider the ansatz

$$
\phi(x) = u(\log x), \ x > 0. \tag{3.32}
$$

Then from the chain rule we get

$$
\phi'(x) = \frac{1}{x}u'(\log x), \ \phi''(x) = \frac{1}{x^2}\left(u''(\log x) - u'(\log x)\right).
$$

Therefore equation (3.31) becomes

$$
u'' + (a - 1)u' + bu = 0, \tag{3.33}
$$

which is a linear ODE with constant coefficients and we have already seen the methods to solve it. Hence ansatz (3.32) where u is a solution to (3.33) indeed provides a solution to (3.31). We solve some problems using the method described now.

Example 3.1.30. Solve $x^2 y'' + 8xy' + 10y = 0, \ x > 0$.

Solution: Let $\phi(x) = u(\log x)$ be a solution to the given ODE. Then we readily see that

$$
\phi'(x) = \frac{1}{x}u'(\log x), \ \phi''(x) = \frac{1}{x^2}\left(u''(\log x) - u'(\log x)\right).
$$

Then we substitute $x\phi'$ and $x^2\phi''$ in the given equation to realize that u satisfies

$$u'' + 7u' + 10u = 0. \tag{3.34}$$

It is easy to find that the solution to (3.34) is

$$u(x) = C_1 e^{-2x} + C_2 e^{-5x}.$$

This in turn gives us that

$$\phi(x) = u(\log x) = C_1 e^{-2\log x} + C_2 e^{-5\log x} = \frac{C_1}{x^2} + \frac{C_2}{x^5}, \quad x > 0,$$

is a solution to the given problem.   □

Example 3.1.31. Solve $x^2 y'' - 5xy' + 9y = 0$, $x > 0$.

Solution: As in the previous example, let $\phi(x) = u(\log x)$ be a solution to the given ODE. Then we obtain that u satisfies

$$u'' - 6u' + 9u = 0.$$

Therefore we have $u(x) = (C_1 + C_2 x)e^{3x}$. Hence we get

$$\phi(x) = u(\log x) = (C_1 + C_2 \log x)e^{3\log x} = (C_1 + C_2 \log x)x^3, \quad x > 0,$$

is a solution to the given problem.   □

Example 3.1.32. Solve $x^2 y'' - 3xy' + 13y = 0$, $x > 0$.

Solution: As before, let $\phi(x) = u(\log x)$ be a solution to the given ODE. Then we readily see that u satisfies
$$u'' - 4u' + 13u = 0.$$

A straightforward calculation gives us that

$$u(x) = (A\cos(3x) + B\sin(3x))\, e^{2x}.$$

Therefore we find that

$$\phi(x) = u(\log x) = x^2 \left(A\cos(3\log x) + B\sin(3\log x)\right), \quad x > 0,$$

is a solution to the given problem.   □

Remark 3.1.33. One can extend this method to solve an n-th order Euler equation given by

$$x^n \frac{d^n y}{dx^n} + a_{n-1}x^{n-1}\frac{d^{n-1}y}{dx^{n-1}} + \cdots + a_0 y = 0, \quad x > 0, \tag{3.35}$$

where $a_0, \ldots, a_{n_1} \in \mathbb{R}$. Assume that a solution to (3.35) is of the form $\phi(x) = u(\log x)$. It is left to the reader as an easy exercise to find the n-th order linear equation with constant coefficients that u satisfies (at least for $n = 3$ and 4).

Remark 3.1.34. Instead of working with the ansatz $\phi(x) = u(\log x)$, one can start with the assumption that $\phi = x^m$ for some $m \in \mathbb{C}$ to be a solution to (3.31). In other words, we want to find m such that $\phi = x^m$ is a solution to (3.31). In order to do that, we substitute $\phi' = mx^{m-1}$ and $\phi'' = m(m-1)x^{m-2}$ in (3.31) to get

$$x^m \left(m(m - 1) + am + b \right) = 0$$

or

$$m^2 + (a - 1)m + b = 0. \tag{3.36}$$

Let m_1 and m_2 be the roots of (3.36). Suppose $m_1 \neq m_2$, then solution to (3.31) is

$$\phi(x) = C_1 x^{m_1} + C_2 x^{m_2}, \ x > 0.$$

Suppose m_1 and m_2 are complex conjugates with $m_1 = \alpha + i\beta$ then from theory of complex functions, we have

$$x^{m_1} = x^\alpha x^{i\beta} = x^\alpha e^{\log x^{i\beta}} = x^\alpha e^{i\beta \log x} = x^\alpha \left(\cos(\beta \log x) + i \sin(\beta \log x) \right).$$

Then solution to (3.31) is

$$\phi(x) = x^\alpha \left(A \cos(\beta \log x) + B \sin(\beta \log x) \right).$$

Suppose we have equal roots for (3.36), i.e., $m_1 = m_2$. Then the solution to (3.31) other than $\phi_1 = x^{m_1}, x > 0$ can easily be found using the the same technique employed to find the solutions to linear ODEs with constant coefficients whose associated polynomials have repeated roots. In fact the solution turns out to be $\phi_2(x) = x^{m_1} \log x, x > 0$. Hence the solution in this case is

$$\phi(x) = C_1 x^{m_1} + C_2 x^{m_1} \log x, \ x > 0.$$

To find ϕ_2 one can also use the Lagrange method (see Section 3.2.3).

The reader is advised to solve problems in Examples 3.1.30–3.1.32 using the choice $\phi = x^m$ and Remark 3.1.34.

3.2   ODEs with variable coefficients

Most of the linear ODEs which arise in the modeling of various physical phenomena have variable coefficients. Henceforth, throughout the chapter we study only linear ODEs with variable coefficients. We begin with the definition of a linear ODE.

Definition 3.2.1. Let $J \subseteq \mathbb{R}$ be an open interval and $a_i \in C(\bar{J})$, $0 \le i \le n$. An equation of the form

$$a_n(x)y^{(n)}(x) + \cdots + a_1(x)y'(x) + a_0(x)y(x) = 0, \quad x \in J, \tag{3.37}$$

is called a homogeneous linear ODE of order n. If $f \in C(\bar{J})$, $f \ne 0$, then the equation

$$a_n(x)y^{(n)}(x) + \cdots + a_1(x)y'(x) + a_0(x)y(x) = f(x), \quad x \in J \tag{3.38}$$

is called a non-homogeneous linear ODE of order n.

Equations (3.2.1)–(3.38) are called linear equations because the associated differential operator $\mathscr{M} : C^n(J) \to C(J)$ given by

$$\mathscr{M}[y] = a_n y^{(n)} + \cdots + a_1 y' + a_0 y$$

is linear, i.e.,

$$\mathscr{M}[ay_1 + by_2] = a\mathscr{M}[y_1] + b\mathscr{M}[y_2], \; y_1, y_2 \in C^n(J), \; a, b \in \mathbb{R}.$$

Throughout the chapter, we assume that $a_n(x) \ne 0$, $x \in \bar{J}$. This is equivalent to the assumption that $a_n \equiv 1$ in (3.37)–(3.38). For, if $a_n(x) \ne 1$ for some $x \in \bar{J}$, then we divide (3.37), (3.38) with a_n to get the ODEs whose coefficient of $y^{(n)}$ is the constant function one. Justification for this assumption is given in the following example.

Example 3.2.2. Show that the problem

$$x^2 y''(x) + 6xy'(x) + 4y(x) = 0, \quad x \in (0,1), \tag{3.39a}$$

$$y(0) = 10, \; y'(0) = 2 \tag{3.39b}$$

does not admit a solution.

Solution. In the given problem, the coefficient of y'' vanishes at $x = 0$ which is in the closure of $(0, 1)$. We first notice that the given ODE is an Euler equation. We seek a solution to (3.39a)–(3.39b) in the form $\phi = x^m$. After substituting this ansatz in (3.39a) to solve for m, we get that solution to (3.39a) as

$$\phi(x) = \frac{c_1}{x} + \frac{c_2}{x^4}, \; x \in (0,1),$$

where c_1 and c_2 are arbitrary constants. The condition $\phi(0) = 10$ never holds for any value of c_1, c_2. Hence we conclude that there exists no solution to (3.39a)–(3.39b). $\qquad \square$

We begin our discussion with the following existence and uniqueness result.

Theorem 3.2.3. Let $a_0, a_1, \ldots, a_{n-1}, f \in C(\bar{J})$ and $x_0 \in J$. Consider the problem

$$\begin{cases} y^{(n)}(x) + a_{n-1}(x)y^{(n-1)}(x) + \cdots + a_1(x)y'(x) + a_0(x)y(x) = f(x), \; x \in J, \\ y(x_0) = \alpha_0, \; y'(x_0) = \alpha_1, \; \ldots, \; y^{(n-1)}(x_0) = \alpha_{n-1}. \end{cases}$$

$$\tag{3.40}$$

Then there exists an interval J_1 such that $x_0 \in J_1 \subseteq J$ and (3.40) has a unique solution in the interval J_1.

The main idea of the proof of Theorem 3.2.3 is to convert (3.40) to a Cauchy problem problem of a system of first order ODEs and use the existence and uniqueness result for the system. An outline of the proof of existence and uniqueness result for Cauchy problem/initial value problem in the context of system of ODEs is given in Chapter 5. Therefore for time being the reader is advised to take Theorem 3.2.3 for granted. In fact, we use this theorem to prove many results in this chapter. Before we move on to the next subsection we state a very useful remark.

Remark 3.2.4. Assume the hypotheses of Theorem 3.2.3. If w is a solution to

$$\begin{cases} w^{(n)}(x) + a_{n-1}(x)w^{(n-1)}(x) + \cdots + a_1(x)w'(x) + a_0(x)w(x) = 0, \ x \in J, \\ w(x_0) = w'(x_0) = \cdots = w^{(n-1)}(x_0) = 0, \end{cases}$$

then $w \equiv 0$ on J.

3.2.1 Dimension of the solution space

We now introduce the notion of linearly dependent/independent subsets of the vector space of continuous functions[2] $C(J)$. Then we discuss the dimension of the solution space of the linear ODE with variable coefficients.

Definition 3.2.5. We say that a nonempty set $S \subset C(J)$ is a linearly dependent set in $C(J)$ if there exists $\{\phi_1, \ldots, \phi_k\} \subseteq X$ and a nonzero vector $(c_1, \ldots, c_k) \in \mathbb{R}^k$ such that

$$c_1\phi_1 + \cdots + c_k\phi_k \equiv \mathbf{0},$$

where $\mathbf{0}$ on the right hand side is the constant function which takes zero everywhere. If S is not a linearly dependent set, then it is said to be a linearly independent set.

We now see some examples of linearly dependent/independent sets.

Example 3.2.6. Let $S = \{f_k \in C(J) : f_k(x) = x^k, k \in \mathbb{N} \cup \{0\}\}$. Show that S is linearly independent.

Solution. Suppose $c_1 f_{k_1} + \cdots + c_n f_{k_n} = \mathbf{0}$. This is the same as

$$c_1 x^{k_1} + c_2 x^{k_2} + \cdots + c_n x^{k_n} = 0, \ x \in J.$$

This implies that the polynomial $P(X) := c_1 X^{k_1} + c_2 X^{k_2} + \cdots + c_n X^{k_n}$ has infinite number of roots, namely all elements of J. Thus $P(X) \equiv 0$, yielding that $c_1 = c_2 = \cdots = c_n = 0$. Therefore S is linearly independent. □

---

[2]Our results regarding the linear independent/dependent subsets of $C(J)$ can be easily extended to the case where J is any interval with positive length.

Example 3.2.7. Assume that α, β, and γ are distinct positive numbers. Show that $\{x^\alpha, x^\beta, x^\gamma\}$ is a linearly independent set in $C(J)$ whenever $J \subset [0, \infty)$.

Solution. Without loss of generality, we assume that $\alpha < \beta < \gamma$. Let

$$c_1 x^\alpha + c_2 x^\beta + c_3 x^\gamma = 0, \ x \in J.$$

Thus for $x \neq 0$, we have

$$c_1 + c_2 x^{\beta - \alpha} + c_3 x^{\gamma - \alpha} = 0. \qquad (3.41)$$

On differentiating (3.41), we get

$$(\beta - \alpha) c_2 x^{\beta - \alpha - 1} + (\gamma - \alpha) c_3 x^{\gamma - \alpha - 1} = 0, \ x \in J\backslash\{0\}.$$

Hence we obtain

$$c_2 + c_3 \left(\frac{\gamma - \alpha}{\beta - \alpha}\right) x^{\gamma - \beta} = 0, \ x \in J\backslash\{0\}. \qquad (3.42)$$

Again on differentiating (3.42), we find

$$c_3(\gamma - \beta) \left(\frac{\gamma - \alpha}{\beta - \alpha}\right) x^{\gamma - \beta - 1} = 0, \ x \in J\backslash\{0\}.$$

Therefore we get $c_3 = 0$. Substituting $c_3 = 0$ in (3.42), we realize that $c_2 = 0$. From (3.41) it follows that $c_1 = 0$. Hence the given set is linearly independent. □

Henceforth, the n-th order differential operator that we consider in chapter is given by

$$\mathscr{L}[y] := y^{(n)} + a_{n-1} y^{(n-1)} + \cdots + a_1 y' + a_0 y, \qquad (3.43)$$

where a_k, $0 \leq k \leq n - 1$, is continuous on $\bar{J}$. The main objective of this subsection is to show that the set of all solutions to $\mathscr{L}[y] = 0$ forms an n-dimensional vector subspace of $C^n(J)$. This is given in the next theorem.

Theorem 3.2.8. *Let $\mathscr{L}$ be as in (3.43). Then the set of all solutions to $\mathscr{L}[y] = 0$ is an n-dimensional vector subspace of $C^n(J)$.*

Proof. If u_1, u_2 are two solutions to $\mathscr{L}[y] = 0$, then for $c_1, c_2 \in \mathbb{R}$, we have

$$\mathscr{L}[c_1 u_1 + c_2 u_2] = c_1 \mathscr{L}[u_1] + c_2 \mathscr{L}[u_2] = 0.$$

Hence $\phi = c_1 u_1 + c_2 u_2$, is also a solution to $\mathscr{L}[y] = 0$. Hence the set of all solutions to $\mathscr{L}[y] = 0$ is a vector subspace of $C^n(J)$.

We now show that it indeed has the dimension n. The proof uses the existence part of Theorem 3.2.3. We first fix $x_0 \in J$. Let $\psi_i(.)$, $0 \leq i \leq n - 1$, be the solution to the following problem

$$\begin{cases} \psi_i^{(n)}(x) + a_{n-1}(x)\psi_i^{(n-1)}(x) + \cdots + a_1(x)\psi_i'(x) + a_0(x)\psi_i(x) = 0, \ x \in J, \\ \psi_i^{(i)}(x_0) = 1, \psi_i^{(k)}(x_0) = 0, \ 0 \leq k \leq n - 1, k \neq i. \end{cases}$$

$$(3.44)$$

Claim 1. The set $S := \{\psi_i : 0 \le i \le n\}$ is linearly independent in $C^n(J)$. For, let $c_0, \ldots, c_{n-1}$ be such that

$$c_0 \psi_0 + \cdots + c_{n-1} \psi_{n-1} = \mathbf{0}$$

or

$$c_0 \psi_0(x) + \cdots + c_{n-1} \psi_{n-1}(x) = 0, \quad x \in J. \tag{3.45}$$

We now differentiate identity (3.45) k times, $1 \le k \le n - 1$, with respect to x to obtain the following identities, $x \in J$,

$$\left\{ \begin{array}{l} c_0 \psi_0'(x) + \cdots + c_{n-1} \psi_{n-1}'(x) = 0, \\ \vdots \\ c_0 \psi_0^{(n-1)}(x) + \cdots + c_{n-1} \psi_{n-1}^{(n-1)}(x) = 0, \end{array} \right. \tag{3.46}$$

We put $x = x_0$ in (3.45) and use the initial conditions in (3.44) to get $c_0 = 0$. Then we put $x = x_0$ in the first equation of (3.46) and use the initial conditions in (3.44) to arrive at $c_1 = 0$. Using the same strategy (by putting $x = x_0$ in the second equation of (3.46)), we conclude $c_2 = 0$. In this way, we show that all the constants $c_0, \ldots, c_{n-1}$ must be zeros. Thus, we obtain that the set S is linearly independent in $C^n(J)$.

Claim 2. We show that S spans the set of all solutions to $\mathscr{L}[y] = 0$.

The important ingredient to prove this claim is the uniqueness part of Theorem 3.2.3. Let ϕ be a solution to $\mathscr{L}[y] = 0$. Define the function

$$z(x) = \phi(x) - \left(\phi(x_0)\psi_0(x) + \cdots + \phi^{(n-1)}(x_0)\psi_{n-1}(x) \right), \quad x \in J.$$

By the linearity of $\mathscr{L}$ and (3.44), we have

$$\mathscr{L}[z](x) = \mathscr{L}[\phi](x) - \left(\phi(x_0).\mathscr{L}[\psi_0](x) + \cdots + \phi^{(n-1)}(x_0).\mathscr{L}[\psi_{n-1}](x) \right) = 0.$$

Using the initial conditions that ψ_k's satisfy, we immediately get that

$$z(x_0) = z'(x_0) = \cdots = z^{(n-1)}(x_0) = 0.$$

In view of Remark 3.2.4, we get $z \equiv 0$. Hence ϕ is a linear combination of $\psi_0, \ldots, \psi_{n-1}$ which in turn implies that the set S spans the set of all solutions of $\mathscr{L}[\psi] = 0$. This completes the proof of the announced result. □

Here onward, the set of all solutions to the linear ODE $\mathscr{L}[y] = 0$ is referred to as *the solution space* of $\mathscr{L}[y] = 0$. We conclude this section with the following definition.

Definition 3.2.9. We say that $\phi = c_1 \phi_1 + \cdots + c_n \phi_n$, $c_i \in \mathbb{R}$, $1 \le i \le n$ is the general solution to the n-th order ODE $\mathscr{L}[y] = 0$, where $\mathscr{L}$ is given in (3.43), if $\{\phi_1, \ldots, \phi_n\}$ forms a basis for the solution space of $\mathscr{L}[y] = 0$.

The reader is advised to check whether the solutions provided in the worked out examples given in Section 3.1.1 are indeed general solutions.

3.2.2   Wronskian and its properties

In this subsection, we introduce a function called the Wronskian of functions and study its properties. In fact, the Wronskian plays a key role in the study of linear ODEs.

Definition 3.2.10. Suppose $\phi_1, \phi_2, \ldots, \phi_n$ are $(n-1)$ times differentiable functions on J. Then we say that the Wronskian of $\phi_1, \phi_2, \ldots, \phi_n$ is the function given by the determinant

$$
W(x; \phi_1, \ldots, \phi_n) := \begin{vmatrix} \phi_1(x) & \phi_2(x) & \cdots & \phi_n(x) \\ \phi_1'(x) & \phi_2'(x) & \cdots & \phi_n'(x) \\ \vdots & \vdots & & \vdots \\ \phi_1^{(n-1)}(x) & \phi_2^{(n-1)}(x) & \cdots & \phi_n^{(n-1)}(x) \end{vmatrix}.
$$

Our next objective is to characterize linearly independent finite sets in $C^n(J)$ in terms of the corresponding Wronskian. To that end, we prove the following result.

Proposition 3.2.11. *If $S := \{\phi_1, \ldots, \phi_n\}$ is a linearly dependent subset of $C^{n-1}(J)$, then $W(x; \phi_1, \ldots, \phi_n) = 0, \ x \in J$.*

Proof. Since S is linearly dependent in $C^{n-1}(J)$, there exists a nonzero vector $(c_1, \ldots, c_n) \in \mathbb{R}^n$ such that we have the following identity

$$
c_1 \phi_1(x) + \cdots + c_n \phi_n(x) = 0, \ x \in J. \tag{3.47}
$$

On differentiating identity (3.47) k-times, $1 \le k \le n-1$, we get

$$
c_1 \phi_1^{(k)}(x) + \cdots + c_n \phi_n^{(k)}(x) = 0, \ x \in J. \tag{3.48}
$$

For better understanding, we write (3.47)–(3.48) in a matrix form given by

$$
\begin{pmatrix} \phi_1(x) & \phi_2(x) & \cdots & \phi_n(x) \\ \phi_1'(x) & \phi_2'(x) & \cdots & \phi_n'(x) \\ \vdots & \vdots & & \vdots \\ \phi_1^{(n-1)}(x) & \phi_2^{(n-1)}(x) & \cdots & \phi_n^{(n-1)}(x) \end{pmatrix} \begin{pmatrix} c_1 \\ c_2 \\ \vdots \\ c_n \end{pmatrix} = \begin{pmatrix} 0 \\ 0 \\ \vdots \\ 0 \end{pmatrix}. \tag{3.49}
$$

Since system (3.49) has a nonzero solution, namely $(c_1, \ldots, c_n)$, the determinant of the square matrix in (3.49) is zero. In other words, we have $W(x; \phi_1, \ldots, \phi_n) = 0, x \in J$. ☐

The converse of Proposition 3.2.11 is not always true. In the next example, we provide two linearly independent functions whose Wronskian is identically zero.

Example 3.2.12. Let $\phi_1, \phi_2 : \mathbb{R} \to \mathbb{R}$ be defined by

$$\phi_1(x) = \begin{cases} x^3, & x \geq 0, \\ 0, & x < 0, \end{cases} \quad \text{and} \quad \phi_2(x) = \begin{cases} 0, & x \geq 0, \\ x^3, & x < 0. \end{cases}$$

Clearly both ϕ_1 and ϕ_2 are C^2-functions.
We now show that $W(x; \phi_1, \phi_2) = 0$, $x \in \mathbb{R}$. For, if $x < 0$, then we observe

$$\phi_1(x) = \phi_1'(x) = 0,$$

which gives $W(x; \phi_1, \phi_2) = 0$, $x < 0$. Similarly, it follows that $W(x; \phi_1, \phi_2) = 0$, $x \geq 0$.
Finally, we prove that ϕ_1 and ϕ_2 are linearly independent. For, let $c_1 \phi_1(x) + c_2 \phi_2(x) = 0$, $x \in \mathbb{R}$. Put $x = 1$ and $x = -1$ to get $c_1 = 0$ and $c_2 = 0$, respectively. Thus ϕ_1, ϕ_2 are linearly independent in $C^1(\mathbb{R})$. □

Moreover if we assume that $\phi_1, \ldots, \phi_n$ are solutions to $\mathscr{L}[y] = 0$ then the converse of Proposition 3.2.11 holds true. The details are given in the following result.

Theorem 3.2.13. *Let $\mathscr{L}$ be as in (3.43). Assume that $\phi_1, \ldots, \phi_n \in C^n(J)$ are solutions to $\mathscr{L}[y] = 0$. If the Wronskian $W(x; \phi_1, \ldots, \phi_n) = 0$, $x \in J$ then the set $S = \{\phi_1, \ldots, \phi_n\}$ is linearly dependent.*

Proof. Let $x_0 \in J$. As the Wronskian of $\phi_1, \ldots, \phi_n$ is given to be the constant zero function, in particular, we have $W(x_0; \phi_1, \ldots, \phi_n) = 0$. Therefore, there exists a nonzero vector $(c_1, \ldots, c_n) \in \mathbb{R}^n$ which is a solution to the following system of equations

$$\begin{pmatrix} \phi_1(x_0) & \phi_2(x_0) & \cdots & \phi_n(x_0) \\ \phi_1'(x_0) & \phi_2'(x_0) & \cdots & \phi_n'(x_0) \\ \vdots & \vdots & & \vdots \\ \phi_1^{(n-1)}(x_0) & \phi_2^{(n-1)}(x_0) & \cdots & \phi_n^{(n-1)}(x_0) \end{pmatrix} \begin{pmatrix} c_1 \\ c_2 \\ \vdots \\ c_n \end{pmatrix} = \begin{pmatrix} 0 \\ 0 \\ \vdots \\ 0 \end{pmatrix}. \quad (3.50)$$

We now define

$$\Phi(x) := c_1 \phi_1(x) + \cdots + c_n \phi_n(x), \quad x \in J.$$

In view of linearity of $\mathscr{L}$, we get $\mathscr{L}\Phi(x) = 0$, $x \in J$. Moreover from (3.50) we have

$$\Phi(x_0) = \Phi'(x_0) = \cdots = \Phi^{(n-1)}(x_0) = 0.$$

From Remark 3.2.4, this readily implies that $\Phi \equiv 0$ on J. Hence we produced a nonzero vector $(c_1, \ldots, c_n)$ such that $c_1 \phi_1(x) + \cdots + c_n \phi_n(x) = 0$, $x \in J$. Therefore S is linearly dependent. □

We now prove another interesting property of the Wronskian which says that if we know that the Wronskian of solutions of $\mathscr{L}[y] = 0$ vanishes at a point in J, then it is identically zero on J. In order to do that, we prove the following technical result.

Proposition 3.2.14. *Let $\mathscr{L}$ be as in (3.43). Moreover, let $\phi_1,\ldots,\phi_n$ be solutions to $\mathscr{L}[y] = 0$ on J. If $x_0 \in J$, then*

$$W(x;\phi_1,\ldots,\phi_n) = W(x_0;\phi_1,\ldots,\phi_n)e^{-\int_{x_0}^x a_{n-1}(t)dt}, \quad x \in J. \tag{3.51}$$

Proof. On differentiating the Wronskian with respect to x and using the derivative of the determinant formula (see Appendix C), we arrive at

$$\frac{d}{dx}W(x;\phi_1,\ldots,\phi_n) = \begin{vmatrix} \phi_1'(x) & \phi_2'(x) & \cdots & \phi_n'(x) \\ \phi_1'(x) & \phi_2'(x) & \cdots & \phi_n'(x) \\ \vdots & \vdots & & \vdots \\ \phi_1^{(n-1)}(x) & \phi_2^{(n-1)}(x) & \cdots & \phi_n^{(n-1)}(x) \end{vmatrix}$$

$$+\cdots+ \begin{vmatrix} \phi_1(x) & \phi_2(x) & \cdots & \phi_n(x) \\ \phi_1'(x) & \phi_2'(x) & \cdots & \phi_n'(x) \\ \vdots & \vdots & & \vdots \\ \phi_1^{(n)}(x) & \phi_2^{(n)}(x) & \cdots & \phi_n^{(n)}(x) \end{vmatrix}.$$

Observe that all the determinants except the last one on the right hand side of above equation vanish. Moreover we have

$$\phi_i^{(n)}(x) = -a_{n-1}(x)\phi_i^{(n-1)}(x) - \cdots - a_0(x)\phi_i(x).$$

Thus, we obtain

$$\frac{d}{dx}W(x;\phi_1,\ldots,\phi_n) = \begin{vmatrix} \phi_1 & \phi_2 & \cdots & \phi_n \\ \phi_1' & \phi_2' & \cdots & \phi_n' \\ \vdots & \vdots & & \vdots \\ -\sum_{j=0}^{n-1} a_j\phi_1^{(j)} & -\sum_{j=0}^{n-1} a_j\phi_2^{(j)} & \cdots & -\sum_{j=0}^{n-1} a_j\phi_n^{(j)} \end{vmatrix}.$$

Using the row operation $R_n \to R_n + a_0(x)R_1 + a_1(x)R_2 + \cdots + a_{n-2}(x)R_{n-1}$, we get

$$\frac{d}{dx}W(x;\phi_1,\ldots,\phi_n) = \begin{vmatrix} \phi_1 & \phi_2 & \cdots & \phi_n \\ \phi_1' & \phi_2' & \cdots & \phi_n' \\ \vdots & \vdots & & \vdots \\ -a_{n-1}\phi_1^{(n-1)} & -a_{n-1}\phi_2^{(n-1)} & \cdots & -a_{n-1}\phi_n^{(n-1)} \end{vmatrix}$$

$$= -a_{n-1}(x)W(x;\phi_1,\ldots,\phi_n). \tag{3.52}$$

On integrating (3.52), we obtain the required formula. $\qquad\square$

Remark 3.2.15. In view of Proposition 3.2.14, it is clear that for arbitrary $x_0, x \in J$ both $W(x;\phi_1,\ldots,\phi_n)$ and $W(x_0;\phi_1,\ldots,\phi_n)$ have the same sign. In other words, if $W(x_0;\phi_1,\ldots,\phi_n)$ is positive (or negative or zero) for some $x_0 \in J$, then $W(x;\phi_1,\ldots,\phi_n)$ remains positive (or negative or zero) for all $x \in J$.

Remark 3.2.16. Let $\mathscr{L}$ be as in (3.43). Assume that $S = \{\phi_1, \ldots, \phi_n\}$ is a set of solutions to $\mathscr{L}[y] = 0$. Then in view of Proposition 3.2.11, Theorem 3.2.13 and Remark 3.2.15, we conclude that the set S linearly dependent in $C^n(J)$ if and only if $W(x_0; \phi_1, \ldots, \phi_n) = 0$ for some $x_0 \in J$. Equivalently, the set S is linearly independent in $C^n(J)$ if and only if the Wronskian $W(x; \phi_1, \ldots, \phi_n)$ never vanishes.

3.2.3 Lagrange's method of reduction of the order

In this subsection, we discuss a method of reducing a linear homogeneous ODE of order n whose solution is known, to an ODE of order $(n-1)$. The advantage of this method is that the other solutions to the original n-th order equation are given in terms of the solutions to the equation of order $(n-1)$. This method is similar to the method of variation of parameters (see Section 3.3.1). Before getting into the general n-th order case, we shall discuss the second order case. To this end, we assume that ϕ_1 is a solution to

$$y'' + a_1(x)y' + a_0(x)y = 0, \ x \in J, \tag{3.53}$$

and ϕ_1 never vanishes in J. Set $\phi_2(x) = u(x)\phi_1(x)$, $x \in J$. We want to find u such that ϕ_2 is a solution to (3.53). By substituting ϕ_2 in (3.53), we get

$$\phi_1 u'' + (2\phi_1' + a_1\phi_1)u' + (\phi_1'' + a_1\phi_1' + a_0\phi_1)u = 0.$$

Since ϕ_1 is a solution to (3.53), we get the following equation for u as

$$\phi_1 u'' + (2\phi_1' + a_1\phi_1)u' = 0, \ x \in J. \tag{3.54}$$

In order to solve (3.54), we put $v = u'$ to arrive at

$$\phi_1 v' + (2\phi_1' + a_1\phi_1)v = 0, \ x \in J. \tag{3.55}$$

Hence in order to find the other solution ϕ_2 to the second order ODE (3.53), it is enough to solve the first order ODE (3.55). This is what we meant by the reduction of the order. It is easy to see that a solution to (3.55) is given by

$$v(x) = e^{-\int^x \left(\frac{2\phi_1'}{\phi_1} + a_1\right)dx} = \frac{e^{-\int^x a_1\,dx}}{\phi_1^2(x)}.$$

Hence another solution to (3.53) is

$$\phi_2(x) = \phi_1(x)\int^x v\,dx = \phi_1(x)\int^x \frac{e^{-\int^x a_1\,dx}}{\phi_1^2(x)}\,dx. \tag{3.56}$$

One can easily observe that ϕ_1 and ϕ_2 are linearly independent. Therefore the general solution to (3.53) is

$$\phi(x) = c_1\phi_1(x) + c_2\phi_2(x).$$

We now give some examples in which we use this method to find the general solutions of homogeneous ODEs.

Example 3.2.17. Find the general solution to

$$(5 + x^2)y'' - 2xy' + 2y = 0, \ x \in [1, \infty).$$

Solution. We first write the given equation into the standard form given in (3.53) to obtain

$$y'' - \frac{2x}{5 + x^2}y' + \frac{2}{5 + x^2}y = 0. \tag{3.57}$$

Comparison with (3.53) gives us $a_1 = -\frac{2x}{5+x^2}$ and $a_0 = \frac{2}{5+x^2}$.

Since $a_1(x) + xa_0(x) = 0$, we deduce that $\phi_1(x) = x$ is a solution to (3.57). Owing to formula (3.56), we get another solution as

$$\phi_2(x) = x \int^x v dx = x \int^x \frac{e^{-\int a_1}}{x^2} dx = x \int^x \frac{x^2 + 5}{x^2} dx = x(x - \frac{5}{x}) = x^2 - 5.$$

Therefore

$$\phi(x) = c_1 x + c_2(x^2 - 5), \ x \in [1, \infty),$$

where c_1 and c_2 are arbitrary real numbers is a general solution to the given differential equation. □

Example 3.2.18. Find the general solution to

$$xy'' - (2x + 2)y' + (x + 2)y = 0, \ x \in [1, \infty).$$

Solution. Write the given equation in the standard form to get

$$y'' - (2 + \frac{2}{x})y' + (1 + \frac{2}{x})y = 0. \tag{3.58}$$

Thus we have $a_1 = -\frac{2}{x} - 2$ and $a_0 = 1 + \frac{2}{x}$. Since $1 + a_1 + a_2 = 0$, we deduce that $\phi_1 = e^x$ is a solution to (3.58). In order to find other solution, we employ formula (3.56) and get

$$\phi_2(x) = e^x \int^x v dx = e^x \int^x \frac{e^{-\int a_1}}{e^{2x}} dx = e^x \int^x \frac{x^2 e^{2x}}{e^{2x}} dx = \frac{x^3}{3} e^x.$$

Hence

$$\phi(x) = c_1 e^x + c_2 x^3 e^x, \ x \in [1, \infty), \ c_1, c_2 \in \mathbb{R},$$

is a general solution to the given equation. □

Example 3.2.19. Find the general solution to

$$(\sin x + \cos x)y'' + 2\sin(x)y' + (\sin x - \cos x)y = 0, \ x \in [0, \frac{\pi}{2}].$$

Solution. As before, we write the given equation as

$$y'' + \frac{2\sin x}{\sin x + \cos x}y' + \frac{\sin x - \cos x}{\sin x + \cos x}y = 0. \tag{3.59}$$

Here we notice that $a_1 = \frac{2\sin x}{\sin x + \cos x}$ and $a_0 = \frac{\sin x - \cos x}{\sin x + \cos x}$.

Since $1 - a_1 + a_2 = 0$, we deduce that $\phi_1 = e^{-x}$ is a solution to (3.58). We first consider

$$\int^x a_1 dx = \int^x \left(1 - \frac{\cos x - \sin x}{\sin x + \cos x}\right) dx = x - \log(\sin x + \cos x).$$

Then the other solution to (3.59) is given by

$$\phi_2(x) = e^{-x} \int^x \frac{e^{-\int a_1}}{e^{-2x}} dx = e^{-x} \int^x e^x (\sin x + \cos x) dx = \sin x.$$

Thus the general solution to equation (3.59) is

$$\phi(x) = c_1 e^{-x} + c_2 \sin x, \ x \in [0, \tfrac{\pi}{2}],$$

where $c_1, c_2 \in \mathbb{R}$. □

We now extend this method to an n-th order ODE given by

$$\mathscr{L}[y](x) := y^{(n)} + a_{n-1}(x)y^{(n-1)} + \cdots + a_1(x)y' + a_0(x)y = 0, \ x \in J. \ (3.60)$$

Let ϕ be a solution to (3.60) and ϕ does not vanish in J. We want to find u such that $\eta(x) := u(x)\phi(x)$, $x \in J$, is a solution to (3.60). On substituting η in (3.60) and using the Leibniz rule of successive differentiation, we get

$$u(x)(\mathscr{L}[\phi](x)) + \phi(x)u^{(n)}(x) + b_{n-1}(x)u^{(n-1)}(x) + \cdots + b_1(x)u'(x) = 0,$$

for some $b_1, \ldots, b_{n-1}$ which depend on ϕ. Since ϕ is a solution to (3.60), u satisfies

$$\phi(x)u^{(n)}(x) + b_{n-1}(x)u^{(n-1)}(x) + \cdots + b_1(x)u'(x) = 0, \ x \in J. \quad (3.61)$$

Now if we put $v(x) = u'(x)$ in (3.61) then v satisfies

$$\phi(x)v^{(n-1)}(x) + b_{n-1}(x)v^{(n-2)}(x) + \cdots + b_1(x)v(x) = 0. \quad (3.62)$$

Since ϕ does not vanish in J, owing to Theorems 3.2.3 and 3.2.8, equation (3.62) has $(n-1)$ linearly independent solutions, say $v_1, \ldots, v_{n-1}$. Therefore

$$u_1 := \int^x v_1 dx, \ldots, u_{n-1} := \int^x v_{n-1} dx,$$

are linearly independent solutions to (3.61). Hence

$$\phi, \phi \int^x v_1 dx, \ldots, \phi \int^x v_{n-1} dx,$$

are solutions to (3.60).

Claim: The set $\{\phi, \phi \int^x v_1 dx, \ldots, \phi \int^x v_{n-1} dx\}$ is linearly independent in $C^n(J)$.
For, we suppose

$$c_1\phi + c_2\phi \int^x v_1 dx + \cdots + c_n\phi \int^x v_{n-1} dx = 0, \ x \in J.$$

As ϕ never vanishes, we have

$$c_1 + c_2 \int^x v_1 dx + \cdots + c_n \int^x v_{n-1} dx = 0, \ x \in J. \qquad (3.63)$$

On differentiating (3.63) we get

$$c_2 v_1 + \cdots + c_n v_{n-1} = 0.$$

Since $v_1, \ldots, v_{n-1}$ are linearly independent, we obtain $c_2 = \cdots = c_n = 0$. In view of (3.63), we get $c_1 = 0$. This completes the proof of the claim. We summarize our discussion in the following Theorem.

Theorem 3.2.20. *Suppose ϕ is a non-vanishing solution to (3.60). Then there exists an $(n-1)$-th order ODE given in (3.62) whose coefficients depend on ϕ such that the following holds. If $v_1, \ldots, v_{n-1}$ are linearly independent solutions to (3.62), then the set*

$$\left\{ \phi, \phi \int^x v_1 dx, \ldots, \phi \int^x v_{n-1} dx \right\}$$

forms a basis for the solution space of (3.60).

3.2.4   Zeros of the solutions to second order ODEs

In this subsection, we restrict ourselves to the second order linear homogeneous ODEs and study the oscillation behavior of the solutions to these equations. Consider

$$y'' + a_1(x)y' + a_0(x)y = 0, \ x \in J, \qquad (3.64)$$

where $a_1 \in C(\bar{J})$, $a_0 \in C^1(\bar{J})$. We provide some sufficient conditions on a_1 and a_0 such that every solution to (3.64) has finite number of zeros. Moreover, we present sufficient conditions for having infinite number of zeros for the solutions. To this end, we first begin with the following elementary observation.

Proposition 3.2.21. *Let ϕ be a solution to (3.64), and $[\alpha, \beta] \subseteq J$. If ϕ is not identically zero in $[\alpha, \beta]$, then there exist at most finite number of zeros of ϕ in $[\alpha, \beta]$.*

Proof. Suppose ϕ is not a constant zero function and there exists a sequence (γ_n) of distinct numbers in $[\alpha, \beta]$ such that $\phi(\gamma_n) = 0$, $n \in \mathbb{N}$. Since $[\alpha, \beta]$ is a closed and bounded interval, there exists a subsequence (γ_{n_k}) of (γ_n) and a number $\gamma \in [\alpha, \beta]$ such that $(\gamma_{n_k}) \to \gamma$. By continuity of ϕ, we get $\phi(\gamma_{n_k}) \to \phi(\gamma)$ and thus

$$\phi(\gamma) = 0. \qquad (3.65)$$

On the other hand, for each index n_k, by Rolle's theorem, there exists a δ_{n_k} between γ_{n_k} and $\gamma_{n_{k-1}}$ such that $\phi'(\delta_{n_k}) = 0$. Moreover, we notice that $\delta_{n_k} \to \gamma$. Owing to continuity of ϕ', we have $\phi'(\delta_{n_k}) \to \phi'(\gamma)$. Hence we get

$$\phi'(\gamma) = 0. \qquad (3.66)$$

In view of (3.65)–(3.66) and Remark 3.2.4, we get $\phi \equiv 0$, which is a contradiction. This completes the proof. $\qquad\square$

Our next objective is to find a second order linear differential equation which appears 'simpler' than (3.64), and whose solutions vanish precisely at the points where solutions to (3.64) vanish. To this end, let ϕ be a solution to (3.64) and $\phi = uv$. On substituting ϕ in (3.64), we get

$$(uv)''(x) + a_1(x)(uv)'(x) + a_0(x)(uv)(x) = 0, \ x \in J$$

or

$$vu'' + (a_1 v + 2v')u' + (v'' + a_1 v' + a_0 v)u = 0, \ x \in J. \tag{3.67}$$

We now choose

$$v(x) = \exp(-\int^x \frac{a_1}{2} dx), \ x \in J,$$

so that the coefficient of u' in (3.67) vanishes. Since v is a positive function, ϕ vanishes precisely at the points where u vanishes. Moreover, u is a solution to the ODE

$$u''(x) + q(x)u(x) = 0, \ x \in J, \tag{3.68}$$

where $q(x) = \dfrac{v''(x) + a_1(x)v'(x) + a_0(x)v(x)}{v(x)}, \ x \in J.$

Therefore in order to study the zeros of the solutions to (3.64), it is enough to study the zeros of (3.68). Thus we restrict ourselves to (3.68) for various choices of q. We refer to (3.68) as the normal form of (3.64).

Next, we consider $y'' - y = 0$. The general solution to this equation is

$$\phi = c_1 e^x + c_2 e^{-x}, \ c_1, c_2 \in \mathbb{R}.$$

It is straightforward to notice that ϕ vanishes at most once in $\mathbb{R}$ for any choice of $c_1, c_2 \in \mathbb{R}$. In fact, this observation can be generalized in the following result.

Proposition 3.2.22. *Assume that* $q \in C(\bar{J})$, $q(x) \leq 0$, $x \in \bar{J}$. *If* ϕ *is a nonzero solution to* (3.68), *then there exists at most one real solution to the equation* $\phi(x) = 0$.

Proof. In view of Proposition 3.2.21, the zeros of ϕ are discrete. Let x_1 and x_2 be two consecutive zeros of ϕ with $x_1 < x_2$. Without loss of generality, we assume that $\phi(\alpha) > 0$, $\alpha \in (x_1, x_2)$. Since $\phi(x_1) = 0$, we have

$$\phi'(x_1) = \lim_{\alpha \to x_1^+} \frac{\phi(\alpha) - \phi(x_1)}{\alpha - x_1} = \lim_{\alpha \to x_1^+} \frac{\phi(\alpha)}{\alpha - x_1} \geq 0.$$

Since ϕ is a nonzero function and in view of Remark 3.2.4, both ϕ and ϕ' cannot vanish simultaneously at any point (in particular at x_1). Therefore we get $\phi'(x_1) > 0$. Similarly one can show that $\phi'(x_2) < 0$. Thus we obtain

$$\phi'(x_1) > 0 > \phi'(x_2). \tag{3.69}$$

On the other hand, we have

$$\phi''(\alpha) = -q(\alpha)\phi(\alpha) \geq 0, \ \alpha \in (x_1, x_2).$$

Hence ϕ' is non-decreasing on $[x_1, x_2]$. This is a contradiction to (3.69). Therefore we conclude that there exists at most one solution to $\phi(x) = 0$. $\qquad\square$

We now consider the case when $q > 0$ in (3.68). In particular, if $q \equiv 1$ then any solution to (3.68) has infinitely many real zeros. A generalization of this is presented in the next result.

Proposition 3.2.23. *Let $J = (x_0, \infty)$ and $q \in C(\bar{J})$. Moreover, assume that $q > 0$ and $\int_{x_0}^{\infty} q(t)dt = \infty$. If ϕ is a nonzero solution to (3.68), then there exist infinitely many positive solutions to $\phi(x) = 0$.*

Proof. On the contrary, suppose that there exist only finite number of positive solutions to $\phi(x) = 0$, say $x_1, \ldots, x_n$. Without loss of generality, we assume that $\phi(\alpha) > 0$, $\alpha > x_n$. Fix $M > x_n$.

Step 1: In this step, we prove that there exists $M_1 > M$ such that $\phi'(M_1) < 0$. For, we define

$$\psi(x) = \frac{\phi'(x)}{\phi(x)}, \ x \geq M.$$

Then we have

$$\psi'(x) = -q(x) - \psi^2(x) \leq -q(x), \ x \geq M.$$

Now by integrating from M to N, we get

$$\psi(N) - \psi(M) \leq -\int_{M}^{N} q(t)dt, \ N > M. \tag{3.70}$$

In view of our hypotheses on q, we can choose $\widetilde{M}$ such that

$$\int_{M}^{\widetilde{M}} q(t)dt > |\psi(M)| + 1. \tag{3.71}$$

By taking $N = \widetilde{M}$ in (3.70) and using (3.71), we estimate

$$\psi(\widetilde{M}) \leq \psi(M) - |\psi(M)| - 1 \leq -1.$$

Since ψ and ϕ' have same signs in $[M, \infty)$, we conclude Step 1 by taking $M_1 = \widetilde{M}$.

Step 2: In this step, we get a contradiction to the assumption that ϕ never vanishes for $x > x_n$, which in turns completes the proof.
To that end, we first observe that

$$\phi''(x) = -q(x)\phi(x) < 0, \ x \geq M.$$

Therefore the Taylor expansion of ϕ about the point $x = M_1$ gives us the inequality

$$\phi(x) = \phi(M_1) + (x - M_1)\phi'(M_1) + \frac{(x - M_1)^2}{2}\phi''(\xi) \qquad (3.72)$$

$$\leq \phi(M_1) + (x - M_1)\phi'(M_1), \quad x \geq M. \qquad (3.73)$$

Since $\phi'(M_1) < 0$, it is possible to choose x 'large enough' such that $\phi(x) < 0$. In particular, if $x = \frac{\phi(M_1)+1}{|\phi'(M_1)|} + M_1$, then $\phi(x) \leq -1$. This is a contradiction to the assumption that $\phi(\alpha) > 0$, $\alpha > M$. Hence there exist infinite number of positive solutions to $\phi(x) = 0$. $\qquad\square$

The reader is advised to go through the proof of the previous proposition carefully and check whether the positivity condition on q can be replaced with $q \geq 0$ in the hypotheses.

Remark 3.2.24. The condition $\displaystyle\int_{x_0}^{\infty} q(t)dt = \infty$, is essential in Proposition 3.2.23. For instance, consider the ODE, $y'' + \dfrac{1}{4x^2}y = 0$, $x \in [1, \infty)$. By comparing with (3.68), we get $q(x) = \frac{1}{4x^2}$ and $\displaystyle\int_1^{\infty} q(t)dt = \frac{1}{4}$. Since it is the Euler equation, we find that one of its solutions to be $\phi(x) = \sqrt{x}$ (see Subsection 3.1.3) which does not have infinite number zeros in $[1, \infty)$.

We now consider $y'' + y = 0$. The functions $\phi_1 = \sin x$ and $\phi_2 = \cos x$ are two linearly independent solutions to the equation under consideration. Some straightforward observations pertaining to the zeros of ϕ_1 and ϕ_2 are as follows.

(*i*) The functions ϕ_1 and ϕ_2 do not vanish simultaneously.

(*ii*) Between any two zeros of ϕ_1 there is a zero of ϕ_2 and vice-versa.

With little more effort, one can show that the functions $\psi_1 = \sin(\alpha x)$ and $\psi_2 = \cos(\alpha x)$ which are linearly independent solutions to the equation $y'' + \alpha^2 y = 0$, also have properties (*i*) and (*ii*). We now ask a question whether these two properties hold for any two linearly independent solutions to any second order linear ODE with variable coefficients. We have an affirmative answer to this question. The details are provided in the following result, known as the Sturm separation theorem.

Theorem 3.2.25 (Sturm separation theorem). *Let $q \in C(\bar{J})$. If ϕ_1 and ϕ_2 are two linearly independent solutions to (3.68). If x_1, x_2 are successive zeros of ϕ_1 and $x_1 < x_2$, there exists a unique zero of ϕ_2 in (x_1, x_2).*

Proof. Without loss of generality, we assume that $\phi_1(\alpha) > 0$ for every $\alpha \in (x_1, x_2)$. Using the same argument as in the proof of Proposition 3.2.22, we can show that $\phi_1'(x_1) > 0$ and $\phi_1'(x_2) < 0$. On the other hand, in view of Remark 3.2.15, we have

$$W(x_1; \phi_1, \phi_2)W(x_2; \phi_1, \phi_2) > 0.$$

Thus we have

$$(\phi_1\phi_2' - \phi_2\phi_1')(x_1)(\phi_1\phi_2' - \phi_2\phi_1')(x_2) > 0,$$

or

$$\phi_2(x_1)\phi_1'(x_1)\phi_2(x_2)\phi_1'(x_2) > 0.$$

Since $\phi_1'(x_1)$ and $\phi_1'(x_2)$ have opposite signs, $\phi_2(x_1)$ and $\phi_2(x_1)$ also have opposite signs. Hence by the intermediate value theorem ϕ_2 has at least one real zero in (x_1, x_2). This completes the existence part of the proof. Uniqueness of a zero of ϕ_2 follows immediately by interchanging the roles of ϕ_1 and ϕ_2 in the existence part of this proof. □

So far, we have seen the behavior of zeros of solutions of a single ODE. In the next example, we consider two ODEs to study the relative position of their zeros. Assume that y and u are solutions to the ODEs $y'' + 4y = 0$ and $u'' + u = 0$, respectively. Then we have

$$y(x) = c_1 \sin(2x) + c_2 \cos(2x), \ u(x) = d_1 \sin(x) + d_2 \cos(x)$$

for some $c_1, c_2, d_1, d_2 \in \mathbb{R}$. It is easy to observe that between any two zeros of u there exists at least one zero of y. This is generalized in the following theorem which is known as the Sturm comparison theorem.

Theorem 3.2.26 (Sturm comparison theorem). *Let* $r, \ q \in C(\bar{J})$. *Assume* $r(x) \geq q(x)$ *in an open interval* J. *Assume that* y *and* u *are solutions to*

$$y''(x) + r(x)y(x) = 0, \ x \in J, \tag{3.74}$$
$$u''(x) + q(x)u(x) = 0, \ x \in J, \tag{3.75}$$

respectively. If $x_1, x_2 \in J$ *are two roots of* u, *then there exists a root of* y *in* (x_1, x_2) *unless* $r(t) \equiv q(t)$ *and* u *is a constant multiple of* y *in* (x_1, x_2).

Proof. Without loss of generality, we assume that x_1, x_2 are successive zeros of u and $u(\alpha) > 0$ for all $\alpha \in (x_1, x_2)$. Using the same argument as in the proof of Proposition 3.2.22, we get $u'(x_1) > 0$ and $u'(x_2) < 0$. Then the values of Wronskian of y and u at $x_1, \ x_2$ are given by

$$W(x_1; y, u) = y(x_1)u'(x_1), \quad W(x_2; y, u) = y(x_2)u'(x_2). \tag{3.76}$$

Moreover, we have

$$\begin{aligned} \frac{d}{dx}W(x; y, u) &= y(x)u''(x) - u(x)y''(x) \\ &= (r(x) - q(x))\, u(x)y(x). \end{aligned} \tag{3.77}$$

Suppose $r(\tilde{x}) > q(\tilde{x})$ for some $\tilde{x} \in (x_1, x_2)$. We now assume that $y > 0$ in (x_1, x_2). Then from (3.77), it follows that $W' \geq 0$ and thus W is non-decreasing. Since $\frac{dW}{dx}(\tilde{x}; y, u) > 0$, there exists $x_3, x_4 \in (x_1, x_2)$ such that

$W' > 0$ in (x_3, x_4). Using the mean value theorem, we arrive at

$$
\begin{aligned}
W(x_1; y, u) &\leq W(x_3; y, u) \\
&= W(x_4; y, u) + (x_3 - x_4)\frac{dW}{dx}(c; y, u) \\
&< W(x_4; y, u) \\
&\leq W(x_2; y, u),
\end{aligned} \tag{3.78}
$$

where $c \in (x_3, x_4)$. But from (3.76), we obtain

$$W(x_1, y, u) \geq 0, \; W(x_2, y, u) \leq 0,$$

which is a contradiction to (3.78).

Using a similar argument, one can get a contradiction when $y < 0$ in (x_1, x_2). Hence y should vanish in (x_1, x_2) at least once.

On the other hand, if $r(t) \equiv q(t)$ and y is not a constant multiple of u, then in view of the Sturm separation theorem y vanishes in (x_1, x_2). □

There are numerous interesting applications of this theorem (see Exercises 3.16–3.21).

3.3   Non-homogeneous ODEs with variable coefficients

In this section, we discuss a method to solve the non-homogeneous equations with variable coefficients. In the constant coefficients case, we have seen some methods when the non-homogeneous term has a special structure (viz., it is a finite linear combination of polynomial, exponential, trigonometric functions). In this section, we develop a method which works in more generic situations. Before we do that, we prove the following result:

Lemma 3.3.1. *Let* $f, a_i \in C(\bar{J})$, $0 \leq i \leq n - 1$. *We assume that* ψ *is a (particular) solution to a linear non-homogeneous equation of order* n *given by*

$$\mathscr{L}[y] := y^{(n)} + a_{n-1}y^{(n-1)} + \cdots + a_0 y = f, \; x \in J. \tag{3.79}$$

Let $\phi_1, \ldots, \phi_n$ *be linearly independent solutions to the associated homogeneous equation* $\mathscr{L}[y] = 0$. *If* ϕ *is any other solution to* $\mathscr{L}[y] = f$, *then there exist* n *constants* $c_1, \ldots, c_n$, *such that* $\phi = \psi + c_1\phi_1 + \cdots + c_n\phi_n$.

Proof. We consider

$$\mathscr{L}[\phi - \psi] = \mathscr{L}[\phi] - \mathscr{L}[\psi] = f - f = 0.$$

Thus $\phi - \psi$ is a solution to the nth order homogeneous equation $\mathscr{L}[y] = 0$. Since $\phi_1, \ldots, \phi_n$ are linearly independent solutions to $\mathscr{L}[y] = 0$, owing to Theorem 3.2.8, there exist constants $c_i \in \mathbb{R}$, $1 \leq i \leq n$, such that

$$\phi - \psi = c_1\phi_1 + \cdots + c_n\phi_n, \; x \in J.$$

This completes the proof. □

Definition 3.3.2. If $\phi_1, \ldots, \phi_n$ and ψ are as in Lemma 3.3.1, then

$$\phi = \psi + c_1\phi_1 + \cdots + c_n\phi_n,$$

where c_i's are arbitrary constants, is called as *the general solution* to (3.79).

Here onward, solving non-homogeneous ODE (3.79) means finding its general solution.

3.3.1 Method of variation of parameters

Before we present the method of variation of parameters for the n-th order ODEs, we demonstrate this method in the case of the first order and the second order linear ODEs.

Let f, $a \in C(\bar{J})$. We first consider

$$y'(x) + a(x)y(x) = f(x), \ x \in J. \tag{3.80}$$

Let ϕ_1 be a nonzero solution to the corresponding homogeneous differential equation, i.e.,

$$\phi_1'(x) + a(x)\phi_1(x) = 0, \ x \in J. \tag{3.81}$$

We observe that ϕ_1 never vanishes in J. We assume that there exists a solution to (3.80) of the form

$$\psi(x) = u(x)\phi_1(x), \ x \in J.$$

where u is suitably chosen later. (This is the vital step in this method.) In other words, we now need to find u such that $\psi = u\phi_1$ is a solution to (3.80). To this end, we substitute $\psi = u\phi_1$ in (3.80) to obtain

$$u'\phi_1 + u(\phi_1' + a\phi_1) = f, \ x \in J. \tag{3.82}$$

Therefore from (3.81)–(3.82) we get $u'\phi_1 = f$ and

$$u(x) = \int^x \frac{f(t)}{\phi_1(t)}\,dt, \ x \in J.$$

Hence we get

$$\psi(x) = \phi_1(x)\int^x \frac{f(t)}{\phi_1(t)}\,dt, \ x \in J,$$

is a solution to (3.80).

Thus the general solution to (3.80) is given by

$$\phi(x) = ce^{-\int^x a(t)dt} + \phi_1(x)\int^x \frac{f(t)}{\phi_1(t)}\,dt, \ x \in J, \tag{3.83}$$

where c is an arbitrary real number.

We next consider the second order non-homogeneous ODE given by

$$y''(x) + a_1(x)y'(x) + a_0(x)y(x) = f(x), \ x \in J. \tag{3.84}$$

Let ϕ_1, ϕ_2 be linearly independent solutions to the corresponding homogeneous equation

$$y''(x) + a_1(x)y'(x) + a_0(x)y(x) = 0, \ x \in J. \tag{3.85}$$

Owing to Remark 3.2.15, the Wronskian $W(x; \phi_1, \phi_2)$ of ϕ_1, ϕ_2 never vanishes in J. As in the case of the first order ODE, we assume that

$$\psi(x) = u_1(x)\phi_1(x) + u_2(x)\phi_2(x), \ x \in J, \tag{3.86}$$

is a solution to (3.84) for some differentiable functions u_1 and u_2. In other words, we find two functions u_1, u_2 such that ψ given by (3.86) is a solution to (3.84). On differentiating (3.86), we obtain

$$\psi' = u_1'\phi_1 + u_1\phi_1' + u_2'\phi_2 + u_2\phi_2'.$$

We set

$$u_1'\phi_1 + u_2'\phi_2 = 0, \tag{3.87}$$

so that

$$\psi' = u_1\phi_1' + u_2\phi_2'. \tag{3.88}$$

It is worth noticing that if u_1, u_2 satisfy (3.87), then ψ' does not contain derivatives of u_1 or u_2 but contains the derivatives of ϕ_1 and ϕ_2. We now differentiate (3.88) to arrive at

$$\psi'' = u_1'\phi_1' + u_1\phi_1'' + u_2'\phi_2' + u_2\phi_2''. \tag{3.89}$$

If we do not assume (3.87), then we would have obtained terms containing the second derivative of u_1, u_2 in ψ''. By substituting ψ, ψ' and ψ'' in (3.84), we get

$$(\phi_1'' + a_1\phi_1' + a_0\phi_1)u_1 + (\phi_2'' + a_1\phi_2' + a_0\phi_2)u_2 + u_1'\phi_1' + u_2'\phi_2' = f \tag{3.90}$$

Since ϕ_1 and ϕ_2 are solutions to (3.85), equation (3.90) becomes

$$u_1'\phi_1' + u_2'\phi_2' = f. \tag{3.91}$$

For simplicity, we write (3.87) and (3.91) in a matrix form

$$\begin{pmatrix} \phi_1 & \phi_2 \\ \phi_1' & \phi_2' \end{pmatrix} \begin{pmatrix} u_1' \\ u_2' \end{pmatrix} = \begin{pmatrix} 0 \\ f \end{pmatrix} \tag{3.92}$$

Since $W(x; \phi_1, \phi_2) \neq 0$, one can easily solve (3.92) to obtain

$$u_1(x) = -\int^x \frac{\phi_2(t)f(t)}{W(t; \phi_1, \phi_2)} dt, \quad u_2(x) = \int^x \frac{\phi_1(t)f(t)}{W(t; \phi_1, \phi_2)} dt. \tag{3.93}$$

Therefore a particular solution to (3.84) is given by

$$\psi(x) = -\phi_1(x) \int^x \frac{\phi_2(t)f(t)}{W(t;\phi_1,\phi_2)} dt + \phi_2(x) \int^x \frac{\phi_1(t)f(t)}{W(t;\phi_1,\phi_2)} dt, \ x \in J. \quad (3.94)$$

Hence the general solution to (3.84) for $x \in J$ is

$$\psi(x) = c_1\phi_1(x)+c_2\phi_2(x)-\phi_1(x) \int^x \frac{\phi_2(t)f(t)}{W(t;\phi_1,\phi_2)} dt+\phi_2(x) \int^x \frac{\phi_1(t)f(t)}{W(t;\phi_1,\phi_2)} dt, \quad (3.95)$$

where c_1, c_2 are arbitrary constants.

To summarize briefly, if ϕ_1, ϕ_2 are solutions to (3.85) and u_1, u_2 satisfy (3.92) then

$$\psi = u_1\phi_1 + u_2\phi_2, \ x \in J,$$

is a solution to (3.84). Owing to this method, finding a solution to the second order non-homogeneous ODEs has reduced to finding the solution to a 2×2 system of linear equations.

Before we move on to a higher order version of this method, we present some examples which are non-homogeneous versions of Examples 3.2.17–3.2.18 that were discussed in Section 3.2.3.

Example 3.3.3. Solve $(5 + x^2)y'' - 2xy' + 2y = (x^2 + 5)^2$, $x \in [1, \infty)$.

Solution: We first write the given equation in the standard form given in (3.84) to get

$$y'' - \frac{2x}{5 + x^2}y' + \frac{2}{5 + x^2}y = x^2 + 5.$$

Comparison with (3.84) gives $a_1 = -\frac{2x}{5+x^2}$, $a_0 = \frac{2}{5+x^2}$ and $f = x^2 + 5$.

From Example 3.2.17, it follows that $\phi_1(x) := x$, $\phi_2(x) := x^2 - 5$, are two independent solutions to the associated homogeneous equation

$$y'' - \frac{2x}{5 + x^2}y' + \frac{2}{5 + x^2}y = 0.$$

Moreover their Wronskian is $W(x;\phi_1,\phi_2) = \phi_1\phi_2' - \phi_1'\phi_2 = x^2 + 5$.

The method of variation of parameters gives us that a particular solution to the given problem is of the form

$$\psi(x) = u_1(x)x + u_2(x)(x^2 - 5),$$

where u_1 and u_2 are given in (3.93). Thus we have

$$u_1(x) = -\int^x \frac{(t^2 - 5)(t^2 + 5)}{t^2 + 5} dt = 5x - \frac{x^3}{3}, \quad u_2(x) = \int^x \frac{t(t^2 + 5)}{t^2 + 5} dt = \frac{x^2}{2}.$$

Therefore

$$\psi(x) = \left(5x - \frac{x^3}{3}\right)x + \frac{x^2}{2}(x^2 - 5) = \frac{x^4}{6} + \frac{5x^2}{2},$$

is a particular solution to the given ODE. The general solution to the given ODE is

$$\phi(x) = \frac{x^4}{6} + \frac{5x^2}{2} + c_1 x + c_2(x^2 - 5), \ x \in [1, \infty),$$

where $c_1, c_2 \in \mathbb{R}$. □

Example 3.3.4. Solve $xy'' - (2x + 2)y' + (x + 2)y = x^2 + x$, $x \in [1, \infty)$.

Solution: As in the previous example, we write the given equation in the standard form as

$$y'' - (2 + \frac{2}{x})y' + (1 + \frac{2}{x})y = x + 1.$$

Here we note that $f(x) = x + 1$. Recall from Example 3.2.18 that $\phi_1 = e^x$ and $\phi_2 = x^3 e^x$ are linearly independent solutions to the associated homogeneous equation. A straightforward calculation gives us that $W(x; \phi_1, \phi_2) = 3x^2 e^{2x}$. We now use (3.93) to compute

$$u_1 = -\int^x \frac{(1+t)t^3 e^t}{3t^2 e^{2t}} dt = \left(1 + x + \frac{x^2}{3}\right) e^{-x},$$

and

$$u_2 = \int^x \frac{(1+t)e^t}{3t^2 e^{2t}} dt = -\frac{e^{-x}}{3x}.$$

Hence a particular solution to the given ODE is

$$\psi = u_1\phi_1 + u_2\phi_2 = 1 + x + \frac{x^2}{3} - \frac{x^2}{3} = 1 + x.$$

Therefore

$$\phi(x) = c_1 e^x + c_2 x^3 e^x + x + 1, \ x \in [1, \infty), c_1, c_2 \in \mathbb{R},$$

is the general solution to the given differential equation. □

We now present the method of variation of parameters for an n-th order linear ODE. Let $f, a_i \in C(\bar{J})$, $0 \leq i \leq n - 1$. Consider

$$\mathcal{L}[y](x) := y^{(n)} + a_{n-1}(x)y^{(n-1)} + \cdots + a_1(x)y' + a_0(x)y = f(x), \ x \in J. \quad (3.96)$$

Let $\phi_1, \ldots, \phi_n$ be linearly independent solutions to the homogeneous equation $\mathcal{L}[y] = 0$. We want to find differentiable functions $u_1, \ldots, u_n$ such that the function

$$\psi(x) = u_1(x)\phi_1(x) + \cdots + u_n(x)\phi_n(x), \ x \in J,$$

is a solution to $\mathcal{L}[y] = f$. As in the case of the second order ODE, we first differentiate ψ to get

$$\psi' = u_1'\phi_1 + \cdots + u_n'\phi_n + u_1\phi_1' + \cdots + u_n\phi_n'. \quad (3.97)$$

If we set

$$u_1'\phi_1 + \cdots + u_n'\phi_n = 0, \quad (3.98)$$

then from (3.97)–(3.98) we have

$$\psi' = u_1\phi_1' + \cdots + u_n\phi_n'. \tag{3.99}$$

Again, if we assume

$$u_1'\phi_1' + \cdots + u_n'\phi_n' = 0, \tag{3.100}$$

then (3.99) gives us

$$\psi'' = u_1\phi_1'' + \cdots + u_n\phi_n''.$$

Owing to (3.100), we do not have the second order derivative terms in ψ''. In the same manner, if we assume that

$$u_1'\phi_1^{(k-1)} + \cdots + u_n'\phi_n^{(k-1)} = 0, \quad 1 \le k \le n-1, \tag{3.101}$$

then by induction we can easily show that

$$\psi^{(k)} = u_1\phi_1^{(k)} + \cdots + u_n\phi_n^{(k)}, \quad 1 \le k \le n-1. \tag{3.102}$$

By putting $k = n-1$ in (3.102) and differentiating once again, we get

$$\psi^{(n)} = u_1'\phi_1^{(n-1)} + \cdots + u_n'\phi_n^{(n-1)} + u_1\phi_1^{(n)} + \cdots + u_n\phi_n^{(n)}.$$

On substituting $\psi, \psi', \ldots, \psi^{(n)}$ in (3.96), we obtain

$$u_1'\phi_1^{(n-1)} + \cdots + u_n'\phi_n^{(n-1)} + u_1\mathscr{L}[\phi_1] + \cdots + u_n\mathscr{L}[\phi_n] = f,$$

or

$$u_1'\phi_1^{(n-1)} + \cdots + u_n'\phi_n^{(n-1)} = f. \tag{3.103}$$

We have a linear system of n equations given in (3.101) and (3.103) for the n unknowns, namely $u_1, \ldots, u_n$. We write the system in a matrix form to get

$$\begin{pmatrix} \phi_1 & \phi_2 & \cdots & \phi_n \\ \phi_1' & \phi_2' & \cdots & \phi_n' \\ \vdots & \vdots & & \vdots \\ \phi_1^{(n-1)} & \phi_2^{(n-1)} & \cdots & \phi_n^{(n-1)} \end{pmatrix} \begin{pmatrix} u_1' \\ u_2' \\ \vdots \\ u_n' \end{pmatrix} = \begin{pmatrix} 0 \\ 0 \\ \vdots \\ f \end{pmatrix}. \tag{3.104}$$

Since $\phi_1, \ldots, \phi_n$ are linearly independent, the square matrix in (3.104), say A, is nonsingular. Hence we can find $u_1', \ldots, u_n'$ uniquely. For instance, one can use the Crammer rule to get $u_i' = \frac{W_i}{W}$, $1 \le i \le n$, where W_i is the determinant of the matrix obtained by replacing the i-th column of A by the column vector $(0, \ldots, 0, f)$ and W is the determinant of A. Therefore a particular solution to (3.96) is

$$\psi(x) = \phi_1(x) \int^x \frac{W_1(t)}{W(t)} dt + \cdots + \phi_n(x) \int^x \frac{W_n(t)}{W(t)} dt, \quad x \in J.$$

Then the general solution to (3.96) is given by

$$\phi(x) = \sum_{i=1}^{n} c_i \phi_i(x) + \sum_{i=1}^{n} \phi_i(x) \int^x \frac{W_i(t)}{W(t)} dt, \quad x \in J. \qquad (3.105)$$

As before, we observe that finding a solution to (3.96) amounts to solving $n \times n$ system of linear equations (3.104). This method was also invented by Lagrange (1775).

Exercise 3.1. Find the general solution to the following homogeneous equations:

(*i*) $y'' - 12y' - 20y = 0$.

(*ii*) $y'' - 14y' + 49y = 0$.

(*iii*) $y'' - 2y' + 10y = 0$.

(*iv*) $y''' - 5y'' + 2y' + 8y = 0$.

(*v*) $y''' - y'' + y' - y = 0$.

(*vi*) $y^{(vi)} - 2y^{(v)} + 3y^{(iv)} - 4y''' + 3y'' - 2y' + y = 0$.

Exercise 3.2. Find the general solution to the following non-homogeneous equations:

(*i*) $y'' - 12y' + 20y = e^{3x}$.

(*ii*) $y'' - 14y' + 49y = e^{7x}$.

(*iii*) $y'' - 2y' + 10y = e^x + \sin x$.

(*iv*) $y''' - 5y'' + 2y' + 8y = e^{-x}$.

(*v*) $y''' - y'' + y' - y = x^4 - 3x^2$.

Exercise 3.3. Show that the following sets are linearly independent in $C(J)$, where J is an interval in $\mathbb{R}$.

(*i*) $\{x^{\alpha} / \alpha > 0\}$.

(*ii*) $\{e^{\alpha x} / \alpha \in \mathbb{R}\}$.

(*iii*) $\{\sin(\alpha x + \beta) / \alpha > 0, \beta \in [0, \pi)\}$.

Hint: (*iii*) Let $\sum\limits_{k=0}^{N} c_k \sin(n_k x + \beta_k) = 0$. On differentiating $4l$ times $0 \le l \le N$, we get $\sum\limits_{k=0}^{N} c_k n_k^{4l} \sin(n_k x + \beta_k) = 0$. Show that this system of $(n+1) \times (n+1)$ linear equations has only trivial solution $c_0 = \cdots = c_N = 0$.

Exercise 3.4. Find the general solution to the following homogeneous equations:

(i) $y'' + 2mxy' + 2my = 0$, if $\phi_1(x) = e^{-mx^2}$ is its solution

(ii) $xy'' - y' - 4x^3y = 0$, if $\phi_1(x) = e^{-x^2}$ is its solution.

(iii) $x^2y'' - 2xy' - 10y = 0$.

(iv) $x^2y'' - xy' + y = 0$.

(v) $(x^2 - 1)y'' - 2xy' + 2y = 0$.

(vi) $xy'' + (x + 5)y' + 5y = 0$.

(vii) $2x(1 - 2x)y'' + (4x^2 + 1)y' - (1 + 2x)y = 0$.

Exercise 3.5. Find the general solution to the following non-homogeneous equations:

(i) $x^2y'' - 2xy' - 10y = 1$.

(ii) $x^2y'' - xy' + y = x$.

(iii) $(x^2 - 1)y'' - 2xy' + 2y = x^2$.

(iv) $xy'' + (x + 5)y' + 5y = 1 + x^2$.

Exercise 3.6. Consider the ODE with constant coefficients given by

$$y^{(n)} + a_{n-1}y^{(n-1)} + \cdots + a_1y' + a_0y = 0, \quad a_i \in \mathbb{R}, \tag{3.106}$$

where $a_i \in \mathbb{R}$ for $0 \leq i \leq n - 1$ and the polynomial

$$P(X) = X^n + a_{n-1}X^{n-1} + \cdots + a_1X + a_0$$

corresponding to (3.106). Then prove the following.

(i) If the real parts of all the roots of the polynomial P are negative, then every solution ϕ to (3.106) satisfies $\phi(x) \to 0$ as $x \to \infty$.

(ii) If the real part of at least one root of the polynomial P is positive, then there exists a solution ϕ to (3.106) and a sequence $(x_n) \to \infty$ such that $|\phi(x_n)| \to \infty$, as $n \to \infty$.

Exercise 3.7. Let $f : (a, b) \to \mathbb{R}$ be a continuous map and $x_0 \in (a, b)$. Then show that the general solution to $y^{(n)}(x) = f(x)$ is

$$\phi(x) = c_0 + \cdots + c_{n-1}x^{n-1} + \frac{1}{(n-1)!}\int_{x_0}^x (x - t)^{n-1}f(t)dt,$$

where $c_1, \ldots, c_{n-1} \in \mathbb{R}$.

Exercise 3.8. Assume that $h, \phi_1, \phi_2 : (a, b) \to \mathbb{R}$ are differentiable functions. Then show the following properties of the Wronskian:

(i) $W(x; h\phi_1, h\phi_2) = h^2 W(x; \phi_1, \phi_2)$, $x \in (a, b)$.

(ii) $W(x; h\phi_1, \phi_1) = -h'(x)\phi_1^2(x)$, $x \in (a, b)$.

Exercise 3.9. Let $a_1, a_0 \in C([a, b])$ and $\phi_1, \phi_2 : (a, b) \to \mathbb{R}$ be solutions to

$$y''(x) + a_1(x)y'(x) + a_0(x)y(x) = 0, \ x \in (a, b).$$

Suppose there exists $x_0 \in (a, b)$ such that both ϕ_1 and ϕ_2 have local maximum (minimum) at x_0. Then show that there exists $k \in \mathbb{R}$ such that $\phi_1 = k\phi_2$.

Exercise 3.10. Suppose $\phi, \psi : (-a, a) \to \mathbb{R}$ are C^2-functions and $W(x; \phi, \psi)$ never vanishes in $(-a, a)$. Then find continuous functions $a_0, a_1 : (-a, a) \to \mathbb{R}$ such that ϕ and ψ are solutions to

$$y''(x) + a_1(x)y'(x) + a_0(x)y(x) = 0, \ x \in (-a, a).$$

Exercise 3.11. Do there exist continuous functions $a_0, a_1 : (-a, a) \to \mathbb{R}$ for some $a > 0$ such that the ODE $y''(x) + a_1(x)y'(x) + a_0(x)y(x) = 0, \ x \in (-a, a)$, has the following solutions?

(i) $(1 - \cos x)h(x)$, where $h \in C^2(\mathbb{R})$.

(ii) $x^n g(x)$, where $n > 2$, $g \in C^2(\mathbb{R})$.

Hint: Observe the behavior of the given functions at $x = 0$.

Exercise 3.12. Find $a_1, a_0 \in C(\mathbb{R})$ such that the equation $y''(x) + a_1(x)y'(x) + a_0(x)y(x) = 0, \ x \in \mathbb{R}$, has a pair of solutions ϕ, ψ such that:

(i) $\phi(x) = x\psi(x), \ x \in \mathbb{R}$.

(ii) $\phi(x)\psi(x) = 10, \ x \in \mathbb{R}$.

Hint: Find ψ in both the cases and substitute in the given equation.

Exercise 3.13. Assume that $\alpha : \mathbb{R} \to \mathbb{R}$ is a C^2-function. Further assume that $\alpha(x)$ is increasing to ∞ as $x \to \infty$. Show that every solution to $y''(x) + \alpha(x)y(x) = 0, \ x \in \mathbb{R}$, is bounded in $[0, \infty)$.

Hint: Multiply the given equation with y', use integration by parts and Gronwall's lemma.

Exercise 3.14. Assume that $\phi : [0, 1] \to \mathbb{R}$ is a solution to $(py')' + qy = 0$, $x \in (0, 1)$, where $p, q \in C^1([0, 1])$. Suppose $p(x) > 0$, $q(x) \neq 0$, $(pq)'(x) > 0$, $x \in [0, 1]$. If $0 < \xi_1 < \xi_2 < 1$ are local maxima of ϕ then show that $\phi^2(\xi_1) \leq \phi^2(\xi_2)$.

Hint: Set $\psi(x) := \phi^2(x) + \dfrac{p^2(x)(\phi')^2(x)}{p(x)q(x)}$ and notice that $\psi' \leq 0$.

Exercise 3.15. For each $a \in \mathbb{R}$, show that any nonzero solution to

$$y''(x) - 2xy'(x) + 2ay(x) = 0, \ x \in \mathbb{R},$$

has at most finitely many zeros in $\mathbb{R}$.

Hint: Convert to the normal form and use Propositions 3.2.21–3.2.22.

Exercise 3.16. Let $h \in C(\mathbb{R})$ be such that $h(x) \to 0$ as $x \to \infty$. Then show that any solution to the equation $y''(x) + (1 + h(x))y(x) = 0$, $x \in \mathbb{R}$, has infinitely many zeros in $\mathbb{R}$.

Exercise 3.17. Show that any solution to the equation $y''(x) + (1+x)y(x) = 0$, $x \in (0, \infty)$, has infinitely many zeros in $(0, \infty)$.

Exercise 3.18. Assume that $\epsilon > 0$ and ϕ is a solution to $y''(x) + q(x)y(x) = 0$, $x \in (0, \infty)$, where $q(x) \geq \dfrac{1 + \epsilon}{4x^2}$, for every $x > 0$. Then show that ϕ has infinitely many solutions in $(0, \infty)$.

Hint: Let u be a solution to the Euler equation $4x^2u''(x) + (1 + \epsilon)u(x) = 0$. Compute u explicitly and use Theorem 3.2.26.

Exercise 3.19. Let $0 < m < M$ and $q : [a, b] \to [m, M]$ be a non-constant continuous function. Let $a \leq x_1 < x_2 < \cdots < x_n \leq b$ be the zeros of a solution ϕ to $y''(x) + q(x)y(x) = 0$, $x \in (a, b)$. Then prove the following statements.

(i) $\dfrac{\pi}{\sqrt{m}} \geq x_{i+1} - x_i \geq \dfrac{\pi}{\sqrt{M}}$, $1 \leq i \leq n - 1$.

(ii) $n + 1 > (b - a)\dfrac{\sqrt{m}}{\pi} \geq n$.

Hint: Use the Sturm comparison theorem (Theorem 3.2.26) for the given equation and $u''(x) + mu(x) = 0$, $v''(x) + Mv(x) = 0$.

Exercise 3.20. Let ϕ be a solution to $y''(x) + \left(1 + \dfrac{1 - 4a^2}{4x^2}\right) y(x) = 0$, where $a \in \mathbb{R}$. Prove the following statements.

(i) The function ϕ has infinitely many zeros in $(0, \infty)$.

(ii) If $0 < x_1 < x_2 < \cdots$ are consecutive zeros of ϕ then the sequence $(x_{n+1} - x_n)$ converges to π as $n \to \infty$.

Hint: Use Exercises 3.16, 3.19.

Exercise 3.21. Let ϕ be a solution to the Airy equation $y''(x) + xy(x) = 0$, where $a \in \mathbb{R}$. Prove the following statements.

(i) The function ϕ has infinitely many zeros in $(0, \infty)$.

(ii) If $0 < x_1 < x_2 < \cdots$ are consecutive zeros of ϕ then the sequence $(x_{n+1} - x_n)$ converges to 0 as $n \to \infty$.

Hint: For $M \in \mathbb{N}$, consider $u''(x) + Mu(x) = 0$ and use Theorem 3.2.26.

Chapter 4

Boundary value problems

4.1 Introduction

Modeling of physical phenomena often leads to ODEs along with the conditions in which the unknown or its derivative or both are specified at more than one point. Such conditions are called the boundary conditions and the ODE along with the boundary conditions are called the boundary value problem (BVP). Throughout the chapter, we consider the boundary conditions prescribed at the end points of an interval where the ODE is posed. Such problems are called two point boundary value problems.

BVPs arise naturally in mathematical physics through partial differential equations. In particular, we get second order ODEs with boundary conditions while solving the Laplace, the heat and the wave equations posed on domains like rectangles, cuboids, cylinders, spheres etc., (see [1, 25]).

A second order linear boundary value problem (BVP) is defined as

$$\begin{cases} a_2(x)y''(x) + a_1(x)y'(x) + a_0(x)y(x) = f(x), \ a < x < b, \\ U_1[y: a, b] = A_1 y(a) + A_2 y'(a) + A_3 y(b) + A_4 y'(b) = \alpha, \\ U_2[y: a, b] = B_1 y(b) + B_2 y'(b) + B_3 y(a) + B_4 y'(a) = \beta. \end{cases} \quad (4.1)$$

where $a_0, a_1, f \in C([a, b])$, $a_2 \in C^1([a, b])$, the constants A_i's, B_i's satisfy the non degeneracy conditions: (i) $(A_1^2 + A_2^2)(B_1^2 + B_2^2) \neq 0$, (ii) (A_1, A_2, A_3, A_4) and (B_1, B_2, B_3, B_4) are linearly independent.

In a Cauchy problem or IVP where the data is prescribed at a point x_0, it is enough to find a solution in an interval containing x_0. In the case of BVP the solution needs to be defined in the entire interval $[a, b]$ and it has to satisfy the ODE along with the boundary conditions. In order to solve (4.1), usually, we first find the general solution to the linear ODE in it which contains two arbitrary constants. Then we substitute the general solution in the boundary conditions to determine those constants. If $f \equiv 0$, then we say that (4.1) is a *homogeneous BVP*, otherwise (4.1) is said to be a *non-homogeneous BVP*. Throughout this chapter, we assume that $a_2(x) \neq 0$, $x \in [a, b]$. The following example motivates us to make this assumption.

Example 4.1.1. Show that the BVP

$$x^2 y'' + 6xy' + 4y = 0, \quad x \in (0,1), \qquad (4.2a)$$
$$y(0) = 10, \ y(1) = 2, \qquad (4.2b)$$

does not admit a solution.

Solution. We first notice that the given ODE is an Euler equation whose the general solution is

$$\phi(x) = \frac{c_1}{x} + \frac{c_2}{x^4}, \ x \in (0,1),$$

where c_1 and c_2 are arbitrary constants. The condition $\phi(0) = 10$ never holds for any choice of c_1, c_2. Hence we conclude that there exists no solution to the given BVP. □

Throughout the chapter, without loss of generality, we assume that either $a_2(x) > 0$, $x \in [a, b]$, or $a_2(x) = 1$, $x \in [a, b]$ (by dividing (4.1) with a_2 if necessary) depending on the context. In particular, for the discussion regarding the existence and the uniqueness of a solution to the BVP (4.1) we assume that $a_2 \equiv 1$. Before we state the existence and uniqueness result, let us consider the following examples.

Example 4.1.2. Consider the boundary value problem

$$\begin{cases} y'' + y = 1, \quad 0 < x < \dfrac{\pi}{2}, \\ y(0) = 2, \ y\left(\dfrac{\pi}{2}\right) = 1. \end{cases} \qquad (4.3)$$

From the previous chapter, we find that the general solution to the given non-homogeneous ODE is

$$\phi(x) = 1 + c_1 \sin x + c_2 \cos x, \ x \in [0, \tfrac{\pi}{2}], \ c_1, c_2 \in \mathbb{R}.$$

From the non-homogeneous boundary conditions given in (4.3), we immediately get that the function

$$\phi(x) = 1 + \cos x, \ x \in [0, \tfrac{\pi}{2}],$$

is a unique solution to (4.3).
We now consider the associated homogeneous ODE with homogeneous boundary conditions, i.e.,

$$\begin{cases} y'' + y = 0, \ 0 < x < \dfrac{\pi}{2}, \\ y(0) = 0, \quad y\left(\dfrac{\pi}{2}\right) = 0. \end{cases} \qquad (4.4)$$

It is straightforward to observe that $\phi_1(x) \equiv 0$ (the trivial solution) is the only solution to the homogeneous boundary value problem (4.4). Thus for both the given non-homogeneous BVP (4.3) and the associated homogeneous BVP (4.4) there exists a unique solution. □

Example 4.1.3. Consider the following boundary value problem

$$y'' + y = 1, \ 0 < x < \pi, \tag{4.5a}$$
$$y(0) = 0, \quad y(\pi) = 1. \tag{4.5b}$$

As in Example 4.1.2, the general solution to non-homogeneous ODE (4.5a) is

$$\phi(x) = 1 + c_1 \sin x + c_2 \cos x, \ x \in [0, \pi], \ c_1, c_2 \in \mathbb{R}.$$

One can easily notice that for no values of $c_1, c_2 \in \mathbb{R}$, the function ϕ satisfies the non-homogeneous boundary conditions given in (4.5b). Hence there is no solution to BVP (4.5a)–(4.5b).
We now consider the corresponding homogeneous boundary value problem, i.e.,

$$y'' + y = 0, \ 0 < x < \pi, \tag{4.6a}$$
$$y(0) = 0, \quad y(\pi) = 0. \tag{4.6b}$$

Then
$$\phi_1(x) = c_1 \sin x, \ x \in [0, \pi], \ c_1 \in \mathbb{R},$$

is a solution to homogeneous ODE (4.6a) with homogeneous boundary conditions (4.6b). Hence there are infinitely many solutions to the associated homogeneous BVP (4.6a)–(4.6b).
Therefore there exists no solution to the given non-homogeneous BVP (non-existence) whereas the associated homogeneous BVP has a non-trivial solution (non-uniqueness of solution). □

Example 4.1.4. Consider the following boundary value problem

$$y'' + y = \cos x, \ 0 < x < \pi, \tag{4.7a}$$
$$y(0) = 5, \quad y(\pi) = -5. \tag{4.7b}$$

The general solution to the non-homogeneous equation is
$$\phi(x) = c_1 \sin x + c_2 \cos x + \frac{x}{2} \sin x, \ x \in [0, \pi], \ c_1, c_2 \in \mathbb{R}.$$

One can clearly observe that
$$\phi(x) = 5 \cos x + c_1 \sin x + \frac{x}{2} \sin x, \ x \in [0, \pi], \ c_1 \in \mathbb{R},$$

is a solution to the given non-homogeneous equation with the non-homogeneous boundary conditions. Hence there are infinitely many solutions to (4.7a)–(4.7b).
Now consider the corresponding homogeneous boundary value problem, i.e.,

$$y'' + y = 0, \ 0 < x < \pi, \tag{4.8a}$$
$$y(0) = 0, \quad y(\pi) = 0. \tag{4.8b}$$

Then
$$\phi_1(x) = c_1 \sin x, \; c_1 \in \mathbb{R}, \; x \in [0, \pi],$$

is a solution to the homogeneous ODE with homogeneous boundary conditions. Hence there are infinitely many solutions to the BVP (4.8a)–(4.8b). Therefore there exist more than one solutions to the given non-homogeneous BVP (non-uniqueness of solution) whereas the associated homogeneous BVP has a non-trivial solution (non-uniqueness of solution). $\square$

The phenomenon observed in Examples 4.1.2–4.1.4 is summarized as follows: If the non-homogeneous BVP has a unique solution, then the associated homogeneous BVP has only trivial solution. On the other hand, if the homogeneous BVP has a non-trivial solution, then the non-homogeneous problem may have infinitely many solutions or no solutions. This is proved in the following theorem for general BVPs.

Theorem 4.1.5. *Assume that $a_1 \in C^1([a,b]), a_0, f \in C([a,b])$. Let $\mathcal{L}$ be a second order linear differential operator given by*

$$\mathcal{L}[y](x) = y''(x) + a_1(x)y'(x) + a_0(x)y(x), \; x \in (a,b).$$
Consider the non-homogeneous BVP

$$\begin{cases} \mathcal{L}[y](x) = f(x), & a < x < b, \\ U_1[y; a, b] = \alpha, \\ U_2[y; a, b] = \beta, \end{cases} \tag{4.9}$$

and the corresponding homogeneous BVP

$$\begin{cases} \mathcal{L}[y](x) = 0, & a < x < b, \\ U_1[y; a, b] = 0, \\ U_2[y; a, b] = 0. \end{cases} \tag{4.10}$$

Then (4.9) has a unique solution if and only if (4.10) has only the trivial solution.

Proof. Let ϕ_1 and ϕ_2 be two linearly independent solutions to $\mathcal{L}[y] = 0$. Then the general solution to $\mathcal{L}[y] = 0$ is

$$\phi(x) = c_1\phi_1(x) + c_2\phi_2(x), \; x \in [a,b], \; c_1, c_2 \in \mathbb{R}.$$

By substituting ϕ in the homogeneous boundary conditions given in (4.10), we obtain that ϕ is a solution to (4.10) if and only if c_1 and c_2 satisfy

$$\begin{aligned} 0 &= A_1(c_1\phi_1(a) + c_2\phi_2(a)) + A_2(c_1\phi_1'(a) + c_2\phi_2'(a)) \\ &\quad + A_3(c_1\phi_1(b) + c_2\phi_2(b)) + A_4(c_1\phi_1'(b) + c_2\phi_2'(b)), \\ 0 &= B_3(c_1\phi_1(a) + c_2\phi_2(a)) + B_4(c_1\phi_1'(a) + c_2\phi_2'(a)) \\ &\quad + B_1(c_1\phi_1(b) + c_2\phi_2(b)) + B_2(c_1\phi_1'(b) + c_2\phi_2'(b)). \end{aligned}$$

After rearranging and writing in the matrix form, we get

$$M \begin{bmatrix} c_1 \\ c_2 \end{bmatrix} = \begin{bmatrix} 0 \\ 0 \end{bmatrix}, \tag{4.11}$$

where M is a 2×2 matrix whose entries are given by

$$M_{11} = A_1\phi_1(a) + A_2\phi_1'(a) + A_3\phi_1(b) + A_4\phi_1'(b),$$
$$M_{12} = A_1\phi_2(a) + A_2\phi_2'(a) + A_3\phi_2(b) + A_4\phi_2'(b),$$
$$M_{21} = B_3\phi_1(a) + B_4\phi_1'(a) + B_1\phi_1(b) + B_2\phi_1'(b),$$
$$M_{22} = B_3\phi_2(a) + B_4\phi_2'(a) + B_1\phi_2(b) + B_2\phi_2'(b).$$

On the other hand, let ψ be a particular solution to $\mathscr{L}[y] = f$. Then the general solution to $\mathscr{L}[y] = f$ is

$$\tilde{\phi}(x) = \tilde{c}_1\phi_1(x) + \tilde{c}_2\phi_2(x) + \psi(x), \ x \in [a,b], \ \tilde{c}_1, \tilde{c}_2 \in \mathbb{R}.$$

The function $\tilde{\phi}$ satisfies the non-homogeneous boundary conditions given in (4.9) if and only if $\tilde{c}_1, \tilde{c}_2$ satisfy

$$\begin{aligned}
\alpha &= A_1\big(\tilde{c}_1\phi_1(a) + \tilde{c}_2\phi_2(a) + \psi(a)\big) \ + \ A_2\big(\tilde{c}_1\phi_1'(a) + \tilde{c}_2\phi_2'(a) + \psi'(a)\big) \\
&\quad + A_3\big(\tilde{c}_1\phi_1(b) + \tilde{c}_2\phi_2(b) + \psi(b)\big) \ + \ A_4\big(\tilde{c}_1\phi_1'(b) + \tilde{c}_2\phi_2'(b) + \psi'(b)\big), \\
\beta &= B_3\big(\tilde{c}_1\phi_1(a) + \tilde{c}_2\phi_2(a) + \psi(a)\big) \ + \ B_4\big(\tilde{c}_1\phi_1'(a) + \tilde{c}_2\phi_2'(a) + \psi'(a)\big) \\
&\quad + B_1\big(\tilde{c}_1\phi_1(b) + \tilde{c}_2\phi_2(b) + \psi(b)\big) + B_2\big(\tilde{c}_1\phi_1'(b) + \tilde{c}_2\phi_2'(b) + \psi'(b)\big).
\end{aligned}$$

After writing in the matrix form, we obtain

$$M \begin{bmatrix} \tilde{c}_1 \\ \tilde{c}_2 \end{bmatrix} = \begin{bmatrix} \alpha - A_1\psi(a) - A_2\psi'(a) - A_3\psi(b) - A_4\psi'(b) \\ \beta - B_3\psi(a) - B_4\psi'(a) - B_1\psi(b) - B_2\psi'(b) \end{bmatrix}. \tag{4.12}$$

We conclude the proof with an observation that the following statements are equivalent.

(i) A unique solution to (4.10) is the constant function $\phi \equiv 0$.

(ii) A unique solution of (4.11) is $(c_1, c_2) = (0,0)$.

(iii) The determinant of the matrix M is nonzero.

(iv) A unique non-trivial solution exists to equation (4.12).

(v) A unique non-trivial solution exists to (4.9). $\qquad \qquad \square$

4.2   Adjoint forms

The main objective of this section is to define the adjoint of a second order linear differential operator and derive a criterion for an operator to be self adjoint. Moreover we want to derive certain type of boundary conditions such that the boundary conditions in the given problem and the associated adjoint problem are the same.

To this end, let $a_2 > 0$, $a_2 \in C^2([a, b])$, $a_1 \in C^1([a, b])$, and $a_0 \in C([a, b])$. We consider the linear operator $\mathscr{M} : C^2([a, b]) \to C([a, b])$ given by

$$\mathscr{M}[y](x) = a_2(x)y''(x) + a_1(x)y'(x) + a_0(x)y(x), \; x \in (a, b).$$

Let $y, z \in C^2([a, b])$. Using the integration by parts formula, we obtain

$$
\begin{aligned}
\int_a^b \mathscr{M}[y]z\,dx &= \int_a^b \left[(-a_2 z)'y' - (a_1 z)'y + a_0 yz\right]dx + \left[a_2 zy' + a_1 zy\right]_{x=a}^{x=b} \\
&= \int_a^b \left[(a_2 z)'' - (a_1 z)' + a_0 z\right]y\,dx \\
&\quad + \left[a_2 zy' - (a_2 z)'y + a_1 zy\right]_{x=a}^{x=b}.
\end{aligned}
\tag{4.13}
$$

We now define the adjoint operator of the operator $\mathscr{M}$ as

$$\mathscr{M}^*[z] := (a_2 z)'' - (a_1 z)' + a_0 z, \; z \in C^2([a, b]).$$

A straightforward computation gives that $\mathscr{M} = \mathscr{M}^*$ if and only if

$$a_1 = 2a_2' - a_1, \; a_0 = a_2'' - a_1' + a_0.$$

Therefore $\mathscr{M}$ is the same as its adjoint operator $\mathscr{M}^*$ if and only if $a_1 = a_2'$ or

$$\mathscr{M}[y] = a_2 y'' + a_2' y' + a_0 y = \left(a_2 y'\right)' + a_0 y, \; y \in C^2([a, b]).$$

Lemma 4.2.1. *If* $\mathscr{M} \neq \mathscr{M}^*$, *then there exists a nonzero function* $s \in C^1([a, b])$ *such that* $s\mathscr{M} = (s\mathscr{M})^*$. *In particular, we can take* $s(x) = \frac{1}{a_2(x)} e^{\int_a^x \frac{a_1}{a_2}\,dt}$, $x \in [a, b]$.

Proof. We first consider

$$\frac{1}{a_2}\mathscr{M}[y] = y'' + \frac{a_1}{a_2}y' + \frac{a_0}{a_2}y.
\tag{4.14}$$

On multiplying equation (4.14) with $e^{\int_a^x \frac{a_1}{a_2}\,dt}$ on both sides, we obtain

$$s\mathscr{M}[y] = \left(e^{\int_a^x \frac{a_1}{a_2}\,dt} y'\right)' + \frac{a_0}{a_2} e^{\int_a^x \frac{a_1}{a_2}\,dt} y.$$

Therefore we have $s\mathscr{M} = (s\mathscr{M})^*$. $\square$

Remark 4.2.2. From the derivation of identity (4.13), it follows that for every $y, z \in C^2([a, b])$, we have the identity

$$\int_a^x (z\mathscr{M}[y] - y\mathscr{M}^*[z])dt = \left[a_2(y'z - yz') + (a_1 - a_2')yz\right]_{t=a}^{t=x}.
\tag{4.15}$$

On differentiating (4.15) on both sides with respect to x, we get another identity

$$z \mathcal{M}[y] - y \mathcal{M}^*[z] = \frac{d}{dx} \left[a_2 (y'z - yz') + (a_1 - a_2') yz \right], \quad (4.16)$$

which is called the Lagrange identity.

Definition 4.2.3. A second order linear differential operator $\mathcal{M} : D(\mathcal{M}) \subseteq C^2([a, b]) \to C([a, b])$ is said to be self-adjoint, if

$$\int_a^b y \mathcal{M}[z] dx = \int_a^b z \mathcal{M}[y] dx, \quad y, z \in D(\mathcal{M}),$$

where $D(\mathcal{M})$ is the domain of $\mathcal{M}$.

Unless specified otherwise, all the differential operators we consider henceforth in this chapter are self-adjoint operators.

4.2.1   Boundary conditions

Let $p(x) > 0$, $x \in [a, b]$, $p \in C^1([a, b])$, $q \in C([a, b])$. Consider a second order linear differential operator $\mathcal{L}$ given by

$$\mathcal{L}[y] = (py')' + qy, \ y \in C^2([a, b]). \quad (4.17)$$

We first notice that $\mathcal{L}$ satisfies $\mathcal{L} = \mathcal{L}^*$. On substituting $x = b$ in (4.15), we obtain

$$\int_a^b \left(z \mathcal{L}[y] - y \mathcal{L}[z] \right) dx = \left[p(x) \left(y'(x) z(x) - y(x) z'(x) \right) \right]_a^b. \quad (4.18)$$

We now want to prescribe the boundary conditions on y, z such that the right side of equation (4.18) becomes zero. This can be achieved in different ways by considering different types of boundary conditions. Among them, we discuss two main types here. They are (*i*) separated boundary conditions, and (*ii*) periodic boundary conditions.

(*i*) *Separated boundary conditions*
We assume that y, z satisfy the boundary conditions, which are called separated boundary conditions, given by

$$\begin{cases} A_1 y(a) + A_2 y'(a) = 0, & A_1 z(a) + A_2 z'(a) = 0, \\ B_1 y(b) + B_2 y'(b) = 0, & B_1 z(b) + B_2 z'(b) = 0. \end{cases}$$

Since we have $(A_1, A_2) \neq (0, 0)$, from the boundary conditions that y, z satisfy at $x = a$, we get that the matrix

$$\begin{pmatrix} y(a) & y'(a) \\ z(a) & z'(a) \end{pmatrix}$$

is singular, i.e., $y(a)z'(a) - z(a)y'(a) = 0$. Similarly from the boundary conditions at $x = b$ we get $y(b)z'(b) - z(b)y'(b) = 0$. Therefore the separated boundary conditions make sure that the right hand side of (4.18) vanishes. We set

$$X = \{y \in C^2([a,b]) \ : \ A_1 y(a) + A_2 y'(a) = 0, \ B_1 y(b) + B_2 y'(b) = 0\}. \quad (4.19)$$

From the discussion we had so far, we can conclude that the operator $\mathscr{L}$ given in equation (4.17) is self-adjoint on X with L^2-inner product[1], i.e.,

$$\langle z, \mathscr{L}[y]\rangle_{L^2} = \langle \mathscr{L}[z], y\rangle_{L^2}, \ y, z \in X. \quad (4.20)$$

(ii) Periodic boundary conditions
Let $p(a) = p(b)$. If $y(a) = y(b)$, $y'(a) = y'(b)$ and $z(a) = z(b)$, $z'(a) = z'(b)$, then we immediately conclude that the right hand side of equation (4.18) vanishes. Therefore if $p(a) = p(b)$, then we define

$$X_p = \left\{y \in C^2([a,b]) : y(a) = y(b), \ y'(a) = y'(b)\right\}.$$

Therefore we have

$$\int_a^b \left(z\mathscr{L}[y] - y\mathscr{L}[z]\right)dx = 0, \ y, z \in X_p.$$

Hence $\mathscr{L}$ is a self-adjoint operator on X_p with L^2- inner product i.e.,

$$\langle z, \mathscr{L}[y]\rangle_{L^2} = \langle \mathscr{L}[z], y\rangle_{L^2}, \ y, z \in X_p.$$

The operator $\mathscr{L}$ defined in (4.17) with the separated boundary conditions or periodic boundary conditions is called as the Sturm-Liouville operator. There are another type of Sturm-Liouville operators called singular Sturm-Liouville operators. We do not discuss these operators in this book.

4.3   Green's function

Throughout this section, unless mentioned otherwise, the only differential operator that we consider is

$$\mathscr{L} = \frac{d}{dx}\left(p(x)\frac{d}{dx}\right) + q(x)I, \ x \in (a,b), \quad (4.21)$$

---

[1] Define $L^2([a,b]) = \{f : [a,b] \to \mathbb{R} : f$ is measurable, $\int_a^b |f(x)|^2 dx < \infty\}$. We define an inner product on $L^2([a,b])$ by $\langle f, g\rangle_{L^2} := \int_a^b f(x)g(x)dx$. Then the norm given by this inner product is $\|f\|_{L^2} = \left(\int_a^b |f^2(x)|dx\right)^{\frac{1}{2}}$. For more details on L^2- spaces and inner product spaces see [31].

where I is the identity operator, $p > 0$, $p \in C^1([a, b])$, $q \in C([a, b])$. Our main objective in this section is to prove that $\mathscr{L} : X \to C([a, b])$ is surjective, where X is given in (4.19). In other words, for every $f \in C([a, b])$ we want to show that there exists a solution to the BVP

$$\begin{cases} \mathscr{L}[y](x) = f(x), \ x \in (a, b), \\ A_1 y(a) + A_2 y'(a) = 0, \\ B_1 y(b) + B_2 y'(b) = 0. \end{cases} \quad (4.22)$$

In order to show this we construct a function $G : [a, b] \times [a, b] \to \mathbb{R}$ such that for each $f \in C([a, b])$, the function

$$\phi(.) := \int_a^b G(., \xi) f(\xi) d\xi \in X$$

is a solution to BVP (4.22). This function G is referred to as a Green's function corresponding to $\mathscr{L}$. We now provide a precise definition of a Green's function.

Definition 4.3.1. A Green's function corresponding to the operator $\mathscr{L}$ given in (4.21) with the homogeneous boundary conditions

$$\begin{cases} A_1 y(a) + A_2 y'(a) = 0, \\ B_1 y(b) + B_2 y'(b) = 0, \end{cases} \quad (4.23)$$

is a function $G : [a, b] \times [a, b] \to \mathbb{R}$ satisfying the following conditions:

(i) The function G is continuous on $[a, b] \times [a, b]$.

(ii) The first and second derivatives of G are continuous except on the line segment $x = \xi$. The first derivative of G has a jump discontinuity on $x = \xi$ with

$$\frac{\partial G}{\partial x}(x, \xi) \Big|_{x=\xi^-}^{x=\xi^+} = \frac{1}{p(\xi)}, \ \xi \in (a, b).$$

(iii) For every $\xi \in (a, b)$, G satisfies the following the ODE and the homogeneous boundary conditions:

$$\begin{cases} \dfrac{\partial}{\partial x}\left(p(x)\dfrac{\partial G}{\partial x}(x, \xi)\right) + q(x)G(x, \xi) = 0, \ x \neq \xi, \\ A_1 G(a, \xi) + A_2 \dfrac{\partial G}{\partial x}(a, \xi) = 0, \\ B_1 G(b, \xi) + B_2 \dfrac{\partial G}{\partial x}(b, \xi) = 0. \end{cases}$$

The next theorem gives us the existence of a Green's function.

Theorem 4.3.2. *Let $\mathscr{L}$ be the operator given in (4.21). Assume that $\mathscr{L}[y] = 0$, with the boundary conditions in (4.23) has only the trivial solution. Then a Green's function exists associated to $\mathscr{L}$ with the boundary conditions given in (4.23).*

Proof. Here we present a constructive proof for the existence of a Green's function.

Let $\phi_1 \neq 0$ and $\phi_2 \neq 0$ be solutions to the problems,

$$\begin{cases} \mathscr{L}[\phi_1] = 0, \ x \in (a, b), \\ A_1\phi_1(a) + A_2\phi_1'(a) = 0, \end{cases} \text{ and } \begin{cases} \mathscr{L}[\phi_2] = 0, \ x \in (a, b), \\ B_1\phi_2(b) + B_2\phi_2'(b) = 0, \end{cases}$$

respectively.

Claim: The functions ϕ_1 and ϕ_2 are linearly independent.
Suppose $\phi_1 = k\phi_2$, then ϕ_1 is a solution to

$$\begin{cases} \mathscr{L}[\phi_1] = 0, \ x \in (a, b), \\ A_1\phi_1(a) + A_2\phi_1'(a) = 0, \\ B_1\phi_1(b) + B_2\phi_1'(b) = 0. \end{cases}$$

This implies $\phi_1 \equiv 0$, due to our hypothesis on $\mathscr{L}$. This is a contradiction. Therefore ϕ_1 and ϕ_2 are linearly independent.

Ansatz : Let

$$G(x, \xi) = \begin{cases} c_1(\xi)\phi_1(x), & a \leq x < \xi, \\ c_2(\xi)\phi_2(x), & \xi \leq x \leq b. \end{cases} \tag{4.24}$$

We will choose $c_1(\xi)$ and $c_2(\xi)$ such that G satisfies the conditions (i)–(iii) in Definition 4.3.1. Since we want G to be continuous at $x = \xi$, we obtain

$$\phi_1(\xi)c_1(\xi) - \phi_2(\xi)c_2(\xi) = 0, \ \xi \in [a, b]. \tag{4.25}$$

From the jump condition in the definition of a Green's function, we get

$$\left.\frac{\partial G}{\partial x}\right|_{x=\xi^-}^{x=\xi^+} = \phi_2'(\xi)c_2(\xi) - \phi_1'(\xi)c_1(\xi) = \frac{1}{p(\xi)}. \tag{4.26}$$

From equations (4.25) and (4.26) we find

$$c_1(\xi) = \frac{\phi_2(\xi)}{p(\xi)W(\xi; \phi_1, \phi_2)}, \quad c_2(\xi) = \frac{\phi_1(\xi)}{p(\xi)W(\xi; \phi_1, \phi_2)}. \tag{4.27}$$

It is worth recalling that ϕ_1, ϕ_2 are linearly independent solutions to $\mathscr{L}[y] = 0$ and thus the Wronskian $W(\xi; \phi_1, \phi_2) \neq 0$, $\xi \in [a, b]$. This proves the existence of a Green's function associated with $\mathscr{L}$ and the boundary conditions given in (4.23). From (4.24) and (4.27) a Green's function is given by

$$G(x, \xi) = \begin{cases} \dfrac{\phi_1(x)\phi_2(\xi)}{p(\xi)W(\xi; \phi_1, \phi_2)}, & a \leq x < \xi, \\[4mm] \dfrac{\phi_1(\xi)\phi_2(x)}{p(\xi)W(\xi; \phi_1, \phi_2)}, & \xi \leq x \leq b. \end{cases} \tag{4.28}$$

It is easy to observe that G satisfies the homogeneous boundary conditions given in (4.23) and

$$\frac{\partial}{\partial x}\left(p(x)\frac{\partial}{\partial x}G(x,\xi)\right) + q(x)G(x,\xi) = 0, \ x \neq \xi.$$

This proves the existence of a Green's function. $\qquad\square$

Remark 4.3.3. If ϕ_1, ϕ_2 are two solutions of $\mathscr{L}[y] = 0$, then $pW(.,\phi_1,\phi_2)$ is a constant function. To prove this claim, we first observe that $\mathscr{L} = \mathscr{L}^*$ and use the Lagrange identity (see (4.16)) for ϕ_1, ϕ_2 to get

$$\frac{d}{dx}\left(p(x)W(x;\phi_1,\phi_2)\right) = \phi_1(x)\mathscr{L}[\phi_2](x) - \phi_2(x)\mathscr{L}[\phi_1](x) = 0, \ x \in (a,b).$$

This readily implies that pW is a constant function.

Remark 4.3.4. In view of Remark 4.3.3 and equation (4.28), we get a simplified formula for Green's function

$$G(x,\xi) = \begin{cases} c\phi_1(x)\phi_2(\xi), & a \leq x < \xi, \\ c\phi_1(\xi)\phi_2(x), & \xi \leq x \leq b, \end{cases} \qquad (4.29)$$

where $\dfrac{1}{c} = p(\xi)W(\xi;\phi_1,\phi_2)$.

Example 4.3.5. Find a Green's function for $\begin{cases} \mathscr{L}[y](x) = y''(x), \ x \in (0,1), \\ y(0) = y'(1) = 0. \end{cases}$

Solution: By comparing the problem given in this example with the standard form (4.21), we get $p \equiv 1$.
Let ϕ_1, ϕ_2 satisfy

$$\begin{cases} \phi_1'' = 0, \\ \phi_1(0) = 0, \end{cases} \quad \text{and} \quad \begin{cases} \phi_2'' = 0, \\ \phi_2'(1) = 0, \end{cases}$$

respectively. It is easy to find that a particular choice of ϕ_1 and ϕ_2 is

$$\phi_1(x) = x, \text{ and } \phi_2(x) = 1, \ x \in [0,1].$$

Now the Wronskian of ϕ_1, ϕ_2 is $W(.,\phi_1,\phi_2) \equiv -1$. Therefore we get

$$p(\xi)W(\xi;\phi_1,\phi_2) = -1, \ \xi \in [0,1].$$

From (4.29), Green's function is given by

$$G(x,\xi) = \begin{cases} -x, & 0 \leq x < \xi, \\ -\xi, & \xi \leq x \leq 1. \end{cases}$$

This completes the construction of Green's function. $\qquad\square$

One can define a Green's function when periodic boundary conditions are given instead of separated boundary conditions (4.23). This can be done by just replacing the phrase the boundary conditions in condition (iii) of Definition 4.3.1 by

$$G(a, \xi) = G(b, \xi), \quad \frac{\partial G}{\partial x}(a, \xi) = \frac{\partial G}{\partial x}(b, \xi), \ \xi \in (a, b).$$

Further, one can use a similar technique to construct a Green's function in the case of periodic boundary conditions. This is demonstrated in the following example.

Example 4.3.6. Find a Green's function for $\begin{cases} \mathscr{L}[y] = y'' + y, \ x \in (0, \pi), \\ y(0) = y(\pi), \ y'(0) = y'(\pi). \end{cases}$

Solution: We first observe that $\mathscr{L}[y] = 0$, with the periodic conditions given in the problem has only the trivial solution. By comparing the problem given in this example with the standard form, we get $p \equiv 1$. Let ϕ_1, ϕ_2 be linearly independent solutions of $y'' + y = 0$. We set

$$G(x, \xi) = \begin{cases} c_1(\xi)\phi_1(x) + c_2(\xi)\phi_2(x), & 0 \le x < \xi, \\ c_3(\xi)\phi_1(x) + c_4(\xi)\phi_2(x), & \xi \le x \le \pi. \end{cases}$$

We need to find c_i, $1 \le i \le 4$, such that G satisfies all the conditions to be a Green's function. In particular, we choose $\phi_1(x) = \cos x$ and $\phi_2(x) = \sin x$, to get

$$G(x, \xi) = \begin{cases} c_1(\xi)\cos x + c_2(\xi)\sin x, & 0 \le x < \xi, \\ c_3(\xi)\cos x + c_4(\xi)\sin x, & \xi \le x \le \pi. \end{cases}$$

We next focus on the boundary conditions, i.e., we need

$$G(0, \xi) = G(\pi, \xi), \quad \frac{\partial G}{\partial x}(0, \xi) = \frac{\partial G}{\partial x}(\pi, \xi), \ \xi \in (0, \pi).$$

Therefore we have

$$c_1(\xi) = -c_3(\xi), \quad c_2(\xi) = -c_4(\xi), \ \xi \in (0, \pi).$$

Thus G is given by

$$G(x, \xi) = \begin{cases} c_1(\xi)\cos x + c_2(\xi)\sin x, & 0 \le x < \xi, \\ -c_1(\xi)\cos x - c_2(\xi)\sin x, & \xi \le x \le \pi. \end{cases} \tag{4.30}$$

As G is continuous at $x = \xi$, we have $G(\xi-, \xi) = G(\xi+, \xi)$, i.e.,

$$c_1(\xi)\cos \xi + c_2(\xi)\sin \xi = -c_1(\xi)\cos \xi - c_2(\xi)\sin \xi,$$

or

$$c_1(\xi)\cos \xi + c_2(\xi)\sin \xi = 0. \tag{4.31}$$

From the jump condition, $\dfrac{\partial G}{\partial x}(x, \xi)\Big|_{x=\xi-}^{x=\xi+} = 1$, we have

$$c_1(\xi)\sin\xi - c_2(\xi)\cos\xi - (-c_1(\xi)\sin\xi + c_2(\xi)\cos\xi) = 1.$$

By rearranging the terms, we get

$$c_1(\xi)\sin\xi - c_2(\xi)\cos\xi = \frac{1}{2}. \tag{4.32}$$

We solve (4.31)–(4.32) to obtain

$$c_1(\xi) = \frac{1}{2}\sin\xi, \quad c_2(\xi) = -\frac{1}{2}\cos\xi, \ \xi \in [0,\pi].$$

On substituting the values of c_1, c_2 in (4.30), we find that Green's function is

$$G(x,\xi) = \begin{cases} \dfrac{1}{2}\sin\xi\cos x - \dfrac{1}{2}\cos\xi\sin x, & 0 \le x < \xi, \\ -\dfrac{1}{2}\sin\xi\cos x + \dfrac{1}{2}\cos\xi\sin x, & \xi \le x \le \pi. \end{cases}$$

This function can be written in a simpler form as

$$G(x,\xi) = \tfrac{1}{2}\sin|x-\xi|, \ x,\xi \in [0,\pi].$$

This completes the solution. □

We now present the main theorem of this section which provides the existence of solutions to $\mathscr{L}[y] = f$ with the homogeneous boundary conditions given in (4.23).

Theorem 4.3.7. (Fundamental Theorem of Green's function)
Assume the hypotheses of Theorem 4.3.2. Further, assume that G is a Green's function for $\mathscr{L}$ with boundary conditions given in (4.23) and $f \in C([a,b])$. Define

$$\phi(x) := \int_a^b G(x,\xi)f(\xi)d\xi, \ x \in [a,b].$$

Then ϕ is a solution to the boundary value problem

$$\begin{cases} \mathscr{L}[y](x) = f(x), \ x \in (a,b), \\ A_1 y(a) + A_2 y'(a) = 0, \\ B_1 y(b) + B_2 y'(b) = 0. \end{cases}$$

Proof. Since G satisfies the boundary conditions given in (4.23), it follows that ϕ also satisfies the same boundary conditions. We consider

$$\frac{d\phi}{dx}(x) = \frac{d}{dx}\left[\int_a^x G(x,\xi)f(\xi)d\xi + \int_x^b G(x,\xi)f(\xi)d\xi\right].$$

We now use the Leibniz rule of differentiation to get

$$\frac{d\phi}{dx}(x) = \int_a^x \frac{\partial G}{\partial x}(x,\xi)f(\xi)d\xi + G(x,x-)f(x-) \tag{4.33}$$

$$+ \int_x^b \frac{\partial G}{\partial x}(x,\xi)f(\xi)d\xi - G(x,x+)f(x+)$$

$$= \int_a^b \frac{\partial G}{\partial x}(x,\xi)f(\xi)d\xi. \tag{4.34}$$

The last equality is due to the continuity of G and f. Again on differentiating we find

$$\frac{d^2\phi}{dx^2}(x) = \int_a^b \frac{\partial^2 G}{\partial x^2}(x,\xi)f(\xi)d\xi + \frac{\partial G}{\partial x}(x,x-)f(x) - \frac{\partial G}{\partial x}(x,x+)f(x)$$

$$= \int_a^b \frac{\partial^2 G}{\partial x^2}(x,\xi)f(\xi)d\xi - f(x)\left[\frac{\partial G}{\partial x}(x,\xi)\right]_{\xi=x-}^{\xi=x+}$$

$$= \int_a^b \frac{\partial^2 G}{\partial x^2}(x,\xi)f(\xi)d\xi + \frac{f(x)}{p(x)}. \tag{4.35}$$

For the equality in the last term see Exercise 4.6.
From (4.34), (4.35) and condition (iii) in the definition of Green's function we get

$$\mathscr{L}[y](x) = \int_a^b \left(\frac{\partial}{\partial x}\left(p(x)\frac{\partial G}{\partial x}(x,\xi)\right) + q(x)G(x,\xi)\right)f(\xi)d\xi + p(x)\frac{f(x)}{p(x)} = f(x).$$

This completes the proof. □

Proposition 4.3.8. *(Uniqueness) Under the hypotheses of Theorem 4.3.2, there exists a unique Green's function to $\mathscr{L}$ with the boundary conditions given in (4.23).*

Proof. Let G_1 and G_2 be two Green's functions corresponding to the given operator and the boundary conditions. From Theorem 4.3.7, for every $f \in C([a,b])$, the functions

$$\phi_i(x) := \int_a^b G_i(x,\xi)f(\xi)d\xi, \ x \in [a,b],$$

satisfy $\mathscr{L}[\phi_i] = f$, $i = 1,2$. Since a solution to $\mathscr{L}[y] = f$, with homogeneous boundary conditions (4.23) is assumed to be unique, we get $\phi_1 \equiv \phi_2$. Let $G := G_1 - G_2$. Then we have

$$\int_a^b G(x,\xi)f(\xi)d\xi = 0, \ x \in [a,b], \ f \in C([a,b]).$$

Let $x \in [a,b]$ be fixed. Owing to the continuity of G, we get $G(x,\xi) = 0$, $\xi \in [a,b]$ (see Exercise 4.5). Since x is arbitrary, one obtains that $G \equiv 0$, establishing the uniqueness of a Green's function. □

Proposition 4.3.9. *Under the hypotheses of Theorem 4.3.7, any Green's function corresponding to the Sturm–Liouville operator $\mathscr{L}$ with homogeneous boundary conditions (4.23) is symmetric.*

Proof. We have constructed a Green's function G in Theorem 4.3.2. We notice from the formula given in (4.29) that G is symmetric . The required result follows from Proposition 4.3.8. $\qquad\square$

Example 4.3.10. Solve $\begin{cases} y'' = x^2, \ x \in (0,1), \\ y(0) = y'(1) = 0, \end{cases}$

using Green's function.

Solution: By comparing with the standard form we get $f(x) = x^2, x \in (0,1)$. We recall from Example 4.3.5, that Green's function associated with the given differential operator along with the boundary conditions is

$$G(x, \xi) = \begin{cases} -x, & 0 \leq x < \xi, \\ -\xi, & \xi \leq x \leq 1. \end{cases}$$

In view of Theorem 4.3.7, the function

$$\phi(x) = \int_0^1 G(x, \xi) f(\xi) d\xi = -\int_0^x \xi^3 d\xi - x \int_x^1 \xi^2 d\xi = \frac{x^4}{12} - \frac{x}{3}, \ x \in [0,1]$$

is the solution to the given BVP. $\qquad\square$

Example 4.3.11. Solve $\begin{cases} y'' - y = 1, \ x \in (0,1), \\ y(0) = y(1) = 0, \end{cases}$

using Green's function.

Solution: Here we have $p \equiv 1$. As usual, we assume that ϕ_1, ϕ_2 are solutions to

$$\begin{cases} \phi_1'' - \phi_1 = 0, \ x \in (0,1), \\ \phi_1(0) = 0, \end{cases} \qquad \begin{cases} \phi_2'' - \phi_2 = 0, \ x \in (0,1), \\ \phi_2(1) = 0, \end{cases}$$

respectively. An easy computation gives us that

$$\phi_1(x) = \sinh(x) \text{ and } \phi_2(x) = \sinh(1-x).$$

The Wronskian of these two solutions is $W(\phi_1, \phi_2) \equiv -\sinh 1$. Now, from (4.29), Green's function is

$$G(x, \xi) = \begin{cases} \dfrac{\sinh(1-\xi)\sinh(x)}{-\sinh 1}, & \text{for } 0 \leq x < \xi, \\[2mm] \dfrac{\sinh(\xi)\sinh(1-x)}{-\sinh 1}, & \text{for } \xi \leq x \leq 1. \end{cases}$$

In view of Theorem 4.3.7, the function

$$\phi(x) = \int_0^1 G(x,\xi)f(\xi)d\xi$$

$$= \frac{\sinh(1-x)}{-\sinh 1}\int_0^x \sinh(\xi)d\xi - \frac{\sinh(x)}{\sinh 1}\int_x^1 \sinh(1-\xi)d\xi,$$

$$= \frac{\sinh(x)+\sinh(1-x)}{\sinh 1} - 1, \quad x \in [0,1]$$

is the solution to the given problem.   □

Example 4.3.12. Solve
$$\begin{cases} y'' + y = -1, \quad x \in (0,\frac{\pi}{2}), \\ y(0) = y(\frac{\pi}{2}) = 0, \end{cases}$$

by finding Green's function.

Solution: We first find Green's function. In order to do that, let ϕ_1 and ϕ_2 be solutions to

$$\begin{cases} \phi_1''(x) + \phi_1(x) = 0, \quad x \in (0,\frac{\pi}{2}), \\ \phi_1(0) = 0, \end{cases} \quad \text{and} \quad \begin{cases} \phi_2''(x) + \phi_2(x) = 0, \quad x \in (0,\frac{\pi}{2}), \\ \phi_2(\frac{\pi}{2}) = 0, \end{cases}$$

respectively. It is easy to see that $\phi_1 = \sin x$ and $\phi_2 = \cos x$. This implies that the Wronskian $W(x;\phi_1,\phi_2) = -1$, $x \in (0,\frac{\pi}{2})$. Since $p \equiv 1$, Green's function is given by

$$G(x,\xi) = \begin{cases} -\sin x \cos \xi, & \text{for } 0 \le x < \xi, \\ -\cos x \sin \xi, & \text{for } \xi \le x \le \frac{\pi}{2}. \end{cases}$$

Therefore from Theorem 4.3.7, we obtain

$$\phi(x) = -\int_0^{\frac{\pi}{2}} G(x,\xi)d\xi$$

$$= \int_0^x \cos x \sin \xi d\xi + \int_x^{\frac{\pi}{2}} \sin x \cos \xi d\xi,$$

$$= -1 + \sin x + \cos x, \quad x \in [0,\frac{\pi}{2}],$$

is the required solution.   □

4.3.1   Non-homogeneous boundary conditions

Consider the following non-homogeneous boundary value problem

$$\begin{cases} \mathscr{L}[y](x) = f(x), \quad x \in (a,b), \\ A_1 y(a) + A_2 y'(a) = \alpha, \\ B_1 y(b) + B_2 y'(b) = \beta. \end{cases} \tag{4.36}$$

Let ψ be the solution to the following homogeneous equation with non-homogeneous boundary conditions

$$\begin{cases} \mathscr{L}[\psi](x) = 0, \ x \in (a,b), \\ A_1 \ \psi(a) + A_2 \ \psi'(a) = \alpha, \\ B_1 \ \psi(b) + B_2 \ \psi'(b) = \beta. \end{cases}$$

Let G be Green's function for $\mathscr{L}$ with homogeneous boundary conditions (4.23). Then from Theorem 4.3.7, the function

$$u(x) := \int_a^b G(x,\xi)f(\xi)d\xi,$$

satisfies

$$\begin{cases} \mathscr{L}[u](x) = f(x), \ x \in (a,b), \\ A_1 \ u(a) + A_2 \ u'(a) = 0, \\ B_1 \ u(b) + B_2 \ u'(b) = 0. \end{cases}$$

From the principle of superposition, the solution to (4.36) is given by

$$\phi(x) = u(x) + \psi(x), \ x \in (a,b). \tag{4.37}$$

For, we observe that

$$\mathscr{L}[\phi](x) = \mathscr{L}[u](x) + \mathscr{L}[\psi](x) = f(x), \ x \in (a,b).$$

Moreover, one can easily verify that ϕ satisfies the boundary conditions given in (4.36).

Example 4.3.13. Solve $\begin{cases} y'' = f(x), \ 0 < x < 1, \\ y(0) = 0, \ y(1) + y'(1) = 2. \end{cases}$

Solution: We first find Green's function corresponding to the operator $\mathscr{L}[y] = y''$, along with the homogeneous boundary conditions $y(0) = 0$, $y(1) + y'(1) = 0$. In order to do that we consider

$$\begin{cases} \phi_1''(x) = 0, \ 0 < x < 1, \\ \phi_1(0) = 0 \end{cases} \quad \text{and} \quad \begin{cases} \phi_2''(x) = 0, \ 0 < x < 1, \\ \phi_2(1) + \phi_2'(1) = 0. \end{cases}$$

It is straightforward to compute that $\phi_1(x) = x$ and $\phi_2(x) = 2 - x$. This implies that the Wronskian $W(x; \phi_1, \phi_2) = -2$, $x \in (0,1)$. As $p \equiv 1$, from equation (4.29), Green's function is given by

$$G(x,\xi) = \begin{cases} \dfrac{x}{2}(\xi - 2), \ \ 0 \leq x < \xi, \\ \dfrac{\xi}{2}(x - 2), \ \ \xi \leq x \leq 1. \end{cases}$$

On the other hand, the solution to

$$\begin{cases} \psi''(x) = 0, \ 0 < x < 1, \\ \psi(0) = 0, \\ \psi(1) + \psi'(1) = 2, \end{cases}$$

is given by $\psi(x) = x$, $x \in (0,1)$. From the principle of superposition, the solution to the given problem is

$$\phi(x) = \psi(x) + \int_0^1 G(x,\xi)f(\xi) \, d\xi.$$

Therefore the function

$$\phi(x) = x + \frac{x-2}{2} \int_0^x \xi f(\xi) d\xi + \frac{x}{2} \int_x^1 (\xi - 2)f(\xi)d\xi, \ x \in [0,1],$$

is the required solution. □

4.4   Sturm-Liouville systems and eigenvalue problems

Consider the following eigenvalue problem corresponding to the second order linear differential equations with variable coefficients

$$a_2(x)y'' + a_1(x)y' + \big(a_0(x) + \lambda\big)y = 0. \tag{4.38}$$

Equation (4.38) can be transformed to (see Lemma 4.2.1)

$$\big(p(x)y'\big)' + \big(q(x) + \lambda s(x)\big)y = 0, \tag{4.39}$$

where $p(x) = e^{\int_a^x \frac{a_1(t)}{a_2(t)} dt}$, $B_1(x) = \frac{a_0(x)}{a_2(x)}p(x)$, $s(x) = \frac{p(x)}{a_2(x)}$.

Definition 4.4.1. Let $p \in C^1\big([a,b]\big)$, $q, s \in C\big([a,b]\big)$, $p(x) > 0, s(x) > 0$, $x \in [a,b]$. Then the system

$$\begin{cases} \big(p(x)y'(x)\big)' + \big(q(x) + \lambda s(x)\big)y(x) = 0, \quad a < x < b, \\ A_1y(a) + A_2y'(a) = 0, \\ B_1y(b) + B_2y'(b) = 0, \end{cases} \tag{4.40}$$

is called a regular Sturm-Liouville (S-L) system.

Definition 4.4.2. The values of $\lambda \in \mathbb{C}$ for which the regular S-L system (4.40) has a non-trivial solution are called the eigenvalues and the corresponding non-trivial solutions are called the eigenfunctions of system (4.40).

Before we prove theorems regarding the properties of eigenvalues and eigenfunctions of (4.40), we present some examples. In the following examples, we use the fact that the eigenvalues of (4.40) are real numbers. This result will be proved later (see Theorem 4.4.9).

Example 4.4.3. Find the eigenvalues and the eigenfunctions of

$$y'' + \lambda y = 0, \quad 0 < x < \pi, \tag{4.41a}$$
$$y(0) = 0, \quad y'(\pi) = 0. \tag{4.41b}$$

Solution: In order to find the eigenvalues, we consider three obvious possibilities given by (i) $\lambda < 0$, (ii) $\lambda = 0$, (iii) $\lambda > 0$.

Case 1. Let $\lambda < 0$.

Then the general solution to (4.41a) is

$$\phi(x) = c_1 e^{-\sqrt{-\lambda}x} + c_2 e^{\sqrt{-\lambda}x}, \quad x \in (0, \pi).$$

On substituting this ϕ in the boundary conditions (4.41b), we get

$$c_1 + c_2 = 0,$$
$$\sqrt{-\lambda}\left(c_1 e^{-\sqrt{-\lambda}\pi} - c_2 e^{\sqrt{-\lambda}\pi}\right) = 0.$$

We now express this system in a matrix form as follows

$$\begin{bmatrix} 1 & 1 \\ e^{-\sqrt{-\lambda}\pi} & -e^{-\sqrt{\lambda}\pi} \end{bmatrix} \begin{bmatrix} c_1 \\ c_2 \end{bmatrix} = \begin{bmatrix} 0 \\ 0 \end{bmatrix}.$$

Since the matrix is nonsingular, $c_1 = c_2 = 0$. Thus we have $\phi \equiv 0$. Hence we cannot have a negative eigenvalue.

Case 2. We suppose $\lambda = 0$.

Then the general solution is given by

$$\phi(x) = c_1 x + c_2, \quad x \in (0, \pi), \ c_1, c_2 \in \mathbb{R}.$$

Then the boundary conditions imply that $c_1 = c_2 = 0$. Therefore we get $\phi \equiv 0$. Thus $\lambda = 0$ is also not an eigenvalue.

Case 3. Let $\lambda > 0$.

The general solution in this case is

$$\phi(x) = A\cos(\sqrt{\lambda}x) + B\sin(\sqrt{\lambda}x), \quad A, B \in \mathbb{R}, \ x \in (0, \pi).$$

The boundary condition $\phi(0) = 0$ holds if and only if $A = 0$ whereas $\phi'(\pi) = 0$ is the same as

$$B\sqrt{\lambda}\cos(\sqrt{\lambda}\pi) = 0.$$

Since we need a nonzero solution, we choose $B \neq 0$. This implies $\cos(\sqrt{\lambda}\pi) = 0$ i.e., $\sqrt{\lambda}\pi = (2n-1)\pi/2$, $n \in \mathbb{N}$. Therefore the eigenvalues are

$$\lambda_n = \frac{(2n-1)^2}{4}, \quad n \in \mathbb{N},$$

and the corresponding eigenfunctions are

$$\phi_n(x) = \sin\left(\frac{(2n-1)x}{2}\right), \ n \in \mathbb{N}.$$

This completes the solution.   □

Example 4.4.4. Let $k > 0$. Find the eigenvalues and the eigenfunctions for

$$\begin{cases} y'' + \lambda y = 0, \ 0 < x < 1, \\ y(0) = 0, \ y(1) + ky'(1) = 0. \end{cases}$$

Solution: As in the previous example we consider three cases.
Case 1. Let $\lambda < 0$.
Then

$$\phi(x) = c_1 e^{-\sqrt{-\lambda}x} + c_2 e^{\sqrt{-\lambda}x}, \ x \in (0, 1).$$

On substituting this ϕ in the boundary conditions, we get

$$c_1 + c_2 = 0,$$
$$(1 + k\sqrt{-\lambda})e^{\sqrt{-\lambda}}c_1 + (1 - k\sqrt{-\lambda})e^{-\sqrt{-\lambda}}c_2 = 0.$$

We can express it in a matrix form as follows

$$\begin{bmatrix} 1 & 1 \\ (1 + k\sqrt{-\lambda})e^{\sqrt{-\lambda}} & (1 - k\sqrt{-\lambda})e^{-\sqrt{-\lambda}} \end{bmatrix} \begin{bmatrix} c_1 \\ c_2 \end{bmatrix} = \begin{bmatrix} 0 \\ 0 \end{bmatrix}.$$

We now prove that this matrix is nonsingular for every $k > 0$. For, one can easily observe that the determinant of the matrix satisfies

$$(1 - k\sqrt{-\lambda})e^{-\sqrt{-\lambda}} - (1 + k\sqrt{-\lambda})e^{\sqrt{-\lambda}} < 0,$$

whenever $\lambda < 0$. Therefore we get $c_1 = c_2 = 0$, and we have $\phi \equiv 0$. Hence we cannot have a negative eigenvalue.

Case 2. Here we assume that $\lambda = 0$.
Using an argument similar to the one that is presented in the previous example, one can easily show that $\lambda = 0$ is also not an eigenvalue.

Case 3. Let $\lambda > 0$.
The solution to the given equation is

$$\phi(x) = A\cos(\sqrt{\lambda}x) + B\sin(\sqrt{\lambda}x), \ x \in (0, 1).$$

From the boundary conditions, we get

$$A = 0, \quad B(\sin\sqrt{\lambda} + k\sqrt{\lambda}\cos\sqrt{\lambda}) = 0.$$

Since we need to seek a non-trivial solution to the problem, we want $\sqrt{\lambda}$ to satisfy

$$\sin\sqrt{\lambda} + k\sqrt{\lambda}\cos\sqrt{\lambda} = 0.$$

If $\lambda > 0$ is a solution of the previous equation, then $\cos\sqrt{\lambda} \neq 0$. For, otherwise both $\cos\sqrt{\lambda} = 0$ and $\sin\sqrt{\lambda} = 0$ which is a contradiction. Hence the eigenvalues are precisely the nonzero solutions of

$$\tan\sqrt{\lambda} = -k\sqrt{\lambda}. \tag{4.42}$$

Finally, we observe that there are countably infinite number of solutions to (4.42), say (λ_n). Moreover, we get $(\lambda_n) \to \infty$. Therefore there exist countable eigenvalues

$$\lambda_1 < \lambda_2 < \lambda_3 < \cdots ,$$

and the corresponding eigenfunctions are given by

$$\sin(\sqrt{\lambda_n}x), \ n \in \mathbb{N}.$$

This completes the solution. □

Definition 4.4.5 (Orthogonality). Let $(X, \langle , \rangle)$ be an inner product space. Let $u, v \in X$ be such that $u \neq v$. We say that u and v are orthogonal, if $\langle u, v \rangle = 0$. We say a set S is orthogonal if any two distinct members of S are orthogonal.

Example 4.4.6. Consider the space $L^2([-\pi, \pi])$ with an inner product defined by $\langle u, v \rangle = \int_{-\pi}^{\pi} u(t)v(t)dt$. It is straight forward to verify that

$$\int_{-\pi}^{\pi} \cos(nx)\cos(mx)dx = 0, \ m \neq n, \ m, n \in \mathbb{Z},$$

$$\int_{-\pi}^{\pi} \sin(nx)\sin(mx)dx = 0, m \neq n, \ m, n \in \mathbb{Z},$$

$$\int_{-\pi}^{\pi} \cos(nx)\sin(mx)dx = 0, \ m, n \in \mathbb{Z}.$$

Hence $\left\{\cos(nx) : n \in \mathbb{Z}\right\} \cup \left\{\sin(nx) : n \in \mathbb{Z}\right\}$ is an orthogonal set in $L^2([-\pi, \pi])$. □

Definition 4.4.7 (Orthogonality with respect to a weight function). We say that ϕ and ψ are orthogonal with respect to a weight function $w > 0$ in $[a, b]$ if

$$\int_a^b w(x)\phi(x)\psi(x)dx = 0.$$

Theorem 4.4.8. *Consider regular S-L system (4.40). For $j, k \in \mathbb{N}$, let ϕ_j and ϕ_k be the eigenfunctions of (4.40) corresponding to distinct eigenvalues λ_j and λ_k, respectively. Then ϕ_j and ϕ_k are orthogonal with respect to the weight function s in $[a, b]$.*

Proof. From the hypotheses, we have the following identities

$$\left(p\phi_j'\right)' + (q + \lambda_j s)\,\phi_j = 0, \tag{4.43}$$

and

$$\left(p\phi_k'\right)' + (q + \lambda_k s)\,\phi_k = 0. \tag{4.44}$$

On multiplying (4.43) and (4.44) with ϕ_k and ϕ_j, respectively, and subtracting one from the other, we get

$$\left[p(\phi_j'\phi_k - \phi_k'\phi_j)\right]' + (\lambda_j - \lambda_k)s\phi_j\phi_k = 0. \tag{4.45}$$

On integrating (4.45) over $[a, b]$, we obtain

$$(\lambda_j - \lambda_k)\int_a^b s(x)\phi_j(x)\phi_k(x)dx = \left[p(x)(\phi_j'\phi_k - \phi_k'\phi_j)(x)\right]_{x=a}^{x=b}. \tag{4.46}$$

On the other hand, we have

$$A_1\phi_j(a) + A_2\phi_j'(a) = 0,$$
$$A_1\phi_k(a) + A_2\phi_k'(a) = 0.$$

Since $(A_1, A_2) \neq (0, 0)$, we obtain

$$\phi_j'(a)\phi_k(a) - \phi_k'(a)\phi_j(a) = 0.$$

Similarly as $(B_1, B_2) \neq (0, 0)$, using the same argument we arrive at

$$\phi_j'(b)\phi_k(b) - \phi_k'(b)\phi_j(b) = 0.$$

Hence the right hand side in equation (4.46) is zero. Therefore

$$(\lambda_j - \lambda_k)\int_a^b s(x)\phi_j(x)\phi_k(x)\,dx = 0.$$

This proves the required result because $\lambda_j \neq \lambda_k$. □

Lemma 4.4.9. *All the eigenvalues of regular S-L system* (4.40) *are real.*

Proof. Let λ be an eigenvalue and ϕ be a corresponding eigenfunction. If λ is not real, then we consider the complex conjugate to the S-L system and equate the real and the imaginary parts to zero to get $\overline{\phi}$ as an eigenfunction corresponding to the eigenvalue $\overline{\lambda}$. From the proof of Theorem 4.4.8, we immediately obtain

$$0 = (\lambda - \overline{\lambda})\int_a^b s(x)\phi(x)\overline{\phi}(x)\,dx = 2\,\text{Im}(\lambda)\int_a^b s(x)|\phi(x)|^2 dx.$$

As $s > 0$ and ϕ is non-trivial, the integral on the right hand side of the above equation is nonzero. Therefore $\text{Im}(\lambda) = 0$, which is a contradiction. Hence all the eigenvalues are real. □

Lemma 4.4.10. *An eigenfunction corresponding to an eigenvalue of regular S-L system (4.40) is unique up to a constant factor. In other words, all eigenvalues of S-L system (4.40) are simple.*

Proof. Let ϕ_1, ϕ_2 be two eigenfunctions corresponding to the eigenvalue λ. From the proof of Theorem 4.4.8, we have $W(a; \phi_1, \phi_2) = 0$. By the properties of the Wronskian, we get $W(x, \phi_1, \phi_2) = 0$, $x \in [a, b]$. Therefore we conclude that ϕ_1, ϕ_2 are linearly dependent. □

Proposition 4.4.11. *Consider the eigenvalue problem*

$$(py')' + (q + \lambda s)y = 0, \quad x \in (a, b), \tag{4.47a}$$

$$y(a) = 0, \ y(b) = 0. \tag{4.47b}$$

where $p(x), s(x) > 0$, $x \in [a, b]$. If $q(x) \leq 0$, $x \in [a, b]$, then there exists no non-positive eigenvalue for eigenvalue problem (4.47a)–(4.47b). In other words, all eigenvalues are positive.

Proof. Let λ be an eigenvalue of (4.47a)–(4.47b) and ϕ be a corresponding eigenfunction. We first multiply (4.47a) with ϕ and integrate on both sides with respect to x to get

$$\int_a^b \left(p(x)\phi'(x)\right)'\phi(x)dx + \int_a^b q(x)\phi^2(x)dx + \int_a^b \lambda s(x)\phi^2(x)dx = 0.$$

Now, by the integration by parts formula, we find

$$-\int_a^b p(x)(\phi'(x))^2 dx + \int_a^b q(x)\phi^2(x)dx + \int_a^b \lambda s(x)\phi^2(x)dx = 0. \tag{4.48}$$

By rearranging the terms we obtain

$$\lambda = \frac{\int_a^b p(x)(\phi'(x))^2 dx - \int_a^b q(x)\phi^2(x)dx}{\int_a^b s(x)\phi^2(x)dx}$$

$$> 0.$$

The last inequality follows from the hypotheses $p > 0$, $q \leq 0$, and $s > 0$. This proves the expected result. □

The boundary conditions given in (4.47b) are called the homogeneous Dirichlet boundary conditions. Proposition 4.4.11 is very useful in the computation of eigenvalues and eigenfunctions of S-L system with homogeneous Dirichlet boundary conditions. In the examples which are presented so far, to find the eigenvalues we have proceeded by considering three cases, viz., $\lambda < 0, \lambda = 0$, and $\lambda > 0$. If the hypotheses of Proposition 4.4.11 is satisfied, then it is enough to consider the case $\lambda > 0$, and discard the other two cases.

Example 4.4.12. Solve $\begin{cases} x^2 y'' + xy' + \lambda y = 0, & 1 < x < e, \\ y(1) = 0 = y(e). \end{cases}$

Solution: We write the given ODE in the standard form as

$$(xy')' + \frac{1}{x}\lambda y = 0, \ x \in (1, e).$$

We first use Proposition 4.4.11 to eliminate the possibility of any non-positive number being an eigenvalue. Therefore the general solution to the Euler equation is

$$\phi(x) = A\cos(\sqrt{\lambda}\log x) + B\sin(\sqrt{\lambda}\log x), \ x \in (1, e).$$

Next, the boundary condition $\phi(1) = 0$ gives $A = 0$ and the other boundary condition $\phi(e) = 0$ yields

$$\sin(\sqrt{\lambda}) = 0.$$

Therefore the eigenvalues are

$$\lambda_n = (n\pi)^2, \ n \in \mathbb{N}$$

and the corresponding eigenfunctions are

$$\phi_n(x) = \sin(n\pi \log x), \ n \in \mathbb{N}.$$

This completes the solution.   □

Our next objective is to show that there exist only countably infinite eigenvalues for the S-L system (4.40) when $s \equiv 1$. There are many approaches to establish this result (see[24, 26, 33, 37]). We adopt the procedure prescribed in [37]. To this end, we first recall the definitions of $\mathscr{L}$ and X from the previous section (see equations (4.21) and (4.19)),

$$\mathscr{L} = (py')' + qy, \ y \in C^2([a, b]),$$

where $p \in C^1([a, b])$, $q \in C([a, b])$, $p(x) > 0$, $x \in [a, b]$, and

$$X = \{y \in C^2([a, b]) \ : \ A_1 y(a) + A_2 y'(a) = 0, \ B_1 y(b) + B_2 y'(b) = 0\}.$$

Let G be Green's function corresponding to $\mathscr{L}$ with homogeneous boundary conditions (HBC) given by

$$\begin{cases} A_1 y(a) + A_2 y'(a) = 0, \\ B_1 y(b) + B_2 y'(b) = 0. \end{cases}$$

We now define $\mathcal{G} : C([a, b]) \to X$ as

$$(\mathcal{G}f)(x) = \int_a^b G(x, \xi) f(\xi) d\xi, \ f \in C([a, b]), \ x \in [a, b]. \tag{4.49}$$

We now prove some of the useful properties of $\mathcal{G}$.

Lemma 4.4.13. *Under the hypotheses of Theorem 4.3.2, both* $\mathscr{L}\mathcal{G} :$ $C([a, b]) \to C([a, b])$ *and* $\mathcal{G}L : X \to X$ *are identity operators.*

Proof. From the fundamental theorem of Green's function (Theorem 4.3.7), we have

$$\mathscr{L}[\mathcal{G}f](x) = \mathscr{L}\left(\int_a^b G(x,\xi)f(\xi)d\xi\right) = f(x), \quad x \in [a,b], \tag{4.50}$$

for $f \in C([a,b])$. On the other hand, for $y \in X$, in view of (4.50), we get

$$\mathscr{L}\big[\mathcal{G}\mathscr{L}[y] - y\big] = \mathscr{L}\mathcal{G}\mathscr{L}[y] - \mathscr{L}[y] = 0.$$

Hence $\mathcal{G}\mathscr{L}[y] - y = 0$, $y \in X$, owing to the hypotheses of Theorem (4.3.2). This completes the proof of the lemma. □

Using the density of $C([a,b])$ in $L^2([a,b])$ and continuity of G, $\mathcal{G}$ can be uniquely extended to $L^2([a,b])$. We still denote the extended operator by $\mathcal{G}$ and henceforth we consider only the extended operator $\mathcal{G} : L^2([a,b]) \to L^2([a,b])$.

Lemma 4.4.14. *Assume that the hypotheses of Theorem 4.3.2 holds. A nonzero real number λ is an eigenvalue of $\mathcal{G}$ if and only if $\frac{1}{\lambda}$ is an eigenvalue of $\mathscr{L}$. Moreover, the eigenfunctions corresponding to the eigenvalue $\frac{1}{\lambda}$ of $\mathscr{L}$ are precisely the eigenfunctions of $\mathcal{G}$ corresponding to the eigenvalue λ.*

Proof. Let $\phi \in L^2([a,b])$ be an eigenfunction corresponding to the eigenvalue λ, i.e.,

$$\mathcal{G}\phi = -\lambda\phi. \tag{4.51}$$

We begin with the observation that $\mathcal{G}\phi \in C([a,b])$ because $\phi \in L^2([a,b])$ (see Exercise 4.16). Since the left hand side of (4.51) is a continuous function, so is the right hand side. Therefore ϕ is a continuous function. Using the same argument, we obtain that $\phi \in C^2([a,b])$. Applying the operator $\mathscr{L}$ on both sides of (4.51), and using Lemma 4.4.13, we obtain that

$$\phi + \lambda\mathscr{L}[\phi]=0.$$

Therefore $\frac{1}{\lambda}$ is an eigenvalue of $\mathscr{L}$ and ϕ is a corresponding eigenfunction. On the other hand, let $\mu \neq 0$ be an eigenvalue of $\mathscr{L}$ and $\psi \in X$ be a corresponding eigenfunction. In other words

$$\mathscr{L}[\psi] + \mu\psi = 0.$$

Applying $\mathcal{G}$ on both sides and using Lemma 4.4.13, we have $\psi + \mu\mathcal{G}\psi = 0$. Thus $\frac{1}{\mu}$ is an eigenvalue of $\mathcal{G}$ and ψ is a corresponding eigenfunction. Hence the theorem is proved. □

Lemma 4.4.15. *The operator $\mathcal{G}$ is a self adjoint operator on $L^2([a,b])$.*

Proof. Let $f, g \in C([a,b])$. We assume that $u, v \in X$ are such that $\mathscr{L}[u] = f$ and $\mathscr{L}[v] = g$. From Lemma 4.4.13, we find that $\mathcal{G}f = u$ and $\mathcal{G}g = v$. Moreover, we recall that (see (4.20))

$$\langle u, \mathscr{L}v \rangle_{L^2} = \langle \mathscr{L}u, v \rangle_{L^2}.$$

We now consider

$$\langle \mathcal{G}f, g \rangle_{L^2} = \langle u, \mathscr{L}v \rangle_{L^2} = \langle \mathscr{L}u, v \rangle_{L^2} = \langle f, \mathcal{G}g \rangle_{L^2}, \ f, g \in C([a,b]). \quad (4.52)$$

Let $F, H \in L^2([a,b])$. Since $C([a,b])$ is dense in $L^2([a,b])$, there exist two sequences (f_n), (g_n) in $C([a,b])$ such that $f_n \to F$ and $h_n \to H$, in $L^2([a,b])$. In view of (4.52), we have

$$\langle \mathcal{G}f_n, h_n \rangle_{L^2} = \langle f_n, \mathcal{G}h_n \rangle_{L^2}.$$

Using Exercise 4.19 and taking limit as $n \to \infty$, we get

$$\langle \mathcal{G}F, H \rangle_{L^2} = \langle F, \mathcal{G}H \rangle_{L^2}.$$

This proves that $\mathcal{G}$ is a self adjoint operator on $L^2([a,b])$. $\quad\square$

Lemma 4.4.16. *The operator* $\mathcal{G} : L^2([a,b]) \to L^2([a,b])$ *is a compact operator[2].*

Proof. The proof is out of the scope of this book. However, an interested reader having knowledge about the Ascoli-Arzela theorem, and Hölder's inequality can prove this result by following the outline given in Exercise 4.16. $\quad\square$

Lemma 4.4.17. *If* $\mathcal{G}f = 0$ *for some* $f \in L^2([a,b])$, *then* $f \equiv 0$ *in* $L^2([a,b])$.

Proof. See [37] for a proof. $\quad\square$

We now recall a classical result from the theory of self adjoint compact operators (For a proof, see for instance [16, 21]).

Theorem 4.4.18. *If* H *is a separable Hilbert space and* $T : H \to H$ *is a compact, self adjoint operator, then the eigenvectors of* T *form a complete orthonormal basis[3] of* H.

Using the results that we have proved so far in this section and Theorem 4.4.18, we draw the following conclusions.

1. The nonzero eigenvalues of $\mathscr{L}$ are precisely the reciprocals of those of the self adjoint compact operator $\mathcal{G}$ (see Lemmas 4.4.14–4.4.16). Therefore there exist at most countably infinite eigenvalues for $\mathscr{L}$.

---

[2]A linear operator $T : L^2([a,b]) \to L^2([a,b])$ is said to be a compact operator if for every bounded sequence (f_n) in $L^2([a,b])$, there exists a subsequence, say (f_{n_k}), such that the sequence (Tf_{n_k}) converges in $L^2([a,b])$.

[3]A set $\{e_i : i \in \mathbb{N}\}$ of mutually orthogonal unit vectors is said to be a complete orthonormal basis of H if $x = \sum_{i=0}^{\infty} \langle x, e_i \rangle e_i, \ x \in H$.

2. Every eigenvalue of $\mathscr{L}$ is simple (see Lemma 4.4.10).

3. Due to Lemma 4.4.17, $\lambda = 0$ is not an eigenvalue of $\mathcal{G}$. In view of Theorem 4.4.18 there exists a complete orthonormal basis of $L^2([a, b])$ consisting of only eigenfunctions corresponding to (nonzero) eigenvalues. Since the set of eigenfunctions of $\mathscr{L}$ is the same as the set of eigenfunctions of $\mathcal{G}$ (see Lemma 4.4.14). it also forms a complete orthonormal basis of $L^2([a, b])$. Moreover there exist countably many eigenvalues of $\mathscr{L}$.

In the case when s is not a constant function in the regular S-L system, we state the following result without providing a proof. For a proof, the reader is advised to refer [24].

Theorem 4.4.19. *A regular S-L system* (4.40) *has an infinite sequence of real eigenvalues,* $\lambda_1 < \lambda_2 < \cdots$ *with* $\lambda_n \to \infty$. *Moreover, all the eigenvalues are simple and the* n^{th} *eigenfunction* ϕ_n *has exactly* n *zeros in* (a, b).

We now turn our attention to the eigenvalue problems with periodic boundary conditions.

Definition 4.4.20 (Periodic S-L system)**.** The system

$$\begin{cases} (p(x)y'(x))' + (q(x) + \lambda s(x)) y(x) = 0, \ a < x < b, \\ p(a) = p(b), \ y(a) = y(b), \ y'(a) = y'(b), \end{cases} \tag{4.53}$$

where $p \in C^1([a, b])$, $q, s \in C([a, b])$, $p > 0$ and $s > 0$, is called a periodic Sturm-Liouville system.

We would like to verify whether the results we have proved in the case of regular S-L system hold true when we have periodic boundary conditions. We begin with the following example.

Example 4.4.21. Find all the eigenvalues and the corresponding eigenfunctions of

$$\begin{cases} y'' + \lambda y = 0, \quad -\pi < x < \pi, \\ y(-\pi) = y(\pi), \quad y'(-\pi) = y'(\pi). \end{cases} \tag{4.54}$$

Solution: In the given problem $p \equiv 1$, and we have periodic boundary conditions. Therefore this is a periodic S-L system.
As in the case of regular S-L system we consider three cases.
Case 1. Let $\lambda < 0$.
The solution to the given ODE in this case is

$$\phi(x) = c_1 e^{-\sqrt{-\lambda}x} + c_2 e^{\sqrt{-\lambda}x}, \ x \in (-\pi, \pi).$$

Substituting this ϕ in the boundary conditions in (4.54), we get $c_1 = c_2 = 0$. Thus we have $\phi \equiv 0$ and hence we cannot have a negative eigenvalue.

Case 2. Let $\lambda = 0$.

In this case $\phi(x) = c_1 x + c_2$ is the general solution. The first boundary condition in the given system implies that $c_1 = 0$. Moreover, the second boundary condition is always satisfied. Therefore

$$\phi(x) = c_2, \quad x \in (-\pi, \pi), \quad c_2 \in \mathbb{R} \backslash \{0\}$$

is an eigenfunction.

Case 3. We consider the situation where $\lambda > 0$. Then

$$\phi(x) = c_1 \cos(\sqrt{\lambda} x) + c_2 \sin(\sqrt{\lambda} x), \quad x \in (-\pi, \pi), \quad c_1, c_2 \in \mathbb{R},$$

is the general solution. From the boundary conditions we have

$$2c_2 \sin(\sqrt{\lambda}\pi) = 0, \quad 2c_1 \sin(\sqrt{\lambda}\pi) = 0.$$

Since $(c_1, c_2) \neq (0, 0)$, we shall have $\sin(\sqrt{\lambda}\pi) = 0$. Hence $\lambda_n = n^2$, $n \in \mathbb{N}$ is an eigenvalue. There are two eigenfunctions, namely $\cos(nx), \sin(nx)$ corresponding to the eigenvalue n^2.

Therefore eigenvalues are $\lambda_n = n^2$, $n \in \mathbb{N} \cup \{0\}$ and for each λ_n, the corresponding eigenfunctions are members of the set

$$S_n = \big\{ a_n \cos(nx) + b_n \sin(nx) \; : \; (a_n, b_n) \neq (0, 0) \big\}.$$

This completes the solution. □

Remark 4.4.22. From Example 4.54, we find that when we have periodic boundary conditions the eigenvalues need not be simple. In particular, their multiplicity can be two. On the other hand, the multiplicity of any eigenvalue of (4.53) is at most two. For, let λ be an eigenvalue of (4.53). For this fixed λ, as the differential equation in (4.53) is of the second order, we cannot have three linearly independent solutions for it (irrespective of the boundary conditions that they satisfy). Hence the multiplicity of the eigenvalue is at most two.

Lemma 4.4.23. *The eigenfunctions of periodic S-L system* (4.53) *are orthogonal with respect to the weight function s in $[a, b]$.*

Proof. Let ϕ_j and ϕ_k be the eigenfunctions corresponding to distinct eigenvalues λ_j and λ_k, respectively. From (4.46) in Theorem 4.4.8 we obtain

$$(\lambda_j - \lambda_k) \int_a^b s(x) \phi_j(x) \phi_k(x) \; dx = \big[p(\phi_j' \phi_k - \phi_k' \phi_j) \big]_a^b = 0,$$

where the last equality is due to the periodic boundary conditions. Hence ϕ_j and ϕ_k are orthogonal with respect to the weight function s. □

Lemma 4.4.24. *All the eigenvalues of a periodic S-L system* (4.53) *are real.*

Proof. One can prove this using the same argument that was presented in Lemma 4.4.9. So the proof is left as an exercise to the reader. □

We now state a result without giving a proof for the periodic S-L system. For a proof see [7].

Theorem 4.4.25. *The periodic S-L system* (4.53) *has an infinite sequence of real eigenvalues (with multiplicity at most 2),* $\lambda_1 < \lambda_2 \leq \cdots$ *with* $(\lambda_n) \to \infty$.

Exercise 4.1. Assume that $a_1 \in C^1([a,b]), a_0, f \in C([a,b])$. Let $\mathscr{L}$ be a second order linear differential operator given by

$$\mathscr{L}[y](x) = y''(x) + a_1(x)y'(x) + a_0(x)y(x), \; x \in (a,b).$$

For $y : [a,b] \to \mathbb{R}$, set $BC_1(y) = A_1 y(a) + A_2 y'(a)$, $BC_2(y) = B_1 y(b) + B_2 y'(b)$, where $(A_1, A_2) \neq (0,0)$ and $(B_1, B_2) \neq (0,0)$. Show that the following are equivalent:

(i) The homogeneous problem $\mathscr{L}[y] = 0, \; BC_1(y) = 0, \; BC_2(y) = 0$ has only trivial solution.

(ii) The equation $\mathscr{L}[y] = 0$ has linearly independent solutions ψ_1 and ψ_2 such that $BC_1(\psi_1) = 0, \; BC_2(\psi_2) = 0$.

(iii) There exist two linearly independent solutions ψ_1 and ψ_2 to $\mathscr{L}[y] = 0$ such that $BC_1(\psi_1)BC_2(\psi_2) - BC_1(\psi_2)BC_2(\psi_1) \neq 0$.

Exercise 4.2. Assume that $a_1 \in C^1([a,b]), a_0 \in C([a,b])$. Let $\mathscr{L}$ be a second order linear differential operator given by

$$\mathscr{L}[y](x) = y''(x) + a_1(x)y'(x) + a_0(x)y(x), \; x \in (a,b).$$

Let $(A_1, A_2) \neq (0,0)$ and $(B_1, B_2) \neq (0,0)$. Let ϕ_1 and ϕ_2 be the solutions to $\mathscr{L}[y] = 0$ satisfying $\phi_1(a) = A_2, \phi_1'(a) = -A_1$ and $\phi_2(b) = -B_2, \phi_2'(b) = B_1$, respectively. Then show that the non-homogeneous boundary value problem

$$\begin{cases} \mathscr{L}[y] = f, \; x \in (a,b) \\ A_1 y(a) + A_2 y'(a) = \alpha, \\ B_1 y(b) + B_2 y'(b) = \beta, \end{cases} \tag{4.55}$$

has a solution for every $f \in C([a,b])$, $\alpha, \beta \in \mathbb{R}$ if and only if $W(b; \phi_1, \phi_2) \neq 0$.

Exercise 4.3. Find Green's functions corresponding to the operators and boundary conditions given below:

(i) $\mathscr{L}[y] = y'', \quad y(0) = y'(0), \; y(1) = -y'(1)$.

(ii) $\mathscr{L}[y] = y'' + y, \quad y(0) + y'(0) = 0, \; y(1) + y'(1) = 0$.

(iii) $\alpha \neq 0, \; \mathscr{L}[y] = y'' - \alpha^2 y, \quad y'(0) = 0, \; y(1) = 0$.

(iv) $\alpha \neq 0, \; \mathscr{L}[y] = y'' + \alpha^2 y, \quad y(0) = y(\pi/\alpha), \; y'(0) = y'(\pi/\alpha)$.

(v) $\alpha \neq 0, \; \mathscr{L}[y] = y'' - \alpha^2 y, \quad y(0) = y(1), \; y'(0) = y'(1)$.

Exercise 4.4. Solve the following boundary value problems by computing the associated Green's functions.

(i) $y'' = x$, $x \in (0,1)$, with the boundary conditions $y'(0) = 0, y(1) = 1$.

(ii) $y'' + y = \sin x$, $x \in (0, \frac{\pi}{2})$, with the boundary conditions $y(0) = 0$, and $y(\frac{\pi}{2}) = 5$.

Exercise 4.5. Let $g : [a,b] \to \mathbb{R}$ be a continuous function. If $\int_a^b g(x)\eta(x)dx = 0$ for every $\eta \in C([a,b])$. Then show that $g(x) = 0$, $x \in [a,b]$.

Exercise 4.6. Let $p \in C^1([a,b])$, $q \in C([a,b])$, Let G be Green's function associated with $\mathscr{L}[y] = (py')' + qy$, homogeneous boundary conditions (HBC) $A_1 y(a) + A_2 y'(a) = 0$ and $B_1 y(b) + B_2 y'(b) = 0$. Show that

$$\frac{\partial G}{\partial x}(x, x^-) = \frac{\partial G}{\partial x}(x^+, x) \text{ and } \frac{\partial G}{\partial x}(x, x^+) = \frac{\partial G}{\partial x}(x^-, x), \ x \in (a,b).$$

Hence, deduce that

$$\frac{\partial G}{\partial x}(x, \xi)\Big|_{\xi=x^-}^{\xi=x^+} = -\frac{1}{p(x)}, \ x \in (a,b).$$

Exercise 4.7. Let $\mathscr{L}$ be as in Exercise 4.6 and ϕ_0 be a solution to $\mathscr{L}[\phi_0] = 0$, with homogeneous boundary conditions given in Exercise 4.6. Moreover, let ϕ be a solution to $\mathscr{L}[\phi] = f$, with the same boundary conditions. Then show that

$$\int_a^b f(x)\phi_0(x)dx = 0.$$

Hint: Apply identity (4.20) to ϕ and ϕ_0.

Exercise 4.8. Let $\mathscr{L}$ be as in Exercise 4.6. Let $\mathscr{L}[\phi_0] = 0$ and ϕ_0 satisfy the homogeneous boundary conditions (HBC) given in Exercise 4.6. Moreover assume that

$$\int_a^b f(x)\phi_0(x)dx = 0,$$

for some continuous function f. Show that the boundary value problem $\mathscr{L}[y] = f$ with HBC given in Exercise 4.6 has a solution.

Hint: Let ϕ be the solution to the initial value problem $\mathscr{L}[y] = f$, $y(a) = \phi_0(a)$, and $y'(a) = \phi_0'(a)$. Apply identity (4.18) to ϕ_0 and ϕ to show that ϕ satisfies the required boundary conditions.

Exercise 4.9. Find the eigenvalues and the corresponding eigenfunctions of the following systems:

(i) $\begin{cases} y'' + \lambda y = 0, \ x \in (0,1), \\ y(0) = 0, \ y(1) = 0. \end{cases}$

$$(ii) \begin{cases} y'' + \lambda y = 0, \ x \in (0,1), \\ y'(0) = 0, \ y'(1) = 0. \end{cases}$$

$$(iii) \begin{cases} y'' + \lambda y = 0, \ x \in (0,1), \\ y'(0) = 0, \ y(1) = 0. \end{cases}$$

Exercise 4.10. Show that all the eigenvalues of

$$\begin{cases} (py')' + (q + \lambda s)y = 0, \ x \in (a,b), \\ y'(a) = 0, \ y'(b) = 0, \end{cases}$$

where $p \in C^1([a,b])$, $q, s \in C([a,b])$, $p(x), s(x) > 0, q(x) \le 0$, $x \in [a,b]$, are nonnegative. Moreover, show that $\lambda = 0$ is an eigenvalue of this problem if and only if $q \equiv 0$.

Hint: Use the same technique presented in Proposition 4.4.11.

Exercise 4.11. Let $p \in C^1([a,b])$, $q, s \in C([a,b])$, $p(x), s(x) > 0$, $x \in [a,b]$. If $q(x) \le 0$, $x \in [a,b]$. Then show that all the eigenvalues of the following systems are positive:

$$(i) \begin{cases} (py')' + (q + \lambda s)y = 0, \ x \in (a,b), \\ y(a) = 0, \ y'(b) = 0. \end{cases}$$

$$(ii) \begin{cases} (py')' + (q + \lambda s)y = 0, \ x \in (a,b), \\ y'(a) = 0, \ y(b) = 0. \end{cases}$$

Exercise 4.12. Assume that $A_1 A_2 \le 0$, $B_1 B_2 \ge 0$ and $A_1 B_1 \ne 0$. Let $p \in C^1([a,b])$, $q, s \in C([a,b])$, $p(x), s(x) > 0$, $x \in [a,b]$. If $q(x) \le 0$, $x \in [a,b]$. Then show that all the eigenvalues of the following system are positive:

$$\begin{cases} (py')' + (q + \lambda s)y = 0, \ x \in (a,b), \\ A_1 y(a) + A_2 y'(a) = 0, \\ B_1 y(b) + B_2 y'(b) = 0. \end{cases}$$

Exercise 4.13. Consider the eigenvalue problem

$$\begin{cases} (py')' + (q + \lambda s)y = 0, \ x \in (a,b), \\ y'(a) = 0, \ y'(b) = 0, \end{cases}$$

where $p \in C^1([a,b])$, $q, s \in C([a,b])$, $p(x), s(x) > 0, q(x) \ge 0$, $x \in [a,b]$. If λ is an eigenvalue of this problem and the corresponding eigenfunction ϕ is positive in $[a,b]$, then show that $\lambda \le 0$.

Hint: Divide the given ODE with ϕ and use the integration by parts formula.

Exercise 4.14. Show that $\displaystyle\int_a^b |f(x)g(x)|dx \le \|f\|_2 \|g\|_2$, for every $f, g \in L^2([a,b])$. This is called Hölder's inequality.

Exercise 4.15.   Assume that $g \in L^2([a,b])$ and $h(x) := \int_a^b K(x,\xi)g(\xi)d\xi$, $x \in [a,b]$, where $K : [a,b] \times [a,b] \to \mathbb{R}$ is a continuous function. Show that h is a continuous function.

Exercise 4.16. Assume that $G : [a,b] \times [a,b] \to \mathbb{R}$ is a continuous function. Consider $\mathcal{G} : L^2([a,b], \|\|_2) \to L^2([a,b], \|\|_2)$ defined by

$$\mathcal{G}[f](x) = \int_a^b G(x,\xi)f(\xi)d\xi, \ f \in L^2([a,b]).$$

(i) Show that $\int_a^b |f(x)|dx \le \sqrt{b-a}\|f\|_2, \ f \in L^2([a,b])$.

(ii) If (f_n) is a bounded sequence in $L^2([a,b])$, then show that $(\mathcal{G}[f_n])$ is bounded in $C([a,b], \|\|_\infty)$ where $\|g\|_\infty = \sup_{x\in[a,b]} |g(x)|, \ g \in C([a,b])$.

(iii) If (f_n) is a bounded sequence in $L^2([a,b])$, then show that $(\mathcal{G}[f_n])$ has a convergent subsequence in $C([a,b], \|\|_\infty)$ using the Ascoli–Arzela theorem.

(iv) If $f_n \to f$ in $C([a,b], \|\|_\infty)$, then show that $f_n \to f$ in $L^2([a,b], \|\|_2)$.

(v) Show that $\mathcal{G}$ is a compact operator.

Exercise 4.17.   Assume that $G : [a,b] \times [a,b] \to \mathbb{R}$ is a continuous function. Then show that $\mathcal{G} : C([a,b], \|\|_\infty) \to C([a,b], \|\|_\infty)$ defined by

$$\mathcal{G}[f](x) = \int_a^b G(x,\xi)f(\xi)d\xi, \ f \in C([a,b]),$$

is a compact operator.

Exercise 4.18.   Let $s > 0$ be a continuous function. Define

$$\langle f,g\rangle_s := \int_a^b s(x)f(x)g(x)dx, \ f,g \in C([a,b]).$$

Show that $\langle \ , \ \rangle_s$ defines an inner product on $C([a,b])$.

Exercise 4.19.   Let $K : [a,b] \times [a,b] \to \mathbb{R}$ be a continuous function. Let $f, h \in L^2([a,b])$. Assume that (f_n), (h_n) are sequences in $C([a,b])$ such that $f_n \to f$ and $h_n \to h$ in $L^2([a,b])$. Then show that

$$\langle \mathcal{G}f_n, h_n\rangle_{L^2} \to \langle \mathcal{G}f, h\rangle_{L^2},$$

where $\mathcal{G}$ is given by

$$(\mathcal{G}f)(x) = \int_a^b K(x,\xi)f(\xi)d\xi, \ f \in C([a,b]), \ x \in [a,b].$$

Chapter 5

Systems of first order ODEs

5.1   Introduction

So far we have studied ODEs where the dependent variable is a scalar function. In particular, we have discussed various methods of finding their solutions. Moreover, we have studied the quantitative and qualitative behavior of the solutions of these equations in case we are unable to find them explicitly. In this chapter, we discuss some methods to solve a system of linear ODEs. Let $J \subseteq \mathbb{R}$ be an open interval, $\Omega \subseteq \mathbb{R}^n$ an open set. We assume that $F : \bar{J} \times \bar{\Omega} \to \mathbb{R}^n$ is a continuous function, $Y(t) = \big(y_1(t), \ldots, y_n(t)\big)$, $t \in J$. Our objective is to study the following system

$$\begin{cases} Y'(t) = F(t, Y(t)), \ t \in J, \\ Y(t_0) = Y_0 \in \mathbb{R}^n. \end{cases} \tag{5.1}$$

As in Chapter 2, we say that (5.1) is a Cauchy problem if $t_0 \in J$ and an initial value problem if $J = (t_0, t_1)$, $t_1 \in \mathbb{R} \cup \{\infty\}$. The notion of left, right, and bilateral solutions to (5.1) can be directly borrowed from Chapter 2. Whenever we deal with the initial value problems, we consider the right solutions. On the other hand, we consider the bilateral solutions in case of Cauchy problems. There are many first order systems of ODEs which model physical/ biological phenomena (see [3, 23]). Often, first order partial differential equations can be solved by solving the corresponding characteristic equations which are nothing but systems of ODEs (see [27]).

It is important to realize that any higher order Cauchy/initial value problem of the form

$$\begin{cases} y^{(n)}(t) = F\big(t, y(t), y'(t), \ldots, y^{(n-1)}(t)\big), \ t \in J, \\ y(t_0) = y_0, \ y'(t_0) = y_1, \ldots, y^{(n-1)}(t_0) = y_{n-1}. \end{cases} \tag{5.2}$$

can be converted into a first order system of the form (5.1). The equivalence of an n-th order ODE and the corresponding $n \times n$ system was first used by D'Alembert. Before we go to the general case, consider the following example.

Example 5.1.1. Convert the following second order initial value problem into a system of first order initial value problem

$$\begin{cases} y''(t) + a_1(t)\sin(y'(t)) + a_0(t)y(t) = f(t), \ t > t_0, \\ y(t_0) = y_0, \ y'(t_0) = y_1. \end{cases}$$

Solution. In order to convert this second order ODE into a system, we first set $z(t) = y'(t)$, $t > t_0$. Then the given ODE reduces to

$$z'(t) = -a_1(t)\sin(z(t)) - a_0(t)y(t) + f(t), \ t > t_0.$$

Hence

$$\begin{cases} y'(t) = z(t), \ t > t_0, \\ z'(t) = -a_1(t)\sin(z(t)) - a_0(t)y(t) + f(t), \ t > t_0, \\ y(t_0) = y_0, \ z(t_0) = y_1, \end{cases}$$

is the required system of equations. $\square$

Similarly, the n-th order initial value problem (5.2) can be converted into a system of n first order equations by introducing $(n-1)$ new variables namely

$$z_1(t) = y'(t), \ z_2(t) = y''(t), \ldots, z_{n-1}(t) = y^{(n-1)}(t), t \geq t_0.$$

Thus the initial value problem associated to (5.2) is

$$\begin{cases} y'(t) = z_1(t), \\ z_2'(t) = z_3(t), \\ \vdots \\ z_{n-1}' = F(t, z_0(t), \ldots, z_{n-1}(t)), \\ y(t_0) = y_0, z_1(t_0) = y_1, \ldots, z_{n-1}(t_0) = y_{n-1}. \end{cases}$$

On the other hand, we need to remember that certain quantitative/qualitative properties that the solutions to the higher order equation posses may not hold for the solutions to the corresponding first order system of ODEs. The following example illustrates this.

Example 5.1.2. Let $a, b \in \mathbb{R}$. Consider the second order ODE

$$y''(t) - (b + 2a)y'(t) + [a(a+b) + b^2 e^{2bt}]y(t) = 0, \ t > 0. \tag{5.3}$$

and the corresponding system of first order ODEs

$$\begin{cases} y' = z, \\ z' = (b + 2a)z - [a(a+b) + b^2 e^{2bt}]y. \end{cases} \tag{5.4}$$

One can easily verify that $\phi(t) = e^{at}\cos(e^{bt})$, $t > 0$, is a solution to (5.3) whereas

$$(\phi(t), \psi(t)) = \left(e^{at}\cos(e^{bt}), -be^{(a+b)t}\sin(e^{at}) + ae^{at}\cos(e^{bt})\right), \ t > 0,$$

is a solution to (5.4). We now make the following observations:

(i) If we choose $a = 0$, $b > 0$, then the solution to (5.3) presented here is bounded and the corresponding solution to (5.4) is unbounded.

(ii) If $-a = b > 0$, then the solution to (5.3) converges to 0 as $t \to \infty$, where as the limit of $(y(t), z(t))$ as $t \to \infty$ does not exist. $\square$

5.2   Existence and uniqueness: Picard's method revisited

We now turn our attention toward the existence and uniqueness of the solutions to systems of first order initial value problems. To this end, we introduce a couple of quantities.

(i) We denote by $\|Y\| := (|y_1|^2 + \cdots + |y_n|^2)^{\frac{1}{2}}$, for every vector $Y = (y_1, \ldots, y_n) \in \mathbb{R}^n$.

(ii) Let K be a closed and bounded interval and $h_i : K \to \mathbb{R}$, $1 \le i \le n$, be continuous functions. If $H : K \to \mathbb{R}^n$ is given by $H(t) = (h_1(t), \ldots, h_n(t))$, $t \in K$, then the integral of H over K is a vector in $\mathbb{R}^n$ defined as

$$\int_K H(t)dt = \left(\int_K h_1(t)dt, \int_K h_2(t)dt, \ldots, \int_K h_n(t)dt\right).$$

Lemma 5.2.1. *Let K and H be as in* (ii). *Then we have*

$$\left\|\int_K H(t)dt\right\| \le \int_K \|H(t)\|dt. \tag{5.5}$$

Proof. In view of the Cauchy–Schwarz inequality[1] it follows that for any fixed $(v_1, \ldots, v_n) \in \mathbb{R}^n$ we have

$$\left|(v_1, \ldots, v_n) \cdot \int_K H(t)dt\right| = \left|\sum_{i=1}^n \int_K v_i h_i(t)dt\right|$$

$$\le \int_K \sum_{i=1}^n |h_i(t)v_i|\, dt$$

$$\le \int_K \|H(t)\|\|(v_1, \ldots, v_n)\|dt \tag{5.6}$$

---

[1]If $(\alpha_1, \ldots, \alpha_n)$ and $(\beta_1, \ldots, \beta_n)$ are any two vectors in $\mathbb{R}^n$, then the Cauchy–Schwarz inequality is given by

$$\sum_{i=0}^n |\alpha_i \beta_i| \le \left(\sum_{i=0}^n \alpha_i^2\right)^{\frac{1}{2}} \left(\sum_{i=0}^n \beta_i^2\right)^{\frac{1}{2}}$$

We now put $(v_1, \ldots, v_n) = \int_K H(t)dt$ in (5.6) to arrive at

$$\left\| \int_K H(t)dt \right\|^2 \leq \int_K \|H(t)\|dt \left\| \int_K H(t)dt \right\|.$$

This completes the proof of (5.5). □

The existence and uniqueness result of (5.1) and the strategy used to prove this is exactly the same as that of the scalar case, i.e., when $n = 1$. In particular, we first define Picard's sequence of successive approximations of solution to system (5.1). This is given by

$$Z_0(t) = Y_0, \tag{5.7}$$

$$Z_k(t) = Y_0 + \int_{t_0}^t F(\tau, Z_{n-1}(\tau))d\tau, \ k \in \mathbb{N}. \tag{5.8}$$

Under the hypotheses similar to that of Theorem 2.2.11, we can show that the sequence of successive approximations (Z_k) indeed converges uniformly to the solution to (5.1) in a neighborhood of t_0. The statement of the existence and uniqueness theorem is given in the following theorem.

Theorem 5.2.2 (Cauchy–Lipschitz). *Let $a, b > 0$ and $R := \{(t, y) : |t - t_0| \leq a, \|Y - Y_0\| \leq b\}$. Let $F : R \to \mathbb{R}^n$ be continuous and $M := \sup_{(t,Y) \in R} |F(t, Y)|$. Furthermore, we assume that F satisfies the Lipschitz condition, i.e., there exists $L > 0$ such that*

$$\|F(t, Y_1) - f(t, Y_2)\| \leq L\|Y_1 - Y_2\|, \ (t, Y_1), \ (t, Y_2) \in R. \tag{5.9}$$

Then initial value problem (5.1) has a unique solution on $[t_0 - \delta, t_0 + \delta]$, where

$$\delta = \min\left\{ a, \frac{b}{M} \right\}. \tag{5.10}$$

Proof. The proof is exactly the same as the one given in the scalar case ($n = 1$). To extend the the proof given in the scalar case to the system the only additional result that we need is (5.5). We leave the details of it to the reader. □

This is a local existence result which can be easily stated for initial value problems by modifying the hypotheses on F. Moreover, one can prove the global existence result also along the lines of Theorem 2.2.20. For completeness, we provide the statement here.

Theorem 5.2.3 (Global existence). *Let $f : [t_0 - a, t_0 + a] \times \mathbb{R}^n \to \mathbb{R}^n$ be continuous and there exists $L > 0$ such that*

$$\|F(t, Y_1) - F(t, Y_2)\| \leq L\|Y_1 - Y_2\|, \ t \in [t_0 - a, t_0 + a], \ Y_1, Y_2 \in \mathbb{R}^n. \tag{5.11}$$

Then there exists a unique solution to

$$Y'(t) = F(t, Y(t)), \ Y(t_0) = Y_0,$$

which is defined on $[t_0 - a, t_0 + a]$.

We conclude this section by stating the Peano existence result.

Theorem 5.2.4 (Peano). *Let* $a, b > 0$ *and* $R := \{(t, Y) : |t - t_0| \leq a, \ \|Y - Y_0\| \leq b\}$. *If* F *is continuous on* R, *then* (5.1) *has a solution on* $[t_0 - \delta, t_0 + \delta]$, *for some* $\delta > 0$.

5.3   Systems of linear ODEs with constant coefficients

Henceforth, throughout the chapter, unless specified otherwise, we assume that the systems of ODEs that we consider are posed in the interval $[0, \infty)$. In this section, we restrict ourselves to the linear equations with constant coefficients.

Let $M_n(\mathbb{R})$ denote the set of $n \times n$ matrices with real entries. Consider the initial value problem

$$Y'(t) = AY(t), \ t > 0, \tag{5.12}$$
$$Y(0) = Y_0, \tag{5.13}$$

where $A \in M_n(\mathbb{R})$.

If we consider a higher order linear ODE with constant coefficients along with the initial data and convert that into a system of linear equations (see Section 5.1), then that linear system is of the form (5.12)–(5.13).

Our objective in this section is to find the solution to (5.12)–(5.13). If $n = 1$, then it is easy to observe that the solution to (5.12)–(5.13) is given by $Y(t) = e^{tA}Y_0, \ t > 0$, where $A \in \mathbb{R}$. If A is a square matrix then we would like to give a meaning for the expression e^{tA} and finally show that $Y(t) = e^{tA}Y_0$ is the solution to (5.12)–(5.13).

Remark 5.3.1. The domain on which IVP (5.12)–(5.13) is posed is $[0, \infty)$. In fact, all the results that we prove henceforth can be easily extended to any interval in $\mathbb{R}$. The readers are advised to understand the statements as well the proofs of the results carefully so that they can suitably modify the statements of the results in any interval in $\mathbb{R}$.

5.3.1   Exponential of a matrix and its properties

In this subsection, we define the exponential of a matrix and prove some of its properties which will be used later. There are many equivalent ways of defining the exponential of a matrix. Most of them are generalizations of different ways of defining the exponential function defined on $\mathbb{R}$. In particular, we begin with the definition of the exponential function (as a power series) given by

$$e^t = 1 + \frac{t}{1!} + \frac{t^2}{2!} + \frac{t^3}{3!} + \cdots + \frac{t^k}{k!} + \cdots , \quad t \in \mathbb{R}.$$

A naive approach to extend this definition to matrices is to replace t with a square matrix A. In that case, we need to discuss the convergence of the resultant series in the space of $n \times n$ square matrices $M_n(\mathbb{R})$. For, we introduce the notion of convergence in $M_n(\mathbb{R})$. In order to do that, we first define two norms on $M_n(\mathbb{R})$. The first one is called the Euclidean norm and the second norm is called the maximum norm. Both the norms are defined by identifying matrices as members of $\mathbb{R}^{n^2}$. The Euclidean norm of a matrix $A = (a_{ij}) \in M_n(R)$ is defined as

$$\|A\| = \left(\sum_{i,j=1}^{n} |a_{ij}|^2 \right)^{\frac{1}{2}},$$

and the maximum norm of A is the maximum of the absolute values of its entries, i.e.,

$$\|A\|_\infty = \max_{i,j} |a_{ij}|.$$

It is straightforward to verify that

$$\|A\|_\infty \le \|A\| \le \sqrt{n}\|A\|_\infty, \quad A \in M_n(R). \tag{5.14}$$

Definition 5.3.2. Let (X_k) be a sequence in $M_n(\mathbb{R})$. We say that (X_k) converges in $M_n(\mathbb{R})$ if there exists $X \in M_n(\mathbb{R})$ such that $\|X_k - X\|_\infty \to 0$ as $k \to \infty$. Moreover, we say that X is the limit of X_k and is denoted by $X_k \to X$ in $M_n(\mathbb{R})$.

We say that a series $\sum_{k=0}^{\infty} X_k$ converges if its sequence of partial sums $(\sum_{k=0}^{N} X_k)$ converges in $M_n(\mathbb{R})$ as $N \to \infty$.

In view of (5.14), we can replace the maximum norm in Definition 5.3.2 with the Euclidean norm. The following lemma immediately follows from the definition of convergence.

Lemma 5.3.3. *Let $(X_{k,ij})$ denote the (i,j)-th entry of $X_k \in M_n(\mathbb{R})$, $k \in \mathbb{N}$. Then the following hold true.*

(i) The sequence $X_k \to X$ in $M_n(\mathbb{R})$ if and only if $X_{k,ij} \to X_{ij}$, $1 \le i, j \le n$.

(ii) The series $\sum_{k=0}^{\infty} X_k$ converges if and only if $\sum_{k=0}^{\infty} X_{k,ij}$ converges for each $1 \le i, j \le n$.

Proof. The proof is left to the reader as an easy exercise. $\qquad\square$

Though we discuss only real matrices in this subsection, it is not difficult to see that all the results that are proved here are still valid in the case of complex matrices.

Lemma 5.3.4. *Let $A := (a_{ij}) \in M_n(\mathbb{R})$ and I be the $n \times n$ identity matrix. Then*

the series $\sum_{k=0}^{\infty} \frac{A^k}{k!}$ converges, where $A^0 = I$. Moreover we have

$$\left\| \sum_{k=0}^{\infty} \frac{A^k}{k!} \right\|_{\infty} \leq \exp(n\|A\|_{\infty}). \tag{5.15}$$

Proof. Let $a_{ij}^{(k)}$ denote the (i,j)-th entry of A^k, $k \geq 1$. In view of Lemma 5.3.3, in order to prove the converge of the series of matrices it is enough to prove that $\sum_{k=0}^{\infty} a_{ij}^{(k)}$ converges for each $1 \leq i,j \leq n$. For, we consider

$$|a_{ij}^{(2)}| = \left| \sum_{l_1=1}^{n} a_{il_1} a_{l_1 j} \right| \leq \sum_{l_1=1}^{n} |a_{il_1}||a_{l_1 j}| \leq \sum_{l_1=1}^{n} \|A\|_{\infty}^2 = n\|A\|_{\infty}^2,$$

$$|a_{ij}^{(3)}| = \left| \sum_{1 \leq l_1,l_2 \leq n} a_{il_1} a_{l_1 l_2} a_{l_2 j} \right|$$

$$\leq \sum_{1 \leq l_1,l_2 \leq n} |a_{il_1}||a_{l_1 l_2}||a_{l_2 j}|$$

$$\leq \sum_{1 \leq l_1,l_2 \leq n} \|A\|_{\infty}^3 = n^2 \|A\|_{\infty}^3.$$

By proceeding in the same way, we obtain

$$|a_{ij}^{(k)}| = \left| \sum_{1 \leq l_1,\ldots,l_{k-1} \leq n} a_{il_1} a_{l_1 l_2} \cdots a_{l_{k-1} j} \right|$$

$$\leq \sum_{1 \leq l_1,\ldots,l_{k-1} \leq n} |a_{il_1}||a_{l_1 l_2}| \cdots |a_{l_{k-1} j}|$$

$$\leq \sum_{1 \leq l_1,\ldots,l_{k-1} \leq n} \|A\|_{\infty}^k$$

$$\leq n^{k-1}\|A\|_{\infty}^k, \quad k \geq 1. \tag{5.16}$$

From (5.16), we obtain

$$\sum_{k=1}^{N} \frac{|a_{ij}^{(k)}|}{k!} \leq \sum_{k=1}^{N} \frac{n^{k-1}\|A\|_{\infty}^k}{k!} \leq \sum_{k=1}^{N} \frac{n^k \|A\|_{\infty}^k}{k!}, \quad N \in \mathbb{N}. \tag{5.17}$$

By the comparison test , we get that the series $\sum_{k=1}^{\infty} \frac{a_{ij}^{(k)}}{k!}$ converges absolutely.

Let (δ_{ij}) denote the identity matrix I. Using (5.17) we immediately have

$$|\delta_{ij}| + \sum_{k=1}^{N} \frac{|a_{ij}^{(k)}|}{k!} \leq 1 + \sum_{k=1}^{N} \frac{n^k \|A\|_{\infty}^k}{k!} \leq \exp(n\|A\|_{\infty}), \ N \in \mathbb{N}. \qquad (5.18)$$

Finally, (5.15) is an immediate consequence of (5.18). $\qquad\qquad\qquad\square$

Lemma 5.3.4 helps us to define the exponential of a matrix.

Definition 5.3.5. Let $A \in M_n(\mathbb{R})$. The exponential of the matrix A is defined as

$$e^A = I + \frac{A}{1!} + \frac{A^2}{2!} + \frac{A^3}{3!} + \cdots + \frac{A^k}{k!} + \cdots .$$

Remark 5.3.6. In view of the proof of Lemma 5.3.4, for $1 \leq i, j \leq n$, the (i,j)-th entry in e^A is an absolutely convergent series and

$$\|e^A\|_{\infty} \leq \exp(n\|A\|_{\infty}).$$

Moreover, from Lemma 5.3.3 it follows that

$$\sum_{k=0}^{N} \frac{A^k}{k!} \to e^A \text{ in } M_n(\mathbb{R}) \text{ as } N \to \infty.$$

Example 5.3.7. Find the exponential of the following matrices:

(i) $D = \begin{pmatrix} \lambda & 0 \\ 0 & \mu \end{pmatrix}$, (ii) $A = \begin{pmatrix} 0 & \beta \\ -\beta & 0 \end{pmatrix}$, (iii) $J = \begin{pmatrix} \lambda t & t \\ 0 & \lambda t \end{pmatrix}$.

Solution. (i) Since $D^k = \begin{pmatrix} \lambda^k & 0 \\ 0 & \mu^k \end{pmatrix}$, $k \in \mathbb{N}$, one can immediately obtain

$$e^D = \sum_{k=0}^{\infty} \frac{D^k}{k!} = \begin{pmatrix} \sum_{k=0}^{\infty} \frac{\lambda^k}{k!} & 0 \\ 0 & \sum_{k=0}^{\infty} \frac{\mu^k}{k!} \end{pmatrix} = \begin{pmatrix} e^{\lambda} & 0 \\ 0 & e^{\mu} \end{pmatrix}.$$

(ii) We first observe that $A^2 = -\beta^2 I$. Then by induction one can easily prove that

$$A^{2k} = (-1)^k \beta^{2k} I, \ k \in \mathbb{N}.$$

Therefore we get $A^{2k+1} = A^{2k} A = (-1)^k \begin{pmatrix} 0 & \beta^{2k+1} \\ -\beta^{2k+1} & 0 \end{pmatrix}$, $k \in \mathbb{N}$.

Hence we have

$$
e^A = \begin{pmatrix} \displaystyle\sum_{k=0}^{\infty} \frac{(-1)^n \beta^{2k}}{(2k)!} & \displaystyle\sum_{k=0}^{\infty} \frac{(-1)^n \beta^{2k+1}}{(2k+1)!} \\ \displaystyle\sum_{k=0}^{\infty} \frac{(-1)^{n+1}\beta^{2k+1}}{(2k+1)!} & \displaystyle\sum_{k=0}^{\infty} \frac{(-1)^n \beta^{2k}}{(2k)!} \end{pmatrix}
$$

$$
= \begin{pmatrix} \cos\beta & \sin\beta \\ -\sin\beta & \cos\beta \end{pmatrix}.
$$

(iii) We set $N_1 = \begin{pmatrix} 0 & t \\ 0 & 0 \end{pmatrix}$ so that $J = \lambda t I + N_1$. Since $N_1^2 = \mathbf{0}$, where $\mathbf{0}$ is the 2×2 zero matrix, we use the binomial theorem to obtain

$$
J^k = \lambda^k t^k I + k\lambda^{k-1} t^{k-1} N_1, \ k \in \mathbb{N}.
$$

Hence we get

$$
e^J = \begin{pmatrix} 1 + \dfrac{\lambda t}{1!} + \cdots + \dfrac{\lambda^k t^k}{k!} + \cdots & t + \dfrac{\lambda t^2}{1!} + \cdots + \dfrac{\lambda^k t^{k+1}}{k!} + \cdots \\ 0 & 1 + \dfrac{\lambda t}{1!} + \cdots + \dfrac{\lambda t^k}{k!} + \cdots \end{pmatrix}
$$

$$
= \begin{pmatrix} e^{\lambda t} & t e^{\lambda t} \\ 0 & e^{\lambda t} \end{pmatrix}.
$$

This completes the solution. $\qquad\square$

We now prove a result which is very useful in the computation of the exponential of matrices.

Theorem 5.3.8. *Let A, B, and P be $n \times n$ matrices. Then the following hold true.*
(i) If $B = P^{-1}AP$, then $e^B = P^{-1}e^A P$.
(ii) If $AB = BA$, then $e^{A+B} = e^A e^B$.
(iii) $\left(e^A\right)^{-1} = e^{-A}$.

Proof. (i) From Remark 5.3.6, and Exercise 5.2, we have

$$
P^{-1}\left(\sum_{k=0}^{N} \frac{A^k}{k!}\right) P \to P^{-1} e^A P \text{ in } M_n(\mathbb{R}) \text{ as } N \to \infty.
$$

By induction it is easy to prove that $B^k = P^{-1}A^k P$, $k \in \mathbb{N}$. Therefore we get

$$
\sum_{k=0}^{N} \frac{B^k}{k!} = \sum_{k=0}^{N} \frac{(P^{-1}A^k P)}{k!} = P^{-1}\left(\sum_{k=0}^{N} \frac{A^k}{k!}\right) P \to P^{-1} e^A P \text{ as } N \to \infty.
$$

On the other hand, $\displaystyle\sum_{k=0}^{N}\frac{B^k}{k!}$ converges to e^B in $M_n(\mathbb{R})$. By the uniqueness of the limit, result (i) follows.

(ii) Since $AB = BA$, we use the binomial theorem to compute $(A + B)^k$, $k \in \mathbb{N}$. Then we have

$$e^{A+B} = \sum_{k=0}^{\infty}\frac{(A+B)^k}{k!} = \sum_{k=0}^{\infty}\frac{1}{k!}\sum_{r=0}^{k}\binom{k}{r}A^r B^{k-r} = \sum_{k=0}^{\infty}\sum_{r=0}^{k}\frac{A^r}{r!}\frac{B^{k-r}}{(k-r)!}.$$

We now set

$$T_{2m} = \sum_{k=0}^{2m}\sum_{r=0}^{k}\frac{A^r}{r!}\frac{B^{k-r}}{(k-r)!}, \quad R_m = \sum_{k=0}^{m}\frac{A^k}{k!}, \quad S_m = \sum_{k=0}^{m}\frac{B^k}{k!}.$$

We would like to prove the following convergence result which is very useful:

$$\|T_{2m} - R_m S_m\|_\infty \to 0 \text{ as } m \to \infty. \tag{5.19}$$

For, we observe that T_{2m} has terms which are not in the product $R_m S_m$ because

$$R_m S_m = \sum_{k=0}^{2m}\left(\sum_{0\le r, k-r\le m}\frac{A^r}{r!}\frac{B^{k-r}}{(k-r)!}\right).$$

Therefore we obtain

$$T_{2m} - R_m S_m = \sum_{k=m+1}^{2m}\sum_{r=m+1}^{k}\frac{A^r}{r!}\frac{B^{k-r}}{(k-r)!} + \sum_{k=m+1}^{2m}\sum_{r=0}^{k-(m+1)}\frac{A^r}{r!}\frac{B^{k-r}}{(k-r)!}$$

$$=: C_m + C'_m.$$

For better understanding, the reader is advised to compute $T_4 - R_2 S_2$ and $T_6 - R_3 S_3$. In order to prove (5.19), we need to estimate C_m and $\tilde{C}_m$. To this end, from Exercise 5.1 we first note that

$$\|X_1 X_2\|_\infty \le n\|X_1\|_\infty\|X_2\|_\infty, \quad X_1, X_2 \in M_n(\mathbb{R}).$$

Using this inequality and the triangle inequality, we get

$$\|C_m\|_\infty = \left\|\sum_{k=m+1}^{2m}\sum_{r=m+1}^{k}\frac{A^r}{r!}\frac{B^{k-r}}{(k-r)!}\right\|_\infty$$

$$\le \sum_{k=m+1}^{2m}\sum_{r=m+1}^{k}n\frac{\|A^r\|_\infty}{r!}\frac{\|B^{k-r}\|_\infty}{(k-r)!}$$

$$\le n\left(\sum_{r=m+1}^{2m}\frac{\|A^r\|_\infty}{r!}\right)\left(\sum_{r=0}^{m}\frac{\|B^r\|_\infty}{r!}\right).$$

In view of the estimate, $\|X^k\|_\infty \leq n^k \|X\|_\infty^k$, $X \in M_n(\mathbb{R})$, $k \in \mathbb{N}$, (see Exercise 5.1), we readily get

$$\|C_m\|_\infty \;\leq\; n\Big(\sum_{r=m+1}^{2m} \frac{n^r \|A\|_\infty^r}{r!}\Big)\Big(\sum_{r=0}^{m} \frac{n^r \|B\|_\infty^r}{r!}\Big)$$

$$\leq\; n\exp(n\|B\|_\infty)\sum_{r=m+1}^{\infty}\frac{n^r\|A\|_\infty^r}{r!} \longrightarrow 0 \text{ as } m \to \infty. \quad (5.20)$$

Since C_m' can be obtained from C_m by interchanging A and B, we immediately obtain

$$\|C_m'\|_\infty \leq n\exp(n\|A\|_\infty)\sum_{k=m+1}^{\infty}\frac{n^r\|B\|_\infty^r}{r!} \longrightarrow 0 \text{ as } m \to \infty. \quad (5.21)$$

Hence (5.20)–(5.21) together prove (5.19).

On the other hand, since $T_{2m} \to e^{A+B}$, $R_m \to e^A$, $S_m \to e^B$ as $m \to \infty$, and from Exercise 5.2, we have

$$T_{2m} - R_m S_m \to e^{A+B} - e^A e^B \text{ as } m \to \infty \text{ in } M_n(\mathbb{R}). \quad (5.22)$$

Result (ii) immediately follows from (5.19) and (5.22).

(iii) Since A and $-A$ commute, owing to (ii) we have

$$e^A e^{-A} = e^{-A} e^A = e^{A-A} = e^{\mathbf{0}} = I,$$

where $\mathbf{0}$ is an $n \times n$ zero matrix. This completes the proof of (iii). $\qquad\square$

Remark 5.3.9. The assumption in (ii) of Theorem 5.3.8 cannot be relaxed. For, let $A = \begin{pmatrix} 1 & 0 \\ 0 & 0 \end{pmatrix}$ and $B = \begin{pmatrix} 0 & 1 \\ 0 & 0 \end{pmatrix}$. Then a straightforward computation gives us $AB \neq BA$, $e^A = \begin{pmatrix} e & 0 \\ 0 & 1 \end{pmatrix}$, $e^B = \begin{pmatrix} 1 & 1 \\ 0 & 1 \end{pmatrix}$ and $e^{A+B} = \begin{pmatrix} e & e-1 \\ 0 & 1 \end{pmatrix}$. In this case $e^A e^B \neq e^{A+B}$.

Remark 5.3.10. From Theorem 5.3.8(iii), it is evident that e^A is nonsingular for every $A \in M_n(\mathbb{R})$.

We now compute exponential of matrices that have some special structure, viz., diagonal matrices, Jordan blocks etc. We begin with the case of diagonal matrices.

Lemma 5.3.11. *Assume that $D(\lambda_1, \lambda_2, \ldots, \lambda_n)$ denotes the $n \times n$ diagonal matrix whose (j,j)-th entry is λ_j, $1 \leq j \leq n$. Then*

$$\exp\big(D(\lambda_1, \lambda_2, \ldots, \lambda_n)\big) = D(e^{\lambda_1}, e^{\lambda_2}, \ldots, e^{\lambda_n}). \quad (5.23)$$

Proof. An easy computation gives us that

$$\left(D(\lambda_1, \lambda_2, \ldots, \lambda_n)\right)^k = D(\lambda_1^k, \lambda_2^k, \ldots, \lambda_n^k), \ \ k \in \mathbb{N}.$$

The required result (5.23) readily follows from the definition of the exponential of a matrix. □

We now turn our attention toward a single Jordan block matrix. A Jordan block matrix is an upper (lower) triangular matrix with the following properties.

(i) All the principle diagonal entries being the same.

(ii) All the entries in the first diagonal above (below) the principle diagonal being equal to one.

(iii) Rest of the elements of the matrix must be zeros.

In particular, we consider the case when the Jordan block matrix B is an upper triangular matrix to compute e^{tB} where $t \in \mathbb{R}$. Recall that we have done this for 2×2 case in Example 5.3.7. A more general result is given in the next lemma.

Lemma 5.3.12. *Let $B = (b_{ij}) \in M_n(\mathbb{R})$ be such that*

$$b_{ij} = \begin{cases} \lambda, & j = i, \ 1 \leq i \leq n, \\ 1, & j = i+1, \ 1 \leq i \leq n-1, \\ 0, & \text{otherwise,} \end{cases} \quad \text{or } B = \begin{pmatrix} \lambda & 1 & 0 & \cdots & 0 & 0 \\ 0 & \lambda & 1 & \cdots & 0 & 0 \\ 0 & 0 & \lambda & \cdots & 0 & 0 \\ \vdots & \vdots & \vdots & \ddots & \vdots & \vdots \\ 0 & 0 & 0 & \cdots & \lambda & 1 \\ 0 & 0 & 0 & \cdots & 0 & \lambda \end{pmatrix}$$

Then the exponential of tB is given by

$$e^{tB} = e^{\lambda t} \begin{pmatrix} 1 & t & \dfrac{t^2}{2!} & \cdots & \dfrac{t^{n-1}}{(n-1)!} \\ 0 & 1 & t & \cdots & \dfrac{t^{n-2}}{(n-2)!} \\ 0 & 0 & 1 & \cdots & \dfrac{t^{n-3}}{(n-3)!} \\ \vdots & \vdots & \vdots & \ddots & \vdots \\ 0 & 0 & 0 & \cdots & 1 \end{pmatrix}. \tag{5.24}$$

Proof. Let N be the matrix such that $B = \lambda I + N$. Since I and N commute, by Theorem 5.3.8(ii), we have

$$e^{tB} = e^{t\lambda I} e^{tN}. \tag{5.25}$$

In view of Lemma 5.3.11, we get $e^{t\lambda I} = e^{t\lambda} I$. Therefore (5.25) reduces to

$$e^{tB} = e^{t\lambda} e^{tN}. \tag{5.26}$$

It remains to compute e^{tN}. To this end, we observe that N is a nilpotent matrix. Furthermore, using the special structure of N, we arrive at

$$N = \begin{pmatrix} 0 & 1 & 0 & \cdots & 0 \\ 0 & 0 & 1 & \cdots & 0 \\ \vdots & \vdots & \vdots & \ddots & \vdots \\ 0 & 0 & 0 & \cdots & 1 \\ 0 & 0 & 0 & \cdots & 0 \end{pmatrix}, \quad N^2 = \begin{pmatrix} 0 & 0 & 1 & 0 & \cdots & 0 \\ 0 & 0 & 0 & 1 & \cdots & 0 \\ \vdots & \vdots & \vdots & \vdots & \ddots & \vdots \\ 0 & 0 & 0 & 0 & \cdots & 1 \\ 0 & 0 & 0 & 0 & \cdots & 0 \\ 0 & 0 & 0 & 0 & \cdots & 0 \end{pmatrix},$$

$$N^3 = \begin{pmatrix} 0 & 0 & 0 & 1 & 0 & \cdots & 0 \\ 0 & 0 & 0 & 0 & 1 & \cdots & 0 \\ \vdots & \vdots & \vdots & \vdots & \vdots & \ddots & \vdots \\ 0 & 0 & 0 & 0 & 0 & \cdots & 1 \\ 0 & 0 & 0 & 0 & 0 & \cdots & 0 \\ 0 & 0 & 0 & 0 & 0 & \cdots & 0 \\ 0 & 0 & 0 & 0 & 0 & \cdots & 0 \end{pmatrix}, \cdots, N^{n-1} = \begin{pmatrix} 0 & 0 & 0 & \cdots & 0 & 1 \\ 0 & 0 & 0 & \cdots & 0 & 0 \\ \vdots & \vdots & \vdots & & \vdots & \vdots \\ 0 & 0 & 0 & \cdots & 0 & 0, \end{pmatrix},$$

and $N^n = \mathbf{0}$, where $\mathbf{0}$ is an $n \times n$ zero matrix. Thus we obtain

$$e^{tN} = \sum_{k=0}^{n-1} \frac{t^k N^k}{k!} = \begin{pmatrix} 1 & t & \dfrac{t^2}{2!} & \cdots & \dfrac{t^{n-1}}{(n-1)!} \\ 0 & 1 & t & \cdots & \dfrac{t^{n-2}}{(n-2)!} \\ 0 & 0 & 1 & \cdots & \dfrac{t^{n-3}}{(n-3)!} \\ \vdots & \vdots & \vdots & \ddots & \vdots \\ 0 & 0 & 0 & \cdots & 1 \end{pmatrix}. \tag{5.27}$$

The required result (5.24) is an immediate consequence of (5.26)–(5.27). $\qquad\Box$

We next consider the matrices whose entries are zeros except in some subblocks. An example of such a matrix is a matrix which is in the Jordan canonical form.

Let $J_1, J_2, \ldots, J_r$ be square matrices of possibly different sizes. We denote by

$$\mathbf{BLOCK}(J_1, J_2, \ldots, J_r) = \begin{pmatrix} J_1 & \mathbf{0} & \mathbf{0} & \cdots & \mathbf{0} \\ \mathbf{0} & J_2 & \mathbf{0} & \cdots & \mathbf{0} \\ \mathbf{0} & \mathbf{0} & J_3 & \cdots & \mathbf{0} \\ \vdots & \vdots & \vdots & \ddots & \vdots \\ \mathbf{0} & \mathbf{0} & \mathbf{0} & \cdots & J_r \end{pmatrix}$$

where $\mathbf{0}$ denotes the zero matrix of appropriate size depending on the location where it appears.

Lemma 5.3.13. *Let $A = \mathbf{BLOCK}(J_1, J_2, \ldots, J_r)$ be a matrix in the Jordan form. Then the exponential of A is given by*

$$e^A = \mathbf{BLOCK}(e^{J_1}, e^{J_2}, \ldots, e^{J_r}). \tag{5.28}$$

Proof. It can be readily seen that
$$\left(\mathbf{BLOCK}(J_1, J_2, \ldots, J_r)\right)^k = \mathbf{BLOCK}(J_1^k, J_2^k, \ldots, J_r^k), \ k \in \mathbb{N}.$$

Now, the required result (5.28) immediately follows from the definition of the exponential of a matrix. $\square$

5.3.1.1 Working rule to find e^A

We present an algorithm to find the exponential of a given matrix in three simple steps.

Step 1. Find a nonsingular matrix P, possibly with complex entries, such that $A = PMP^{-1}$, where M is a Jordan matrix. In other words, reduce the given matrix A to a Jordan canonical form.
Let $M = \mathbf{BLOCK}(J_1, J_2, \ldots, J_r)$.

Step 2. Use Lemmas 5.3.12 and 5.3.13 to compute e^M.

Step 3. Finally, compute P^{-1} and use Theorem 5.3.8(i) to get $e^A = Pe^M P^{-1}$.

In some special cases, one can avoid computation of P, M, and P^{-1} which can take a lot of time and effort. One such case is presented in the following example.

Example 5.3.14. Suppose $A \in M_n(\mathbb{R})$ and all its eigenvalues are equal to λ. Then show that

$$e^A = e^\lambda \sum_{k=0}^{n-1} \frac{(A - \lambda I)^k}{k!}. \tag{5.29}$$

Solution. Since $(A - \lambda I)$ and λI commute, by Theorem 5.3.8, we get

$$e^A = e^{(A-\lambda I) + \lambda I} = e^{(A-\lambda I)} e^{\lambda I}. \tag{5.30}$$

On the other hand, since all the eigenvalues of A are equal to λ, the characteristic polynomial of A is $(x - \lambda)^n$. By the Cayley–Hamilton theorem we have $(A - \lambda I)^n = \mathbf{0}$, where $\mathbf{0}$ is the $n \times n$ zero matrix. Hence we get

$$e^{(A-\lambda I)} = \sum_{k=0}^{\infty} \frac{(A - \lambda I)^k}{k!} = \sum_{k=0}^{n-1} \frac{(A - \lambda I)^k}{k!}. \tag{5.31}$$

In view of Lemma 5.3.11, we obtain

$$e^{\lambda I} = e^{\lambda} I. \tag{5.32}$$

The required result, i.e., (5.29) follows from (5.30)–(5.32). □

5.3.2 Solution to $Y' = AY$

In this subsection, we introduce the notion of derivative of the vector valued and the matrix valued functions. The objective of this section is to prove that the solution to $Y' = AY$ is $Y(t) = e^{At} Y(0)$ which is analogous to the situation where A is a real number. We then give some examples where this formula is used to find the explicit solutions to some 2×2, 3×3 systems of ODEs. To this end, we first assume that J is an open interval in $\mathbb{R}$ and $M_{m \times n}(\mathbb{R})$ denotes the set of $m \times n$ real matrices.

Definition 5.3.15. A function $B : J \to M_{m \times n}(\mathbb{R})$ is said to be differentiable at $t_0 \in J$ if the limit

$$\lim_{h \to 0} \frac{B(t_0 + h) - B(t_0)}{h}$$

exists in $M_{m \times n}(\mathbb{R})$.

In view of Lemma 5.3.3 it is straightforward to prove that the function $B(t) = (B_{ij}(t))$, $t \in J$, is differentiable if and only if $B_{ij} : J \to \mathbb{R}$ is differentiable for each $1 \le i \le m$, $1 \le j \le n$.

Lemma 5.3.16. *Let $A(t)$ and $B(t)$ be matrices of sizes $m \times n$ and $n \times r$, respectively, for each $t > 0$. If each entry of $A(t)$ and $B(t)$ is a differentiable function then*

$$\frac{d}{dt}\big(A(t)B(t)\big) = \big(\frac{d}{dt}A(t)\big)B(t) + A(t)\big(\frac{d}{dt}B(t)\big), \quad t > 0.$$

Proof. The proof is left to the reader as a simple exercise. □

We are now ready to prove the main theorem of this section.

Theorem 5.3.17. *Suppose $A \in M_n(\mathbb{R})$, $Y_0 \in \mathbb{R}^n$. Consider the homogeneous system of linear ODEs with constant coefficients*

$$\begin{cases} Y'(t) = AY(t), \ t > 0, \\ Y(0) = Y_0. \end{cases} \tag{5.33}$$

Then

$$Y(t) = e^{tA}Y_0, \ t \geq 0. \tag{5.34}$$

is a unique solution to (5.33).

Proof. We first prove that Y in (5.34) is a solution to (5.33) and later we establish its uniqueness. The main step in the proof is to prove the following identity

$$\frac{d}{dt}e^{tA} = e^{tA}A, \ t > 0. \tag{5.35}$$

For, using the results proved in the previous section, we obtain

$$
\begin{aligned}
\frac{d}{dt}e^{tA} &= \lim_{h \to 0} \frac{e^{(t+h)A} - e^{tA}}{h} & (5.36)\\
&= \lim_{h \to 0} \frac{e^{tA}e^{hA} - e^{tA}}{h} \\
&= \lim_{h \to 0} \left(\frac{e^{hA} - I}{h} \right)e^{tA} \\
&= \lim_{h \to 0} \left(\sum_{k=0}^{\infty} \frac{h^k A^{k+1}}{(k+1)!} \right)e^{tA} \\
&= \left(\lim_{h \to 0} \sum_{k=0}^{\infty} \frac{h^k A^{k+1}}{(k+1)!} \right)e^{tA}. & (5.37)
\end{aligned}
$$

Our objective is to interchange the limit and the summation in (5.37). To this end, we show that each entry in the matrix series in the right hand side of (5.37) is a uniformly convergent series[2] of functions of h in $[-1, 1]$. As in the proof of Lemma 5.3.4, let $a_{ij}^{(k)}$ denote the (i, j)-th entry of A^k, $k \geq 1$. For $1 \leq i, j \leq n$, define a sequence of functions $f_{k,ij} : [-1, 1] \to \mathbb{R}$ as

$$f_{k,ij}(h) = \frac{h^k a_{ij}^{(k+1)}}{(k+1)!}, \ k \in \mathbb{N}.$$

From (5.16), we have

$$\max_{h \in [-1,1]} |f_{k,ij}(h)| \leq \frac{n^k \|A\|_{\infty}^{k+1}}{(k+1)!} \leq \frac{n^{k+1} \|A\|_{\infty}^{k+1}}{(k+1)!} =: M_k.$$

Since the series $\sum_{k=0}^{\infty} M_k$ converges, owing to the Weierstrass M-test[3] the series

---

[2] A series of functions $\sum_{k=0}^{\infty} f_k$ is said to converge uniformly (to f) if the sequence of partial sums ($\sum_{k=0}^{N} f_k$) converges uniformly (to f).

[3] Let $f_k : [a, b] \to \mathbb{R}$ be continuous for each $k \in \mathbb{N}$ and $M_k = \sup_{x \in [a,b]} |f_k(x)|$. If $\sum_{k=0}^{\infty} M_k$ converges then $\sum_{k=0}^{\infty} f_k$ converges uniformly.

$\sum_{k=0}^{\infty} f_{k.ij}$ converges uniformly on $[-1,1]$, $1 \leq i,j \leq n$. Thus we have proved that each entry in the matrix $\sum_{k=0}^{\infty} \frac{h^k A^{k+1}}{(k+1)!}$ is uniformly convergent in $[-1,1]$. Therefore we can interchange the limit and the summation in (5.37) to get

$$\frac{d}{dt} e^{tA} = \left(\sum_{k=0}^{\infty} \lim_{h \to 0} \frac{h^k A^{k+1}}{(k+1)!} \right) e^{tA} = A e^{tA}, \ t > 0.$$

On differentiating Y which is given in (5.34), using (5.35), we arrive at

$$\frac{d}{dt} Y(t) = \frac{d}{dt} (e^{tA} Y_0) = \left(\frac{d}{dt} e^{tA} \right) Y_0 = A e^{tA} Y_0 = AY(t).$$

Moreover, we put $t = 0$ in (5.34) to obtain

$$Y(0) = e^{0A} Y_0 = e^0 Y_0 = Y_0.$$

Hence Y given in (5.34) is indeed a solution to (5.33).
We now turn our attention toward the uniqueness of the solution. It is straightforward to verify that equation (5.33) satisfies the hypotheses of the Cauchy–Lipschitz theorem. Hence it has a unique solution. However, we give another proof of uniqueness by exploiting the linear structure of the problem. For, let Y_1, Y_2 be two solutions to (5.33) and $Z(t) = Y_1(t) - Y_2(t)$, $t \geq 0$. Then Z satisfies

$$\begin{cases} Z'(t) - AZ(t) = \mathbf{0}, \ t > 0, \\ Z(0) = \mathbf{0} \in \mathbb{R}^n. \end{cases} \tag{5.38}$$

On multiplying the first equation in (5.38) with e^{-tA} and using (5.35), Exercise 5.3, we obtain

$$e^{-tA} (Z'(t) - AZ(t)) = \frac{d}{dt} (e^{-tA} Z(t)) = \mathbf{0}.$$

Hence $e^{-tA} Z(t)$ is a constant independent of t, i.e., $e^{-tA} Z(t) = e^0 Z(0) = \mathbf{0}$. In other words, for every fixed t, the vector $Z(t)$ is a solution to the system of linear equations $e^{-tA} Z(t) = \mathbf{0}$. From Remark 5.3.10, the matrix e^{-tA} is nonsingular. Hence $Z(t) = \mathbf{0}$, or $Y_1(t) = Y_2(t)$, $t > 0$. $\qquad \square$

We next present some examples in which we solve some systems of first order linear ODEs using Theorem 5.3.17.

Example 5.3.18. Solve
$$\begin{cases} \dfrac{dx}{dt} = x + y, \\ \dfrac{dy}{dt} = 4x - 2y. \end{cases}$$

Solution. We set $Y(t) = (x(t), y(t))$ and $A = \begin{pmatrix} 1 & 1 \\ 4 & -2 \end{pmatrix}$, so that the given system is of the form $Y' = AY$. Owing to Theorem 5.3.17 the solution to the given system is $Y(t) = e^{tA} Y_0$, where $Y_0 = (x(0), y(0))$. Therefore it is enough to compute e^{tA}. For, we recall the working rule to compute the exponential of a matrix in Section 5.3.1.1.

We first compute the eigenvalues and eigenvectors of A. The characteristic polynomial of A is

$$\det(A - \lambda I) = \det \begin{pmatrix} 1 - \lambda & 1 \\ 4 & -2 - \lambda \end{pmatrix} = \lambda^2 + \lambda - 6.$$

The eigenvalues of A are precisely the roots of $\lambda^2 + \lambda - 6 = 0$, which are given by $\lambda = -3$ and 2.

Let (α, β) be an eigenvector corresponding to $\lambda = -3$, i.e.,

$$(A + 3I) \begin{pmatrix} \alpha \\ \beta \end{pmatrix} = \begin{pmatrix} 4 & 1 \\ 4 & 1 \end{pmatrix} \begin{pmatrix} \alpha \\ \beta \end{pmatrix} = \begin{pmatrix} 0 \\ 0 \end{pmatrix}.$$

Therefore $(1, -4)$ is an eigenvector corresponding to $\lambda = -3$. Similarly, for $\lambda = 2$, an eigenvector is $(1, 1)$. Assume that P is the matrix whose columns are eigenvectors of A and D is the corresponding diagonal matrix, i.e.,

$$P = \begin{pmatrix} 1 & 1 \\ -4 & 1 \end{pmatrix} \text{ and } D = \begin{pmatrix} -3 & 0 \\ 0 & 2 \end{pmatrix}$$

From matrix theory (see [34]), it is follows that

$$A = PDP^{-1}.$$

Thus we have

$$e^{tA} = Pe^{tD}P^{-1} = \frac{1}{5} \begin{pmatrix} 1 & 1 \\ -4 & 1 \end{pmatrix} \begin{pmatrix} e^{-3t} & 0 \\ 0 & e^{2t} \end{pmatrix} \begin{pmatrix} 1 & -1 \\ 4 & 1 \end{pmatrix}$$

Therefore the required solution is

$$\begin{pmatrix} x(t) \\ y(t) \end{pmatrix} = \begin{pmatrix} e^{-3t} + 4e^{2t} & -e^{-3t} + e^{2t} \\ -4e^{-3t} + 4e^{2t} & 4e^{-3t} + e^{2t} \end{pmatrix} \begin{pmatrix} x(0) \\ y(0) \end{pmatrix}.$$

This completes the solution.   □

Example 5.3.19. Solve $\begin{cases} \dfrac{dx}{dt} = 3x - 4y, \\[2mm] \dfrac{dy}{dt} = x - y. \end{cases}$

Solution. We set $Y(t) = (x(t), y(t))$ and $A = \begin{pmatrix} 3 & -4 \\ 1 & 1 \end{pmatrix}$. The characteristic equation of A is $(\lambda - 1)^2 = 0$. Hence the eigenvalues of A are $\lambda = 1, 1$. Let (α, β) be an eigenvector corresponding to $\lambda = 1$. Then we have

$$\begin{pmatrix} 2 & -4 \\ 1 & -2 \end{pmatrix} \begin{pmatrix} \alpha \\ \beta \end{pmatrix} = \begin{pmatrix} 0 \\ 0 \end{pmatrix}.$$

Hence $p_1 := (2, 1)$ is an eigenvector of A and there are no other eigenvectors which are independent of $(2, 1)$. In this case, the matrix A is not diagonalizable and its Jordan canonical form is

$$J := \begin{pmatrix} 1 & 1 \\ 0 & 1 \end{pmatrix}.$$

Our objective is to find a nonsingular matrix P such that $AP = PJ$. By comparing the columns on both sides, it is very easy to observe that the first column of P is an eigenvector of A corresponding to $\lambda = 1$. In particular, we take it as p_1.

Let the second column of P be $p_2 := (\alpha', \beta')$. Then p_2 satisfies

$$Ap_2 = p_2 + p_1.$$

In other words, (α', β') is a solution to

$$\begin{pmatrix} 2 & -4 \\ 1 & -2 \end{pmatrix} \begin{pmatrix} \alpha' \\ \beta' \end{pmatrix} = \begin{pmatrix} 2 \\ 1 \end{pmatrix}.$$

It is not difficult to obtain a particular solution $(\alpha', \beta') = (1, 0)$. (In fact, the general solution is given by $p_2 = (1,0) + \gamma(2,1)$, $\gamma \in \mathbb{R}$.)

Thus we get

$$A = PJP^{-1}, \text{ where } P = \begin{pmatrix} 2 & 1 \\ 1 & 0 \end{pmatrix}.$$

From Lemma 5.3.12, we readily have $e^{tJ} = e^t \begin{pmatrix} 1 & t \\ 0 & 1 \end{pmatrix}$.

Hence we obtain

$$e^{tA} = \begin{pmatrix} 2 & 1 \\ 1 & 0 \end{pmatrix} e^t \begin{pmatrix} 1 & t \\ 0 & 1 \end{pmatrix} \begin{pmatrix} 0 & 1 \\ 1 & -2 \end{pmatrix} = e^t \begin{pmatrix} 2t+1 & -4t \\ t & 1-2t \end{pmatrix}.$$

Therefore the required solution is

$$\begin{pmatrix} x(t) \\ y(t) \end{pmatrix} = e^t \begin{pmatrix} 2t+1 & -4t \\ t & 1-2t \end{pmatrix} \begin{pmatrix} x(0) \\ y(0) \end{pmatrix}. \tag{5.39}$$

Another method: Since both eigenvalues of A are equal, we can use the formula given in Example 5.3.14. Therefore we have

$$e^{tA} = \left(I + (A - I)t \right) e^t = \begin{pmatrix} 2t+1 & -4t \\ t & 1-2t \end{pmatrix} e^t.$$

Once we have e^{tA}, we can write the solution to the given system as in (5.39). $\square$

Example 5.3.20. Solve $\begin{cases} \dfrac{dx}{dt} = y, \\ \dfrac{dy}{dt} = -4x. \end{cases}$

Solution. Let $A = \begin{pmatrix} 0 & 1 \\ -4 & 0 \end{pmatrix}$. Then a straightforward computation gives us the eigenvalues of A are $-2i$ and $2i$, and the corresponding eigenvectors are $(1, -2i)$ and $(1, 2i)$, respectively. By setting

$$P = \begin{pmatrix} 1 & 1 \\ -2i & 2i \end{pmatrix}, \quad D = \begin{pmatrix} -2i & 0 \\ 0 & 2i \end{pmatrix},$$

and using the formula $e^{tA} = Pe^{tD}P^{-1}$, we obtain

$$e^{tA} = \begin{pmatrix} 1 & 1 \\ -2i & 2i \end{pmatrix} \begin{pmatrix} e^{-2it} & 0 \\ 0 & e^{2it} \end{pmatrix} \begin{pmatrix} 2i & -1 \\ 2i & 1 \end{pmatrix} \frac{1}{4i} = \begin{pmatrix} \cos(2t) & \frac{1}{2}\sin(2t) \\ -2\sin(2t) & \cos(2t) \end{pmatrix}.$$

Therefore the required solution is

$$\begin{pmatrix} x(t) \\ y(t) \end{pmatrix} = \begin{pmatrix} \cos(2t) & \frac{1}{2}\sin(2t) \\ -2\sin(2t) & \cos(2t) \end{pmatrix} \begin{pmatrix} x(0) \\ y(0) \end{pmatrix}.$$

This completes the solution. □

Example 5.3.21. Solve
$$\begin{cases} \dfrac{dx}{dt} = 2x, \\[2mm] \dfrac{dy}{dt} = 3x + 2y, \\[2mm] \dfrac{dz}{dt} = 5x - 2y - z, \end{cases}$$

using the exponential of the matrix corresponding to this system.

Solution. Let
$$A = \begin{pmatrix} 2 & 0 & 0 \\ 3 & 2 & 0 \\ 5 & -2 & -1 \end{pmatrix}.$$

Since A is a lower triangular matrix, the eigenvalues of A are its diagonal entries, i.e., $\lambda = 2, 2$, and -1.

One can easily compute that an eigenvector corresponding to $\lambda = -1$ is $p_3 := (0, 0, 1)$. Similarly, for $\lambda = 2$, an eigenvector is $p_1 := (0, 3, -2)$. Moreover, there is no other eigenvector for $\lambda = 2$ which is independent of p_1. Hence a Jordon canonical form for A is given by

$$J = \begin{pmatrix} 2 & 1 & 0 \\ 0 & 2 & 0 \\ 0 & 0 & -1 \end{pmatrix}.$$

We shall find a nonsingular matrix P such that $AP = PJ$. By comparing the first and the third columns on both sides of $AP = PJ$, we obtain that the first and third columns of P are precisely the eigenvectors of A corresponding to the eigenvalues $\lambda = -1$ and $\lambda = 2$, respectively. In particular, we take p_1 and p_3 as the first and the third columns of P. Let $p_2 := (a, b, c)$ be the second column of P. Then by comparing the second columns on both sides of $AP = PJ$ we get

$$Ap_2 = p_1 + 2p_2.$$

In other words, (a, b, c) is a solution to

$$\begin{pmatrix} 0 & 0 & 0 \\ 3 & 0 & 0 \\ 5 & -2 & -3 \end{pmatrix} \begin{pmatrix} a \\ b \\ c \end{pmatrix} = \begin{pmatrix} 0 \\ 3 \\ -2 \end{pmatrix}. \tag{5.40}$$

In particular, $(a, b, c) = (1, 2, 1)$ is a solution to (5.40). Hence the required nonsingular matrix is

$$P = \begin{pmatrix} 0 & 1 & 0 \\ 3 & 2 & 0 \\ -2 & 1 & 1 \end{pmatrix}.$$

Thus $e^{tA} = P e^{tJ} P^{-1}$, where e^{tJ} is computed using Lemma 5.3.13 as

$$e^{tJ} = \begin{pmatrix} e^{2t} & te^{2t} & 0 \\ 0 & e^{2t} & 0 \\ 0 & 0 & e^{-t} \end{pmatrix}.$$

The required solution is

$$\begin{pmatrix} x(t) \\ y(t) \\ z(t) \end{pmatrix} = \begin{pmatrix} 0 & 1 & 0 \\ 3 & 2 & 0 \\ -2 & 1 & 1 \end{pmatrix} \begin{pmatrix} e^{2t} & te^{2t} & 0 \\ 0 & e^{2t} & 0 \\ 0 & 0 & e^{-t} \end{pmatrix} \begin{pmatrix} -\frac{2}{3} & \frac{1}{3} & 0 \\ 1 & 0 & 0 \\ -\frac{7}{3} & \frac{2}{3} & 1 \end{pmatrix} \begin{pmatrix} x(0) \\ y(0) \\ z(0) \end{pmatrix}$$

or

$$\begin{pmatrix} x(t) \\ y(t) \\ z(t) \end{pmatrix} = \begin{pmatrix} e^{2t} & 0 & 0 \\ 3te^{2t} & e^{2t} & 0 \\ \frac{7}{3}e^{2t} - 2te^{2t} - \frac{7}{3}e^{-t} & -\frac{2}{3}e^{2t} + \frac{2}{3}e^{-t} & e^{-t} \end{pmatrix} \begin{pmatrix} x(0) \\ y(0) \\ z(0) \end{pmatrix}.$$

This completes the solution. $\square$

5.4 Systems of linear ODEs with variable coefficients

In this section, we study the case when the coefficient matrix in the system of equations is a function of the independent variable. Let $A : [0, \infty) \to M_n(\mathbb{R})$ be continuous with $A(t) := (a_{ij}(t))$, $t \geq 0$. We are interested in studying the following homogeneous linear system of ODEs

$$Y'(t) = A(t)Y(t), \ t > 0, \tag{5.41}$$

$$Y(0) = Y_0 \in \mathbb{R}^n. \tag{5.42}$$

It can be readily seen that (5.41)–(5.42) satisfies the hypotheses of Theorem 5.2.3 and hence it posses the global solution.

We shall exploit the linear structure that is present in equation (5.41) to prove many interesting properties of its solutions.

Lemma 5.4.1. *If* $\Phi_1, \Phi_2 \in \left(C^1((0,\infty)) \right)^n$ *are two solutions to* (5.41) *then*

$$\Phi = C_1 \Phi_1 + C_2 \Phi_2, \ \ C_1, C_2 \in \mathbb{R}$$

is also a solution to (5.41). *In other words, the solution space (the set of solutions) to* (5.41) *is a vector subspace of* $\left(C^1((0,\infty)) \right)^n$.

Proof. The proof is left to the reader as a trivial exercise. $\qquad\qquad \square$

Henceforth, the set of solutions of (5.41) is said to be the *solution space* of (5.41).

Our next result deals with the linear dependence of initial data and linear dependence of the corresponding solutions. Before we state the result, we consider three vector valued C^1-functions that are defined on an interval J, namely $U(t) = \big(u_1(t), u_2(t), u_3(t) \big)$, $V(t) = \big(v_1(t), v_2(t), v_3(t) \big)$, and $W(t) = \big(w_1(t), w_2(t), w_3(t) \big)$.

We say that U, V, and W are linearly dependent in $\left(C^1(J) \right)^3$, if there exists a constant vector $(c_1, c_2, c_3) \neq (0, 0, 0)$, such that

$$c_1 U(t) + c_2 V(t) + c_3 W(t) = \mathbf{0}, \ t \in J,$$

where $\mathbf{0}$ is the constant function which takes zero. On the other hand, when we say $U(t), V(t)$, and $W(t)$ are linearly dependent at $t \in J$, we mean that there exists a nonzero vector $(\tilde{c}_1, \tilde{c}_2, \tilde{c}_3)$, which may depend on t, such that

$$\tilde{c}_1 U(t) + \tilde{c}_2 V(t) + \tilde{c}_3 W(t) = \mathbf{0},$$

where $\mathbf{0} = (0, 0, 0) \in \mathbb{R}^3$ (with the abuse of notation). In this case, when we fix t, we have considered $U(t), V(t)$, and $W(t)$ as vectors in $\mathbb{R}^3$. Hence if U, V, and W are linearly dependent in $\left(C^1(J) \right)^3$, then $U(t), V(t)$, and $W(t)$ are linearly dependent in $\mathbb{R}^3$ for every $t \in J$.

Equivalently, if there exists $t \in J$ such that $U(t), V(t)$, and $W(t)$ are linearly independent in $\mathbb{R}^3$ then U, V, and W are linearly independent in $\left(C^1(J) \right)^3$. The following example demonstrates that the converse need not be true.

Example 5.4.2. Let $U(t) = (\sin t, \cos t, 0)$, $V(t) = (\sin^2 t, \cos^2 t, 0)$, $W(t) = (\sin^3 t, \cos^3 t, 0)$, $t \in \mathbb{R}$. Then clealy the vectors $U(t), V(t)$, and $W(t)$ are linearly dependent in $\mathbb{R}^3$ for each fixed $t \in \mathbb{R}$.

On the other hand, let (c_1, c_2, c_3) be such that

$$c_1 U(t) + c_2 V(t) + c_3 W(t) = \mathbf{0}, \ t \in \mathbb{R},$$

i.e.,

$$c_1(\sin t, \cos t, 0) + c_2(\sin^2 t, \cos^2 t, 0) + c_3(\sin^3 t, \cos^3 t, 0) = (0, 0, 0). \quad (5.43)$$

We now show that $c_1 = c_2 = c_3 = 0$ which implies that U, V, and W are linearly independent in $\left(C^1(\mathbb{R}) \right)^3$. For, we put $t = 0$ and $t = \frac{\pi}{3}$ in (5.43) to get the system of equations

$$
\begin{pmatrix} 1 & 1 & 1 \\ \frac{\sqrt{3}}{2} & \frac{3}{4} & \frac{3\sqrt{3}}{8} \\ \frac{1}{2} & \frac{1}{4} & \frac{1}{8} \end{pmatrix} \begin{pmatrix} c_1 \\ c_2 \\ c_3 \end{pmatrix} = \begin{pmatrix} 0 \\ 0 \\ 0 \end{pmatrix}
$$

Since the square matrix in the previous equation is nonsingular, we immediately obtain that $c_1 = c_2 = c_3 = 0$. □

The concept presented in Example 5.4.2 can be extended to functions in $\left(C^1(J)\right)^n$, where $n \in \mathbb{N}$, J is any open interval in $\mathbb{R}$.
If we further assume that the vector valued C^1-functions are solutions to a system of linear ODEs, then the linear dependence of those functions at any fixed t is the same as the linear dependence of the functions in $\left(C^1(J)\right)^n$. The details are given in the following result.

Theorem 5.4.3. *Assume that $A : [0, \infty) \to M_n(\mathbb{R})$ is a continuous function. Let $\Phi_i \in \left(C^1((0, \infty))\right)^n$ be the solution to*

$$Y'(t) = A(t)Y(t), \ t > 0, \tag{5.44}$$
$$Y(0) = Y_{0,i}, \tag{5.45}$$

for $1 \le i \le n$. Then the set $\{\Phi_i : 1 \le i \le n\}$ is linearly independent in $\left(C^1((0, \infty))\right)^n$ if and only if the set $\{Y_{0,i} : 1 \le i \le n\}$ is linearly independent in $\mathbb{R}^n$.

Proof. One side of the theorem follows immediately from the discussion we had before the statement of the Lemma. However, we write down the details here as follows.
We first assume that $\{Y_{0,i} : 1 \le i \le n\}$ is a linearly independent set in $\mathbb{R}^n$. Let $(c_1, \ldots, c_n) \in \mathbb{R}^n$ be such that

$$c_1\Phi_1(t) + c_2\Phi_2(t) + \cdots + c_n\Phi_n(t) = \mathbf{0}, \ t \ge 0. \tag{5.46}$$

where $\mathbf{0}$ denotes the constant zero function. On substituting $t = 0$ in (5.46) and using the fact that $\{Y_{0,i} : 1 \le i \le n\}$ is linearly independent, we obtain

$$C_1 = 0, C_2 = 0, \ldots, C_n = 0.$$

This shows that $\{\Phi_i : 1 \le i \le n\}$ is linearly independent in $\left(C^1((0, \infty))\right)^n$. Conversely, we assume that $\{\Phi_i : 1 \le i \le n\}$ is linearly independent. Let $(c_1, \ldots, c_n) \in \mathbb{R}^n$ be such that

$$c_1 Y_{0,1} + c_2 Y_{0,2} + \cdots + c_n Y_{0,n} = \mathbf{0}. \tag{5.47}$$

We set $\Phi := c_1\Phi_1(t) + c_2\Phi_2(t) + \cdots + c_n\Phi_n(t)$. Then from Lemma 5.4.1 and equation (5.47), we observe that Φ satisfies

$$\Phi'(t) = A(t)\Phi(t), \ t > 0, \tag{5.48}$$
$$\Phi(0) = \mathbf{0} \in \mathbb{R}^n. \tag{5.49}$$

Since the constant zero function is also a solution to (5.48)–(5.49), from the uniqueness result we have $\Phi \equiv 0$. Thus we get

$$c_1\Phi_1(t) + \cdots + c_n\Phi_n(t) = \mathbf{0}, \ t \geq 0. \tag{5.50}$$

As $\{\Phi_i : 1 \leq i \leq n\}$ is linearly independent, (5.50) gives us

$$c_1 = 0, c_2 = 0, \ldots, c_n = 0.$$

This shows that $\{Y_{0,i} : 1 \leq i \leq n\}$ is a linearly independent set in $\mathbb{R}^n$. $\qquad\square$

As an immediate consequence of Theorem 5.4.3, we can find the dimension of the solution space of (5.41). This is given in the following result.

Theorem 5.4.4. *Assume that $A : [0, \infty) \to M_n(\mathbb{R})$ is a continuous function. Then the dimension of the solution space of (5.41) is n.*

Proof. Let $\mathcal{B} = \{\Phi_\alpha : \alpha \in \Lambda\}$, Λ an indexed set, be a basis of the solution space of (5.41). We define

$$S := \{\Phi_\alpha(0) : \Phi_\alpha \in \mathcal{B}\} \subset \mathbb{R}^n.$$

Let the cardinalities of $\mathcal{B}$ and S be $|\mathcal{B}|$ and $|S|$, respectively. From the uniqueness part of the solution to (5.41), (see the Cauchy–Lipschitz theorem), it follows that

$$|\mathcal{B}| = |S|.$$

Since $\mathcal{B}$ is linearly independent in $\left(C^1\big((0, \infty)\big)\right)^n$, in view of Theorem 5.4.3, S is linearly independent in $\mathbb{R}^n$. Thus S is a finite set and $|S| \leq n$. Hence we have proved that $|\mathcal{B}| \leq n$.

On the other hand, we suppose $|S| < n$. Hence there exists a vector $\mathbf{u} \in \mathbb{R}^n \backslash S$ such that $S \cup \{\mathbf{u}\}$ is linearly independent in $\mathbb{R}^n$. Let Φ be the solution to (5.41), with $\Phi(0) = \mathbf{u}$. Then we notice that $\Phi \notin \mathcal{B}$. Again, due to Theorem 5.4.3 we have $\mathcal{B} \cup \{\Phi\}$ is linearly independent in $\left(C^1((0, \infty))\right)^n$. This is a contradiction, because $\mathcal{B}$ is a basis of the solution space. Therefore we get

$$|\mathcal{B}| = |S| = n.$$

This completes the proof of the theorem. $\qquad\square$

Remark 5.4.5. From the proof of Theorem 5.4.4, it is evident that, if $\mathcal{B} = \{\Phi_\alpha : \alpha \in \Lambda\}$, Λ an indexed set, is a basis of the solution space of (5.41), then $S := \{\Phi_\alpha(0) : \Phi_\alpha \in \mathcal{B}\}$ is a basis of $\mathbb{R}^n$. Conversely, if $\{v_i : 1 \leq i \leq n\}$ is a basis of $\mathbb{R}^n$ then the set

$$\{\Phi_i : \Phi_i \text{ is the solution to (5.41) with } \Phi_i(0) = \mathbf{v_i}, \ 1 \leq i \leq n\}$$

is a basis of the solution space of (5.41).

5.4.1 Solution matrix and fundamental matrix

In this subsection, we define and study two special types of matrices corresponding to (5.41), namely 'solution matrix' and 'fundamental matrix.' These matrices play a role similar to that of e^{tA} in Section 5.3.

Definition 5.4.6. Let Φ_j, $1 \le j \le n$ be solutions to (5.41). The matrix Ψ whose j-th column is Φ_j is called a solution matrix of (5.41).

If Ψ is a solution matrix of (5.41), then

$$\Psi'(t) = A(t)\Psi(t), \ t > 0. \tag{5.51}$$

For, if we fix $1 \le j \le n$, then the j-th columns on the left and right hand side of (5.51) are Φ'_j and $A\Phi_j$, respectively.

We now present a theorem due to Liouville which is very useful in computation of the determinant of the solution matrix of (5.41).

Theorem 5.4.7 (Liouville). *Let $\Psi(t)$ be a solution matrix of (5.41), whose j-th column is Φ_j, $1 \le j \le n$. Then it follows that*

$$\det(\Psi(t)) = \det(\Psi(0))\exp(\int_0^t \sum_{i=1}^n a_{ii}(\tau)d\tau). \tag{5.52}$$

Proof. Since Φ_j is a solution to (5.41), we have

$$\Phi'_{ij} = \sum_{k=1}^n a_{ik}\Phi_{kj}, \ 1 \le i, j \le n. \tag{5.53}$$

From Theorem C.1.3 and (5.53), we readily obtain

$$\frac{d}{dt}\det(\Psi(t)) = \det\begin{pmatrix} \Phi'_{11} & \cdots & \Phi'_{1n} \\ \Phi_{21} & \cdots & \Phi_{2n} \\ \vdots & \cdots & \vdots \\ \Phi_{n1} & \cdots & \Phi_{nn} \end{pmatrix} + \cdots + \det\begin{pmatrix} \Phi_{11} & \cdots & \Phi_{1n} \\ \Phi_{21} & \cdots & \Phi_{2n} \\ \vdots & \cdots & \vdots \\ \Phi'_{n1} & \cdots & \Phi'_{nn} \end{pmatrix}$$

$$= \det\begin{pmatrix} \sum_{k=1}^n a_{1k}\Phi_{k1} & \cdots & \sum_{k=1}^n a_{1k}\Phi_{kn} \\ \Phi_{21} & \cdots & \Phi_{2n} \\ \vdots & \cdots & \vdots \\ \Phi_{n1} & \cdots & \Phi_{nn} \end{pmatrix} + \cdots$$

$$+ \det\begin{pmatrix} \Phi_{11} & \cdots & \Phi_{1n} \\ \Phi_{21} & \cdots & \Phi_{2n} \\ \vdots & \cdots & \vdots \\ \sum_{k=1}^n a_{nk}\Phi_{k1} & \cdots & \sum_{k=1}^n a_{nk}\Phi_{kn} \end{pmatrix}. \tag{5.54}$$

$$=: \alpha_1(t) + \cdots + \alpha_n(t), \tag{5.55}$$

where α_i denotes the i-th determinant in (5.54), $1 \leq i \leq n$. For $1 \leq i, m \leq n$, let $R_{i,m}$ denote the i-th row of the m-th matrix in (5.54). To compute α_m, we apply the row transformation $R_{m,m} \longrightarrow R_{m,m} - \sum\limits_{k \neq m} a_{mk} R_{k,m}$ to find

$$\alpha_m = \det \begin{pmatrix} \Phi_{11} & \cdots & \Phi_{1n} \\ \vdots & \cdots & \vdots \\ \Phi_{(m-1)1} & \cdots & \Phi_{(m-1)n} \\ a_{mm}\Phi_{m1} & \cdots & a_{mm}\Phi_{mn} \\ \Phi_{(m+1)1} & \cdots & \Phi_{(m+1)n} \\ \vdots & \cdots & \vdots \\ \Phi_{n1} & \cdots & \Phi_{nn} \end{pmatrix} = a_{mm} \det \Psi(t). \tag{5.56}$$

In view of (5.55)–(5.56), we have

$$\frac{d}{dt} \det\left(\Psi(t)\right) = \left(\sum_{k=0}^{n} a_{kk}(t)\right) \det\left(\Psi(t)\right), \ t > 0. \tag{5.57}$$

The solution to (5.57) is given by

$$\det\left(\Psi(t)\right) = \det(\Psi(0)) \exp\left(\int_0^t \sum_{i=1}^{n} a_{ii}(\tau) d\tau\right), \ t \geq 0.$$

Hence the theorem is proved. ☐

From formula (5.52) we immediately arrive at the following result.

Corollary 5.4.8. *The function* $\det\left(\Psi(t)\right)$ *is either identically zero or never vanishes in* $[0, \infty)$. ☐

In the definition of the solution matrix Ψ, we have considered that each column is a solution to (5.41). So it is possible that all rows of Ψ are equal. (In that case, both the sides of (5.52) are equal to zero.) We would like to avoid this type of a situation and hence consider the following special type of a solution matrix.

Definition 5.4.9. A solution matrix is said to be a fundamental matrix if its columns are linearly independent in $\left(C^1([0, \infty))\right)^n$.

Example 5.4.10. If A is a constant $n \times n$ matrix, then show that e^{tA} is a fundamental matrix of (5.41).
Solution. Let $\{e_i : 1 \leq i \leq n\}$ be the standard basis in $\mathbb{R}^n$. For $1 \leq i \leq n$, we assume that Φ_i is the solution to

$$\begin{cases} Y' = AY, \ t > 0, \\ Y(0) = e_i. \end{cases}$$

From Theorem 5.3.17, we readily have $\Phi_i(t) = e^{tA}\mathbf{e}_i$. Therefore Φ_i is indeed the i-th column of e^{tA}. In view of Theorem 5.4.3, the set $\{\Phi_i : 1 \leq i \leq n\}$ is linearly independent in $\left(C^1([0,\infty))\right)^n$. Thus each column of e^{tA} is a solution to (5.41) and the columns are linearly independent. Hence e^{tA} is a fundamental matrix of (5.41). □

Lemma 5.4.11 (General solution). *Assume that Ψ is a fundamental matrix of (5.41) whose j-th column is Φ_j, $1 \leq j \leq n$. If $\tilde{\Phi}$ is any other solution to (5.41), then there exists $C = (c_1, \ldots, c_n)$ such that $\tilde{\Phi} = c_1\Phi_1 + \cdots + c_n\Phi_n = \Psi C$.*

Proof. We first observe that the set $S := \{\Phi_j : 1 \leq j \leq n\}$ is a linearly independent subset of the solution space of (5.41). Moreover, from Theorem 5.4.4, the dimension of the solution space is n. Hence S spans the solution space. This completes the proof of the theorem. □

Corollary 5.4.12. *If Ψ is a fundamental matrix of (5.41), then the solution to (5.41)–(5.42) is $Y(t) = \Psi(t)(\Psi(0))^{-1}Y(0)$, $t > 0$.* □

If Ψ is a fundamental matrix of (5.41), then we say that $\Phi = \Psi C$, where C is an arbitrary vector in $\mathbb{R}^n$, is the *general solution* to (5.41). We next present a simple criteria to determine whether the solution matrix is a fundamental matrix.

Theorem 5.4.13. *Assume that Ψ is a solution matrix of (5.41). Then Ψ is a fundamental matrix if and only if*

$$\det\left(\Psi(t)\right) \neq 0, \ t \in [0,\infty). \tag{5.58}$$

Proof. We assume that Ψ is a fundamental matrix of (5.41). Let the j-th column of Ψ be Ψ_j, $1 \leq j \leq n$. From Theorem 5.4.3, the set $\{\Psi_j(0) : 1 \leq j \leq n\}$ is linearly independent in $\mathbb{R}^n$. Hence we get $\det\left(\Psi(0)\right) \neq 0$. From Corollary 5.4.8, we obtain (5.58).

Conversely, let us assume (5.58). In particular, we have $\det\left(\Psi(0)\right) \neq 0$. Then we get that the columns of $\Psi(0)$ are linearly independent in $\mathbb{R}^n$. Again, from Theorem 5.4.3, we get that the columns of Ψ are linearly independent in $\left(C^1([0,\infty))\right)^n$. This proves that Ψ is a fundamental matrix. □

5.4.2 Non-homogeneous ODEs: method of variation of parameters revisited

In this subsection, our objective is to solve the system of non-homogeneous linear ODEs given by

$$Y'(t) = A(t)Y(t) + H(t), \ t > 0, \tag{5.59}$$
$$Y(0) = Y_0, \tag{5.60}$$

where $A : [0, \infty) \to M_n(\mathbb{R})$ and $H : [0, \infty) \to \mathbb{R}^n$ are continuous. In fact, we give an explicit formula for the solution to (5.59)–(5.60) in terms of a fundamental matrix of the associated homogeneous system and H. This is given in the next result.

Theorem 5.4.14. *Let Ψ be a fundamental matrix of $Y'(t) = A(t)Y(t)$, $t > 0$. Then the solution of non-homogeneous system (5.59)–(5.60) is*

$$\Phi(t) = \Psi(t)\big(\Psi(0)\big)^{-1}Y_0 + \Psi(t)\int_0^t \big(\Psi(\tau)\big)^{-1}H(\tau)d\tau, \ t \geq 0. \tag{5.61}$$

Proof. From Lemma 5.4.11, the general solution to $Y'(t) = A(t)Y(t)$ is

$$Y(t) = \Psi(t)C, \ t > 0, \ C \in \mathbb{R}^n.$$

We now wish to find a particular solution to (5.59) using the method of variation of parameters. For, we assume that a particular solution Φ_p to (5.59) is of the form

$$\Phi_p(t) = \Psi(t)U(t), \ t > 0, \tag{5.62}$$

where $U : [0, \infty) \to \mathbb{R}^n$ has to be determined such that (5.62) is a solution to (5.59).

On differentiating (5.62) (see Lemma 5.3.16), we obtain

$$\Phi_p'(t) = \Psi'(t)U(t) + \Psi(t)U'(t), \ t > 0. \tag{5.63}$$

On the other hand, since Φ_p is a solution to (5.59), we readily have

$$\Phi_p'(t) = A(t)\Psi(t)U(t) + H(t), \ t > 0. \tag{5.64}$$

In view of (5.51), (5.63)–(5.64), we immediately get

$$H(t) = \Psi(t)U'(t), \ t > 0.$$

From Theorem 5.4.13, the matrix $\Psi(t)$ is invertible for each $t > 0$ and thus we have

$$U(t) = \int_0^t \big(\Psi(\tau)\big)^{-1}H(\tau)d\tau, \ t > 0. \tag{5.65}$$

Therefore a particular solution which is in the form given in (5.62) is

$$\Phi_p(t) = \Psi(t)\int_0^t \big(\Psi(\tau)\big)^{-1}H(\tau)d\tau, \ t > 0. \tag{5.66}$$

Let Φ be an arbitrary solution to (5.59). Set $Y := \Phi - \Phi_p$. Then Y satisfies the associated homogeneous equation, i.e., $Y'(t) = A(t)Y(t)$, $t > 0$. From Lemma 5.4.11, there exists $C \in \mathbb{R}^n$ such that $Y(t) = \Psi(t)C$, $t > 0$. Therefore we have

$$\Phi(t) = \Phi_p(t) + \Psi(t)C, \ t > 0.$$

We now use (5.66) to conclude that any solution to (5.59) can be written as

$$\Phi(t) = \Psi(t)C + \Psi(t) \int_0^t \left(\Psi(\tau)\right)^{-1} H(\tau)d\tau, \ t > 0, \tag{5.67}$$

for some $C \in \mathbb{R}^n$.

In particular, to solve (5.59)–(5.60), we put $t = 0$ in (5.67). Then we get $Y_0 = \Psi(0)C$ or $C = \left(\Psi(0)\right)^{-1} Y_0$. This gives us the required formula (5.61) $\square$

Remark 5.4.15. In the case of constant coefficients, i.e., when A is independent of t, one can take the fundamental matrix to be e^{tA}. Then formula (5.61) reduces to

$$\Phi(t) = e^{tA}Y_0 + \int_0^t e^{(t-\tau)A} H(\tau)d\tau, \ t > 0. \tag{5.68}$$

For, we substitute $\Psi(t) = e^{tA}$ in formula (5.61), use the facts $e^{0A} = I$ and $\left(e^{tA}\right)^{-1} = e^{-tA}$ to obtain (5.68).

Example 5.4.16. Solve
$$\begin{cases} x' = x + y + t, \\ y' = x - y + 1, \\ x(0) = 1, y(0) = 3. \end{cases}$$

Solution. In order to solve this initial value problem we first find a fundamental matrix of the associated homogeneous system. For, we write the given equation in the matrix form as

$$\begin{pmatrix} x'(t) \\ y'(t) \end{pmatrix} = \begin{pmatrix} 1 & 1 \\ 1 & -1 \end{pmatrix} \begin{pmatrix} x(t) \\ y(t) \end{pmatrix} + \begin{pmatrix} t \\ 1 \end{pmatrix}. \tag{5.69}$$

Let A denote the square matrix in (5.69). We now consider the associated homogeneous system

$$\begin{pmatrix} x'(t) \\ y'(t) \end{pmatrix} = \begin{pmatrix} 1 & 1 \\ 1 & -1 \end{pmatrix} \begin{pmatrix} x(t) \\ y(t) \end{pmatrix}. \tag{5.70}$$

A fundamental matrix of (5.70) is e^{At} where $A = \begin{pmatrix} 1 & 1 \\ 1 & -1 \end{pmatrix}$. We now compute e^{At}. A straightforward computation gives that the eigenvalues of A to be $\pm\sqrt{2}$. Eigenvectors corresponding to $\sqrt{2}$ and $-\sqrt{2}$ are $(1, \sqrt{2} - 1)$ and $(1, -\sqrt{2} - 1)$, respectively. Moreover we have

$$A = P \begin{pmatrix} \sqrt{2} & 0 \\ 0 & -\sqrt{2} \end{pmatrix} P^{-1} \text{ where } P = \begin{pmatrix} 1 & 1 \\ \sqrt{2} - 1 & -\sqrt{2} - 1 \end{pmatrix}$$

Thus a fundamental matrix of (5.70) is

$$
\begin{aligned}
\Psi(t) &= e^{At} \\
&= \frac{1}{2\sqrt{2}} \begin{pmatrix} 1 & 1 \\ \sqrt{2}-1 & -\sqrt{2}-1 \end{pmatrix} \begin{pmatrix} e^{\sqrt{2}t} & 0 \\ 0 & e^{-\sqrt{2}t} \end{pmatrix} \begin{pmatrix} \sqrt{2}+1 & 1 \\ \sqrt{2}-1 & -1 \end{pmatrix} \\
&= \frac{1}{\sqrt{2}} \begin{pmatrix} \sqrt{2}\cosh(\sqrt{2}t)+\sinh(\sqrt{2}t) & \sinh(\sqrt{2}t) \\ \sinh(\sqrt{2}t) & \sqrt{2}\cosh(\sqrt{2}t)-\sinh(\sqrt{2}t) \end{pmatrix}.
\end{aligned}
$$

Let the non-homogeneous term in the given system be $H(t) = (t,1)$, $t > 0$. Then substituting e^{tA} and H in (5.68) one gets the solution to the given IVP to be

$$
\begin{aligned}
\Phi(t) &= \frac{1}{\sqrt{2}} \begin{pmatrix} \sqrt{2}\cosh(\sqrt{2}t)+4\sinh(\sqrt{2}t) \\ 3\sqrt{2}\cosh(\sqrt{2}t)-2\sinh(\sqrt{2}t) \end{pmatrix} \\
&\quad + \frac{1}{\sqrt{2}} \int_0^t \begin{pmatrix} \sqrt{2}\tau\cosh(\sqrt{2}(t-\tau))+\tau\sinh(\sqrt{2}(t-\tau)) \\ \tau\sinh(\sqrt{2}(t-\tau)) \end{pmatrix} d\tau \\
&\quad + \frac{1}{\sqrt{2}} \int_0^t \begin{pmatrix} \sinh(\sqrt{2}(t-\tau)) \\ \sqrt{2}\cosh(\sqrt{2}(t-\tau))-\sinh(\sqrt{2}(t-\tau)) \end{pmatrix} d\tau \\
&= \frac{1}{2\sqrt{2}} \begin{pmatrix} 4\sqrt{2}\cosh(\sqrt{2}t)+9\sinh(\sqrt{2}t)-2\sqrt{2}-\sqrt{2}t \\ 5\sqrt{2}\cosh(\sqrt{2}t)-\sinh(\sqrt{2}t)-\sqrt{2}(1-t) \end{pmatrix}.
\end{aligned}
$$

This completes the solution. $\qquad\square$

For more results on systems of linear ODEs, the reader can refer to [6].

Exercise 5.1. Let $A, B \in M_n(\mathbb{R})$. Then prove the following:

(i) $\|AB\|_\infty \le n\|A\|_\infty\|B\|_\infty$.

(ii) $\|A^k\|_\infty \le n^{k-1}\|A\|_\infty^k$, $k \in \mathbb{N}$.

(iii) the sequence of partial sums $\left(\sum_{k=0}^{m} \frac{A^k}{k!}\right)$ is a Cauchy sequence in $(M_n(\mathbb{R}), \|\|_\infty)$.

(iv) $\left\|\sum_{k=0}^{\infty} \frac{A^k}{k!}\right\|_\infty \le \exp(n\|A\|_\infty)$.

Exercise 5.2. Let $M_{m\times l}(\mathbb{R})$ denote the space of $m \times l$ matrices with real entries. Assume that (X_k) is a sequence in $M_n(\mathbb{R})$, $Y \in M_{m\times n}(\mathbb{R})$ and $Z \in M_{n\times l}(\mathbb{R})$. Then show that $YX_kZ \to YXZ$ in $M_{m\times l}(\mathbb{R})$ whenever $X_k \to X$ in $M_n(\mathbb{R})$. (For $A = (a_{ij}) \in M_{m\times l}(\mathbb{R})$, the norm of A is defined as $\|A\|_\infty = \max_{i,j} |a_{ij}|$.)

Exercise 5.3. Show that $Ae^A = e^A A$, for every $A \in M_n(\mathbb{R})$.

Exercise 5.4. Let A^T denote the transpose of $A \in M_n(\mathbb{R})$. Then show that $e^{(A^T)} = (e^A)^T$, for every $A \in M_n(\mathbb{R})$. Hence deduce that if $A + A^T = \mathbf{0}$, where $\mathbf{0}$ is the zero matrix in $M_n(\mathbb{R})$, then $e^A(e^A)^T = I$.

Exercise 5.5. Let $A \in M_n(\mathbb{R})$. Let λ be an eigenvalue of A and v be the corresponding eigenvector. Then show that e^λ is an eigenvalue of e^A, with the same v being the corresponding eigenvector. Conclude that e^A is never a singular matrix even if A is singular (see Remark 5.3.10).

Exercise 5.6. Let $A \in M_n(\mathbb{R})$, then show that $\det(e^A) = e^{Tr(A)}$.

Exercise 5.7. Find the solution to each of the following systems by computing the exponential of the associated matrix.

(i) $\begin{cases} x' = x - y, \\ y' = x + 2y, \\ x(0) = x_0, \ y(0) = y_0. \end{cases}$
$\qquad$
(ii) $\begin{cases} x' = x + y, \\ y' = -9x + 7y, \\ x(0) = x_0, \ y(0) = y_0. \end{cases}$

(iii) $\begin{cases} x' = 5x - y, \\ y' = 2x + 2y, \\ x(0) = x_0, \ y(0) = y_0. \end{cases}$
$\qquad$
(iv) $\begin{cases} x' = 5y + 5z, \\ y' = 5x + 5z, \\ z' = 5x + 5y, \\ x(0) = x_0, \ y(0) = y_0, \ z(0) = z_0. \end{cases}$

(v) $\begin{cases} x' = z, \\ y' = x - 3z, \\ z' = y + 3z, \\ x(0) = 1, \ y(0) = 1, \ z(0) = 1. \end{cases}$
$\qquad$
(vi) $\begin{cases} x' = 4x + y, \\ y' = -x + 6y, \\ z' = -x + y + 5z, \\ x(0) = 0, \ y(0) = 1, \ z(0) = 0. \end{cases}$

Exercise 5.8. Assume that Ψ is a fundamental matrix of $Y'(t) = A(t)Y(t)$. Set $\mathcal{Y}(t) = \Psi(t)(\Psi(0))^{-1}$.

(i) Show that the matrix valued function $\mathcal{Y}(t)$ is independent of the choice of Ψ. The matrix $\mathcal{Y}$ is called as the *evolution matrix*.

(ii) Let $H(t) := (h_{ij}(t))$, $t > 0$ be a continuous function. Denote by $\int_0^t H(\tau)d\tau$, the matrix whose (i, j)-th entry is $\int_0^t h_{ij}(\tau)d\tau$. Consider the sequence $(\mathcal{Y}_m)$ defined recursively by

$$\mathcal{Y}_0 = I, \ \mathcal{Y}_{m+1} = I + \int_0^t A(\tau)\mathcal{Y}_m(\tau)d\tau, \ m = 0, 1, 2, \ldots$$

Show that $(\mathcal{Y}_m)$ converges to the evolution matrix $\mathcal{Y}$.

(iii) Hence deduce that the evolution matrix is

$$\mathcal{Y}(t) = I + \int_0^t A(t_1)dt_1 + \int_0^t \int_0^{t_1} A(t_1)A(t_2)dt_2dt_1 + \cdots.$$

The above series is called the Peano–Baker series.

(iv) Find the evolution matrix when A is a constant matrix.

Hints: For *(i)*, observe that $\mathcal{Y}'(t) = A(t)\mathcal{Y}(t)$ and $\mathcal{Y}(0) = I$. For *(ii)*, use the arguments in the proof of Cauchy–Lipschitz theorem.

Exercise 5.9. Assume that $A(t) = (a_{ij}(t))$ be such that $a_{ij}(t) > 0$, $t \geq 0$, $1 \leq i, j \leq n$. Then show the following results:
(i) For every $t \geq 0$, every element in the evolution matrix $\mathcal{Y}(t)$ of $Y'(t) = A(t)Y(t)$ is nonnegative.

(ii) If Y and $\tilde{Y}$ are two solutions of $Y'(t) = A(t)Y(t)$ such that $Y_i(0) \leq \tilde{Y}_i(0)$ for $1 \leq i \leq n$, then $Y_i(t) \leq \tilde{Y}_i(t)$, $t \geq 0$, $1 \leq i \leq n$.

Hints: For *(i)*, use Peano–Baker series. For *(ii)*, observe that $Y(t) = \mathcal{Y}(t)Y(0)$ and $\tilde{Y}(t) = \mathcal{Y}(t)\tilde{Y}(0)$, $\forall t \geq 0$.

Exercise 5.10. Prove the following statements:

(i) Let $A(t)$, $t \geq 0$ be continuous. Assume that Ψ is a fundamental matrix of the system $Y'(t) = A(t)Y(t)$, $t \geq 0$. If $B \in M_n(\mathbb{R})$ is nonsingular, then ΨB is also a fundamental matrix of the same system.

(ii) Conversely, if Ψ_1 and Ψ_2 are fundamental matrices of $Y'(t) = A(t)Y(t)$, then there exists a nonsingular matrix B such that $\Psi_2(t) = \Psi_1(t)B$, $t > 0$.

(iii) There exist a system of ODEs $Y'(t) = A(t)Y(t)$, its fundamental matrix Ψ and a nonsingular matrix B such that $B\Psi$ is not a solution matrix of the same system of ODEs.

Exercise 5.11. Prove the following statements:

(i) Assume that Ψ is a fundamental matrix of the system $Y'(t) = A(t)Y(t)$. Then show that $(\Psi^{-1})^T$ is a fundamental matrix of $Z'(t) = -A^T(t)Z(t)$. This is called *the system of adjoint equations* for the given system.
(ii) If Ψ_1 and Ψ_2 are fundamental matrices of the systems $Y'(t) = A(t)Y(t)$ and $Z'(t) = -A^T(t)Z(t)$, respectively, then show that $\Psi_2^T \Psi_1 = B$, where B is a constant matrix.

Exercise 5.12. Prove the following statements:

(i) Assume that $A(t)$ and $\int_{t_0}^t A(\tau)d\tau$ commute for every $t > 0$. Then show that the function $U(t) = \exp(\int_{t_0}^t A(\tau)d\tau)Y_0$ is the solution to $Y' = A(t)Y(t)$, $t > t_0$, $Y(t_0) = Y_0$.

(ii) Use *(i)* of this exercise to solve $\begin{cases} x'(t) = -x(t) + 2ty(t), \\ y'(t) = -2tx(t) - y(t), \\ x(0) = 1,\ y(0) = 1. \end{cases}$

Exercise 5.13. Assume that $a : [0, \infty) \to \mathbb{R}$ is continuous and $\displaystyle\int_0^\infty |a(t)| dt$ is finite. Consider the system $\begin{cases} x'(t) = y(t), \ t > 0, \\ y'(t) = a(t)x(t), \ t > 0. \end{cases}$

(i) If x is bounded in $[0, \infty)$, then show that $y(t) \to 0$ as $t \to \infty$.

(ii) Show that there exists at least one unbounded solution to the system.

Exercise 5.14. Solve the following equations using the method of variation of parameters.

(i) $\begin{cases} x'(t) = y(t), \\ y'(t) = -x(t) + \cos(5t), \\ x(0) = x_0, \ y(0) = y_0. \end{cases}$ $\quad (ii)$ $\begin{cases} x'(t) = 2x(t) - y(t) + e^{2t}, \\ y'(t) = 3x(t) - 2y(t) + e^{3t}, \\ x(0) = x_0, \ y(0) = y_0. \end{cases}$

Exercise 5.15. A function $h : [0, \infty) \to \mathbb{R}$ is said to be of the exponential order if there exists $M, \alpha > 0$ such that $|h(t)| \le Me^{\alpha t}, \ t > 0$. Let Φ be a solution to the system of non-homogeneous equations $Y'(t) = AY(t) + H(t), t > 0$ where $A \in M_n(\mathbb{R})$. Show that if $\|H\|_\infty$ is of the exponential order then so is $\|\Phi\|_\infty$.

Exercise 5.16. Let $\|\ \|_2$ denote the Euclidean norm for vectors and matrices, i.e., $\|(v_1, \ldots, v_n)\|_2 = \sqrt{v_1^2 + \cdots + v_n^2}$, $\|(b_{ij})\|_2 = \sqrt{\displaystyle\sum_{1 \le i,j \le n} b_{ij}^2}$. Then prove the following statements:

(i) Assume that $A : [0, \infty) \to M_n(\mathbb{R})$ satisfies $V^T A(t)V \le 0, \ t \ge 0, \ V \in \mathbb{R}^n$. If $Y'(t) = A(t)Y(t)$, then $\|Y(t)\|_2 \le \|Y(0)\|_2, \ t \ge 0$.

(ii) If there exists $M > 0$ such that $\|A(t)\|_2 \le M, \ t \ge 0$, then the solution to $Y'(t) = A(t)Y(t)$ satisfies $\|Y(t)\|_2 \le e^{Mt}\|Y(0)\|_2, \ t \ge 0$.

Hints: (i) $\frac{d}{dt}(\|Y\|_2)^2 \le 0$.

(ii) Put $Z(t) = e^{-Mt}Y(t)$ then $Z'(t) = B(t)Z(t)$ where $B(t) = A(t) - MI$. Now use (i) to complete the proof.

Chapter 6

Qualitative behavior of the solutions

6.1 Introduction

In this chapter, our objective is to understand the behavior of the solutions to 2×2 systems of ordinary differential equations. In particular, we consider only those systems of ODEs for which global solutions exist and ask questions like: Is the solution bounded? Does the limit of the solution as $t \to \infty$ exist? If it exists, then what is it? Is the solution periodic? and so on.

Therefore in this chapter, our focus is to find the qualitative behavior of the solution for large values of t and not actually on finding solutions to the given ODEs. Hence we do not learn any new method to solve them here. Though many notions and theorems that we study in this chapter can be easily extended to $n \times n$ systems of ODEs (n ODEs in n involving unknowns), we restrict our discussion to 2×2 systems of ODEs.

Notation: From Chapter 5, recall that $\|\cdot\|$ denotes the Euclidean norm of vectors, i.e.,

$$\|(u, v)\| = \sqrt{u^2 + v^2}, \ (u, v) \in \mathbb{R}^2.$$

For 2×2 matrices also we use $\|\cdot\|$ to denote the Euclidean norm, i.e.,

$$\|(b_{ij})\| := \sqrt{\sum_{i,j=1}^{2} b_{ij}^2}, \ (b_{ij}) \in M_2(\mathbb{R}).$$

We begin with some definitions. Let $g_1, g_2 : \mathbb{R}^2 \to \mathbb{R}$ be C^1-functions. Then the system of ODEs, for $t > 0$

$$\begin{cases} x'(t) = g_1(x(t), y(t)), \\ y'(t) = g_2(x(t), y(t)), \end{cases} \tag{6.1}$$

is called an *autonomous* system of ODEs. In an autonomous system, the right hand side does not depend on the independent variable t explicitly. If the right hand side of a system of ODEs depends on t explicitly, then that system is called a non-autonomous systems. Therefore the non-autonomous system is in the following form, $t > 0$,

$$\begin{cases} x'(t) = h_1(t, x(t), y(t)), \\ y'(t) = h_2(t, x(t), y(t)), \end{cases}$$

where $h_1, h_2 : [0, \infty) \times \mathbb{R}^2 \to \mathbb{R}$ are C^1-functions. For example, the system

$$\begin{cases} x'(t) = 2x(t) + \sin(x(t)y(t)), \\ y'(t) = x(t) + 5y(t) + e^{x(t)+y(t)}, \end{cases} \qquad (6.2)$$

is a 2×2 autonomous system with $g_1(u, v) = 2u + \sin(uv)$, $g_2(u, v) = u + 5v + e^{u+v}$. On the other hand, for $t > 0$,

$$\begin{cases} x'(t) = 2x(t) + \sin(x(t)y(t)) + t^4, \\ y'(t) = x(t) + 5y(t) + e^{x(t)+y(t)} + t, \end{cases} \qquad (6.3)$$

is a 2×2 non-autonomous system of ODEs with $h_1(u, v, t) = 2u + \sin(uv) + t^4$, $h_2(u, v, t) = u + 5v + e^{u+v} + t$. We now convert non-autonomous system (6.3) to a 3×3 autonomous system by setting $z(t) = t$. Then (6.3) becomes

$$\begin{cases} x'(t) = 2x(t) + \sin(x(t)y(t)) + z(t)^4, \\ y'(t) = x(t) + 5y(t) + e^{x(t)+y(t)} + z(t), \\ z'(t) = 1. \end{cases}$$

which is a 3×3 autonomous system. In fact this method can be used to convert any $n \times n$ non-autonomous system to an $(n + 1) \times (n + 1)$ autonomous system.

Regarding the domain of definition of (6.1), we have a couple of comments. Firstly, to define autonomous/non-autonomous systems of ODEs the domain need not be unbounded. It could be any interval of positive length. Secondly, to study the long time behavior, the domain of definition of (6.1) should be of the form (t_0, ∞). Any 2×2 autonomous system on (t_0, ∞) can be converted to a system of the form (6.1) by introducing new variables $\tilde{x}(t) = x(t + t_0)$, $\tilde{y}(t) = y(t + t_0)$. That is why without loss of generality, we have taken the domain of (6.1) to be $(0, \infty)$.

The Euclidean space $\mathbb{R}^2$ where every solution $(x(t), y(t))$ to (6.1) lies for each $t \geq 0$ is referred to as the *phase plane*. Since every solution $(x(t), y(t))$, $t \geq 0$, to (6.1) is a curve[1], we refer it as an *integral curve* or a *trajectory* of (6.1) in the phase plane $\mathbb{R}^2$.

In view of the Cauchy–Lipschitz existence theorem, it follows that through every point (namely (x_0, y_0)) in the phase space, there is a unique trajectory $(x(t), y(t))$, $t \geq 0$, of (6.1) such that $(x(0), y(0)) = (x_0, y_0)$. Moreover, at every point (x_1, y_1) in the phase plane, the direction ratios of the tangent vector of the solution through that point are given by $(g_1(x_1, y_1), g_2(x_2, y_2))$.

We now give a physical interpretation of the solution to (6.1). Suppose there

---

[1]Let J be an interval. A continuous function $\gamma : J \to \mathbb{R}^2$ is said to be a curve in $\mathbb{R}^2$ and the image of γ is called the trace of the curve. However, in this chapter we do not distinguish between these two.

is a particle in the xy-plane at $P_0 = (x_0, y_0)$ at time $t = 0$, whose equations of motion for $t > 0$ are given by (6.1). If $(x(t), y(t))$, $t \geq 0$, is the solution to (6.1) with the initial condition $(x(0), y(0)) = (x_0, y_0)$, then the position of the particle at any time $t_1 > 0$ is $P_1 = (x(t_1), y(t_1))$ and the velocity vector at time t_1 is $(x'(t_1), y'(t_1)) = (g_1(P_1), g_2(P_1))$. Since the equations of motion that we have here form an autonomous system, the velocity vector at any time depends only on the position of the particle but not on the time at which the particle is at that position. This is a typical feature of the autonomous systems.

Definition 6.1.1. We say that $(\hat{x}, \hat{y}) \in \mathbb{R}^2$ is a critical point or stationary point of (6.1) if $(x(t), y(t)) = (\hat{x}, \hat{y})$, $t > 0$, is the solution to (6.1) whenever the initial condition $(x(0), y(0)) = (\hat{x}, \hat{y})$ holds.

In the Cartesian coordinate system, the critical points of (6.1) are characterized as the zeros of the system of equations

$$\begin{cases} g_1(\hat{x}, \hat{y}) = 0, \\ g_2(\hat{x}, \hat{y}) = 0. \end{cases} \tag{6.4}$$

For, if $(\hat{x}, \hat{y})$ is such that (6.4) holds, then it is easy to verify that $(x(t), y(t)) = (\hat{x}, \hat{y})$, $t \geq 0$, is the solution to (6.1) satisfying the initial condition $(x(0), y(0)) = (\hat{x}, \hat{y})$.
Conversely, if a constant vector $(\hat{x}, \hat{y})$ is a solution to (6.1), then by substituting $(x(t), y(t)) = (\hat{x}, \hat{y})$, $t \geq 0$ in (6.1) we get that (6.4) holds. The reader is advised to observe the analogy between the notions of critical points of 2×2 systems and scalar equations (see Remark 2.5.7).

Thus as long as we work in the Cartesian coordinate system, the critical points are defined as the points $(\hat{x}, \hat{y})$ such that $g_1(\hat{x}, \hat{y}) = 0$, $g_2(\hat{x}, \hat{y}) = 0$. On the other hand, in the polar coordinates, we do not have such equivalence (see Example 6.5.4).
For a physical interpretation, consider a particle whose equations of motion are given by (6.1). Suppose initially the particle is at a position whose Cartesian coordinates are $(\hat{x}, \hat{y}) \in \mathbb{R}^2$ which satisfy (6.4). Then the particle does not move from that position, i.e., $(x(t), y(t)) = (\hat{x}, \hat{y})$, $t \geq 0$. This is a reason for using the term stationary point.

Notation: Let $(x(t), y(t); x_0, y_0)$ denote the solution to (6.1) with $(x(0), y(0)) = (x_0, y_0)$. Sometimes we write merely $(x(t), y(t))$ instead of $(x(t), y(t); x_0, y_0)$ or $(x(t), y(t); x(0), y(0))$ when there is no ambiguity regarding the initial data. On the other hand, if we want to emphasize the importance of the location of initial data, then we denote the solution to (6.1) by $(x(t), y(t); x_0, y_0)$.
From this notation, it follows that

$$(x(t), y(t); \hat{x}, \hat{y}) = (\hat{x}, \hat{y}; \hat{x}, \hat{y}), \ t \geq 0.$$

We now introduce the concept of a stable critical point. We first formally explain the idea behind the definition before defining it rigorously. We assume that $(\hat{x}, \hat{y})$ is a critical point of (6.1). We 'roughly' say that $(\hat{x}, \hat{y})$ is a stable critical point if the following holds. 'If (x_0, y_0) is "close" to the critical point $(\hat{x}, \hat{y})$, then $(x(t), y(t); x_0, y_0)$ remains "close" to $(x(t), y(t); \hat{x}, \hat{y})$, $t \geq 0$ (or $(\hat{x}, \hat{y}; \hat{x}, \hat{y})$).' This is analogous to the notion of 'point of continuity' because there also we say (roughly) that x_0 is a point of continuity of a function h if $h(x)$ is remains 'close' to $h(x_0)$ whenever x is 'close' to x_0. We now give the precise definition of a stable critical point.

Definition 6.1.2. A critical point $(\hat{x}, \hat{y})$ of (6.1) is said to be stable if for a given $\epsilon > 0$, there exists $\delta > 0$ such that

$$\|(x(t), y(t)) - (\hat{x}, \hat{y})\| < \epsilon, \ t > 0,$$

whenever $\|(x(0), y(0)) - (\hat{x}, \hat{y})\| < \delta$.

Definition 6.1.3. A stable critical point $(\hat{x}, \hat{y})$ of (6.1) is said to be asymptotically stable critical point if there exists $\delta > 0$ such that

$$\lim_{t \to \infty} x(t) = \hat{x}, \quad \lim_{t \to \infty} y(t) = \hat{y}, \tag{6.5}$$

whenever $\|(x(0), y(0)) - (\hat{x}, \hat{y})\| < \delta$.

Definition 6.1.4. A critical point $(\hat{x}, \hat{y})$ of (6.1) is said to be unstable if it is not stable.

In Figure 6.1, we present a situation where $(0,0)$ is stable critical point of (6.1). In particular, for a given $\epsilon > 0$, we show a typical trajectory whose initial point is inside the circle S_δ with center $(0,0)$ radius δ. The trajectory lies inside the circle S_ϵ with center $(0,0)$ radius ϵ.

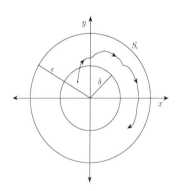

FIGURE 6.1: The trajectory starting inside S_δ cannot cross S_ϵ.

We now give an example of a stable critical point which is not asymptotically stable.

Example 6.1.5. Consider the system for $t > 0$,

$$\begin{cases} x'(t) = -y(t), \\ y'(t) = x(t). \end{cases} \tag{6.6}$$

Clearly, $(0,0)$ is the critical point of the given system. Moreover, from Example 5.3.7 we obtain the solution to (6.6) as

$$\begin{pmatrix} x(t) \\ y(t) \end{pmatrix} = \exp\left\{ t \begin{pmatrix} 0 & -1 \\ 1 & 0 \end{pmatrix} \right\} \begin{pmatrix} x(0) \\ y(0) \end{pmatrix} = \begin{pmatrix} \cos t & -\sin t \\ \sin t & \cos t \end{pmatrix} \begin{pmatrix} x(0) \\ y(0) \end{pmatrix}.$$

A straightforward calculation gives us

$$x^2(t) + y^2(t) = x^2(0) + y^2(0), \ t \geq 0,$$

or

$$\|(x(t), y(t))\| = \|(x(0), y(0))\|, \ t \geq 0. \tag{6.7}$$

For a given $\epsilon > 0$, we choose $\delta = \epsilon$. For this choice of δ, it immediately follows that $\|(x(t), y(t))\| < \epsilon$, $t \geq 0$ whenever $\|(x(0), y(0))\| < \delta$. Hence $(0,0)$ is a stable critical point.

On the other hand, for every nonzero $(x(0), y(0)) \in \mathbb{R}^2$, the limits

$$\lim_{t \to \infty} x(t), \ \lim_{t \to \infty} y(t)$$

do not exist. Hence $(0,0)$ is not an asymptotically stable critical point. □

We conclude this section with a very important remark.

Remark 6.1.6. Let $(\hat{x}, \hat{y}) \in \mathbb{R}^2$ be a critical point of (6.1). In order to study the stability of $(\hat{x}, \hat{y})$, we first set new variables $\tilde{x}(t) = x(t) - \hat{x}$ and $\tilde{y}(t) = y(t) - \hat{x}$. Then we observe that $(0,0)$ is a critical point of the new system

$$\begin{cases} \tilde{x}'(t) = g_1(\tilde{x}(t) + \hat{x}, \tilde{y}(t) + \hat{y}), \\ \tilde{y}'(t) = g_2(\tilde{x}(t) + \hat{x}, \tilde{y}(t) + \hat{y}). \end{cases}$$

The nature of the critical point $(\hat{x}, \hat{y})$ of (6.1) is the same as the nature of $(0,0)$ of this new system.

Motivated by the discussion we had so far, throughout this chapter, unless specified otherwise we:

(i) consider only autonomous systems.

(ii) assume that g_1, g_2 are Lipschitz functions[2] so that there exists a unique global solution to (6.1).

---

[2] A function $h : \Omega \subseteq \mathbb{R}^2 \to \mathbb{R}$ is said to be a Lipschitz function if there exists $L > 0$ such that

$$|h(Y_1) - h(Y_2)| \leq L\|Y_1 - Y_2\|, \ Y_1, Y_2 \in \Omega.$$

(*iii*) consider only initial value problems but not the Cauchy problems.

(*iv*) consider only those systems for which all critical points are isolated. In other words, we assume that for every critical point $(\hat{x}, \hat{y})$ of the system of ODEs there exists $r_0 > 0$ such that there is no critical point inside the circle with the center $(\hat{x}, \hat{y})$, radius r_0.

(*v*) assume (for theoretical purposes) that $(0,0)$ is a critical point of (6.1).

6.2   Linear systems with constant coefficients

Let A be a 2×2 invertible matrix with real entries. This section is devoted to the study of the stability of the critical points of the following homogeneous system of linear ODEs

$$\begin{pmatrix} x'(t) \\ y'(t) \end{pmatrix} = A \begin{pmatrix} x(t) \\ y(t) \end{pmatrix}, \quad t > 0. \tag{6.8}$$

Clearly $(0,0)$ is a unique critical point of the system (6.8). Let $Y(t) = (x(t), y(t))$, $t \geq 0$ (or $Y(t; Y(0))$, $t \geq 0$) denote the solution to (6.8). Then from Theorem 5.3.17 we get

$$Y(t) = e^{At} Y(0), \quad t \geq 0.$$

Using the structure of the exponential matrix we analyze the behavior of $Y(t)$ for large t.

To that end, we begin with the following useful observation. Let $B \in M_2(\mathbb{R})$, and the first and second columns of B be $\mathbf{b_1}$ and $\mathbf{b_2}$, respectively. Then, for every $Z = (z_1, z_2) \in \mathbb{R}^2$, from the triangle inequality and the Cauchy–Schwarz inequality it follows that

$$\begin{aligned} \|BZ\| &= \|z_1 \mathbf{b_1} + z_2 \mathbf{b_2}\| \\ &\leq |z_1| \|\mathbf{b_1}\| + |z_2| \|\mathbf{b_2}\| \\ &\leq \|B\| \|Z\|. \end{aligned} \tag{6.9}$$

Depending on the nature of A, we have three possibilities. They are:

Case–I. The matrix A has real eigenvalues and two linearly independent eigenvectors.

Case–II. Eigenvalues of A are equal and A has a unique eigenvector (up to multiplicity).

Case–III. Eigenvalues of A are complex numbers (whose imaginary part is nonzero).

We now discuss each case in detail.

Case–I. Suppose A has only real eigenvalues, say λ_1 and λ_2, with linearly independent eigenvectors. Hence A is diagonalizable (though $\lambda_1 = \lambda_2$). Let $\mathbf{v_1} = (v_{11}, v_{21})$ and $\mathbf{v_2} = (v_{12}, v_{22})$ be eigenvectors of A corresponding to λ_1 and λ_2, respectively. Let the first and the second columns of matrix P be $\mathbf{v_1}$ and $\mathbf{v_2}$, respectively. Then one can write

$$A = PDP^{-1}, \text{ where } D = \begin{pmatrix} \lambda_1 & 0 \\ 0 & \lambda_2 \end{pmatrix}.$$

By setting $C := (c_1, c_2) = P^{-1}Y(0)$, we get that the solution to (6.8) is

$$
\begin{aligned}
Y(t) &= e^{At}Y(0) \\
&= Pe^{Dt}P^{-1}Y(0) \\
&= \begin{pmatrix} v_{11} & v_{12} \\ v_{21} & v_{22} \end{pmatrix} \begin{pmatrix} e^{\lambda_1 t} & 0 \\ 0 & e^{\lambda_2 t} \end{pmatrix} \begin{pmatrix} c_1 \\ c_2 \end{pmatrix} \\
&= c_1 e^{\lambda_1 t}\mathbf{v_1} + c_2 e^{\lambda_2 t}\mathbf{v_2}. \qquad (6.10)
\end{aligned}
$$

Some immediate consequences of (6.10) are in order.
Firstly, using the triangle inequality, we obtain

$$\left| \|c_1\|\|\mathbf{v_1}\|e^{\lambda_1 t} - |c_2|\|\mathbf{v_2}\|e^{\lambda_2 t} \right| \leq \|Y(t)\| \leq |c_1|\|\mathbf{v_1}\|e^{\lambda_1 t} + |c_2|\|\mathbf{v_2}\|e^{\lambda_2 t}. \quad (6.11)$$

On the other hand, one can easily show that if $Y(0) = \mathbf{v_1}$ (resp. $Y(0) = \mathbf{v_2}$), then $C = (1,0)$ (resp. $C = (0,1)$). Therefore if $Y(0)$ is an eigenvector of A corresponding to the eigenvalue λ_1, i.e., $Y(0) = k\mathbf{v_1}$ for some $k \in \mathbb{R}$, then

$$Y(t) = e^{\lambda_1 t}Y(0), \ t \geq 0. \qquad (6.12)$$

Similarly if $Y(0) = k\mathbf{v_2}$, $k \in \mathbb{R}$, then

$$Y(t) = e^{\lambda_2 t}Y(0), \ t \geq 0. \qquad (6.13)$$

Finally, a simple calculation yields that

$$c_1\mathbf{v_1} + c_2\mathbf{v_2} = PC = PP^{-1}Y(0) = Y(0).$$

Therefore if $\lambda = \lambda_1 = \lambda_2$, then (6.10) implies

$$Y(t) = e^{\lambda t}Y(0), \ t \geq 0. \qquad (6.14)$$

Further, Case-1 can be broadly divided into three sub cases. They are:

Sub case I (a). Both λ_1 and λ_2 are negative.
Sub case I (b). Both λ_1 and λ_2 are positive.
Sub case I (c). λ_1 and λ_2 have opposite signs.

Sub case–I(a). Let λ_1, λ_2 be negative. This sub case is also further subdivided into two more sub cases based on whether λ_1 and λ_2 are different or the same. We first study the case when λ_1 and λ_2 are distinct. For, let $\lambda_1 < \lambda_2 < 0$.

In this case we prove that $(0,0)$ is asymptotically stable. In order to do that, we first prove that $(0,0)$ is stable. From (6.9) and (6.10), it follows that

$$
\begin{aligned}
\|Y(t)\| &\leq \|P\|\|e^{tD}P^{-1}Y(0)\| \\
&\leq \|P\|\|e^{tD}\|\|P^{-1}Y(0)\| \\
&\leq 2\|P\|\|P^{-1}Y(0)\| \\
&\leq 2\|P\|\|P^{-1}\|\|Y(0)\|,
\end{aligned}
\tag{6.15}
$$

for every $t > 0$. Let $\epsilon > 0$ be given. We choose

$$
\delta = \frac{\epsilon}{2\|P\|\|P^{-1}\|},
$$

so that if $\|Y(0)\| < \delta$, then

$$
\|Y(t)\| < 2\|P\|\|P^{-1}\|\delta = \epsilon, \ t \geq 0.
$$

This shows that $(0,0)$ is stable. Next, in view of (6.11) we readily have

$$
\|Y(t)\| \to 0, \text{ as } t \to \infty, \ (c_1, c_2) \in \mathbb{R}^2.
\tag{6.16}
$$

Since $(c_1, c_2)(= P^{-1}Y(0))$ is arbitrary in (6.16), we find that

$$
\|Y(t)\| \to 0, \text{ as } t \to \infty, \ Y(0) \in \mathbb{R}^2.
$$

This proves that $(0,0)$ is asymptotically stable.

We now want to find the direction along which $Y(t) = (x(t), y(t))$ converges to $(0,0)$ as t tends to ∞. From (6.12)–(6.13), we know that if $Y(0)$ is an eigenvector of A, then the trajectory lies on a straight line and converges to $(0,0)$. If $Y(0)$ is not an eigenvector of A, i.e., $c_1 \neq 0$ and $c_2 \neq 0$, then in view of (6.10) we have,

$$
\lim_{t \to \infty} \frac{y'(t)}{x'(t)} = \lim_{t \to \infty} \frac{\lambda_1 c_1 v_{21} e^{\lambda_1 t} + \lambda_2 c_2 v_{22} e^{\lambda_2 t}}{\lambda_1 c_1 v_{11} e^{\lambda_1 t} + \lambda_2 c_2 v_{12} e^{\lambda_2 t}} = \frac{v_{22}}{v_{12}}.
$$

In other words, the straight line $v_{12}y - v_{22}x = 0$ is the asymptote to the trajectory (as $t \to \infty$) that starts from $Y(0)$ which is not an eigenvector of A. Hence $(x(t), y(t))$ converges to $(0,0)$ along the eigenvector corresponding to the larger eigenvalue λ_2. A schematic representation of the trajectories of (6.8) when A has distinct negative eigenvalues is given in Figure 6.2(a). In this figure, we notice that every trajectory is moving toward the origin as t tends to infinity. The trajectories are straight lines whenever $Y(0)$ is an eigenvectors of A (see (6.12)–(6.13)). We now turn our attention to the case when we have equal eigenvalues, i.e., $\lambda_1 = \lambda_2 =: \lambda$. In view of (6.14) we have

$$
\|Y(t)\| \leq \|Y(0)\|, \ t \geq 0.
$$

Given any $\epsilon > 0$, we take $\delta = \epsilon$. Then we have

$$
\|Y(t)\| < \epsilon, \ t \geq 0, \text{ whenever } \|Y(0)\| < \delta.
$$

From this, we conclude that the critical point $(0,0)$ is stable. Furthermore,

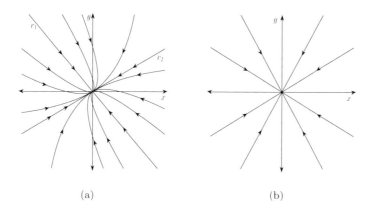

(a)　　　　　　　　　　　　　　　　(b)

FIGURE 6.2: (a) The trajectories of (6.8) when the eigenvalues of A are negative and distinct; (b) The trajectories of (6.8) when both the eigenvalues of A are negative and equal. Here every trajectory is a ray.

from (6.14) it is evident that $\|Y(t)\| \to 0$, as $t \to \infty$, for every choice of $Y(0)$. Therefore $(0,0)$ is asymptotically stable when $\lambda_1 = \lambda_2$. Moreover from (6.14), we find that the trajectory through $(x(0), y(0))$ lies on the straight line $y(0)x - x(0)y = 0$ and moves toward the origin as t increases. The trajectories of (6.8) when A has negative and equal eigenvalues are shown in Figure 6.2(b). In this sub case the critical point $(0,0)$ is called an *asymptotically stable node*.

Sub case–I(b). As in Sub case–I(a), we first consider the case when λ_1, λ_2 are distinct and later we move on to the case when they are equal.
Let $0 < \lambda_1 < \lambda_2$. We show that the $(0,0)$ is unstable. For, we fix $\epsilon = 1$. We show that for given $\delta > 0$, there exists $Y(0)$ such that $\|Y(0)\| < \delta$ and $\|Y(\tilde{t}; Y(0))\| \geq 1$ for some $\tilde{t} > 0$. Let $k > 0$ be such that $k\|\mathbf{v_1}\| < \delta$. We now choose $Y(0) = k\mathbf{v_1}$. Then from (6.12) we obtain

$$\|Y(t)\| = e^{\lambda_1 t}\|Y(0)\| \to \infty \text{ as } t \to \infty.$$

Therefore we can find $\tilde{t} > 0$ such that $\|Y(\tilde{t}; Y(0))\| > \epsilon = 1$. This shows that $(0,0)$ is unstable. Moreover from (6.12)–(6.13) it is easy to observe that $\|Y(t; Y(0))\| \to \infty$, whenever $Y(0)$ is an eigenvector of A.
Furthermore we prove that $\|Y(t; Y(0))\| \to \infty$ as $t \to \infty$ whenever $Y(0) \neq (0,0)$. Without loss of generality, we assume that $Y(0)$ is not an eigenvector of A. From (6.11) we observe

$$\|Y(t)\| \geq e^{\lambda_2 t}\left(|c_2|\|\mathbf{v_2}\| - e^{(\lambda_1 - \lambda_2)t}|c_1|\|\mathbf{v_1}\|\right), \ t \geq 0.$$

As $\lambda_1 < \lambda_2$, we can find $M > 0$ such that

$$|c_1|\|\mathbf{v_1}\|e^{(\lambda_1-\lambda_2)t} < \frac{|c_2|\|\mathbf{v_2}\|}{2}, \ t > M.$$

Therefore we have

$$\|Y(t)\| \geq \frac{|c_2|\|\mathbf{v_2}\|}{2}e^{\lambda_2 t}, \ t > M.$$

Thus $\|Y(t)\| \to \infty$ as $t \to \infty$. Using the same argument as in Sub case I(a), we find the slope of the asymptote of $Y(t)$ to be

$$\lim_{t\to\infty} \frac{y'(t)}{x'(t)} = \frac{v_{22}}{v_{12}}.$$

Hence $Y(t)$ tends to infinity in the direction of the eigenvector corresponding to the larger eigenvalue.

We now assume that $\lambda_1 = \lambda_2 = \lambda$. As in Sub case I(a) we use (6.14) to conclude that

$$\|Y(t)\| = e^{\lambda t}\|Y(0)\| \to \infty, \ Y(0) \in \mathbb{R}^2\backslash\{(0,0)\}.$$

Using the same argument employed to prove instability of $(0,0)$ in the case of distinct eigenvalues one can show that $(0,0)$ is an unstable critical point when $\lambda_1 = \lambda_2$.

In this sub case, the critical point $(0,0)$ is called an *unstable node*. In Figure 6.3(a), we depict the trajectories of (6.8) when A has distinct positive eigenvalues. All non-trivial solutions tend to infinity as t tends to infinity. Moreover, if $Y(0)$ is an eigenvalue of A then the corresponding trajectory is a ray starting from $Y(0)$ which moves along the direction of $Y(0)$. In Figure 6.3(b), the trajectories of (6.8) are shown when the eigenvalues are positive and equal.

Sub case–I(c). We assume that $\lambda_1 < 0 < \lambda_2$.

In this case also we show that $(0,0)$ is an unstable critical point. The argument is the same as that in the previous sub case. We first fix $\epsilon = 1$. For given $\delta > 0$, let $k > 0$ be such that $\|k\mathbf{v_2}\| < \delta$. We now choose $Y(0) = k\mathbf{v_2}$ and use (6.13) to obtain

$$\|Y(t)\| = e^{\lambda_2 t}\|Y(0)\|, \ t > 0.$$

Therefore there exists $t_1 > 0$ such that $\|Y(t_1)\| > 1$. Since δ is arbitrary, $(0,0)$ is unstable. Moreover for this choice $Y(0)$, we have $\|Y(t;Y(0))\| \to \infty$ as $t \to \infty$.

On the other hand, let $m > 0$ be such that $\|m\mathbf{v_1}\| < \delta$. If we take $Y(0) = m\mathbf{v_1}$, then from (6.12) it follows that

$$\|Y(t;Y(0))\| \leq \|Y(0)\|, \ t \geq 0, \text{ and } \lim_{t\to\infty} \|Y(t;Y(0))\| = 0.$$

The main difference between this sub case and the previous sub case is the following. In the previous sub case, we have shown that $\|Y(t;Y(0))\| \to \infty$ as $t \to \infty$ for every $Y(0) \neq (0,0)$, whereas here we have shown that for every $\delta > 0$, there exists $Y(0)$ such that $\|Y(0)\| < \delta$ and $\|Y(t;Y(0))\| \to 0$ as $t \to \infty$.

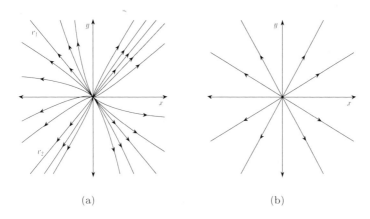

(a)   (b)

FIGURE 6.3: (a) The trajectories of (6.8) when the eigenvalues of A are positive and distinct; (b) The trajectories of (6.8) when both the eigenvalues of A are positive and equal. Here every trajectory is a ray.

In Figure 6.4, the trajectories of (6.8) are depicted when the eigenvalues of A have opposite signs.

In this sub case, the critical point $(0,0)$ is called a *saddle point* or simply *saddle*.

Case–II. Let λ be a repeated eigenvalue of A and $\mathbf{v_1} = (v_{11}, v_{21})$ be a unique eigenvector (up to multiplicity) corresponding to λ. We assume that $\mathbf{v_2} = (v_{12}, v_{22})$ denotes a generalized eigenvector. Let $P \in M_2(\mathbb{R})$, the first and second columns of P be $\mathbf{v_1}$ and $\mathbf{v_2}$, respectively. Then one can write

$$A = PJP^{-1}, \text{ where } J := \begin{pmatrix} \lambda & 1 \\ 0 & \lambda \end{pmatrix}.$$

As in Case-1, we denote $C = (c_1, c_2) = P^{-1}Y(0)$. This immediately gives that the solution to (6.8) as

$$Y(t) = e^{At}Y(0) = Pe^{Jt}P^{-1}Y(0) = e^{\lambda t}PTC, \ t \geq 0, \tag{6.17}$$

where $T = \begin{pmatrix} 1 & t \\ 0 & 1 \end{pmatrix}$. This can be rewritten as

$$Y(t) = e^{\lambda t}\big(c_1\mathbf{v_1} + c_2(t\mathbf{v_1} + \mathbf{v_2})\big), \ t \geq 0. \tag{6.18}$$

Using the triangle inequality, we get

$$\|Y(t)\| \leq e^{\lambda t}\big(|c_1|\|\mathbf{v_1}\| + |c_2|(t\|\mathbf{v_1}\| + \|\mathbf{v_2}\|)\big), \ t \geq 0. \tag{6.19}$$

Suppose $Y(0) = k\mathbf{v_1}$, then we have (see Case–1)

$$Y(t) = e^{\lambda t}Y(0), \ t \geq 0. \tag{6.20}$$

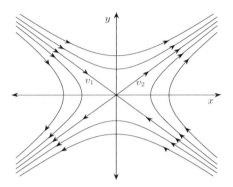

FIGURE 6.4: The trajectories when the eigenvalues of A have opposite signs. If $\lambda_1 < 0 < \lambda_2$, only those trajectories for which $(x(0), y(0))$ are constant multiples of $\mathbf{v_1}$ move toward the origin. All other trajectories are unbounded.

As in the previous case, the behavior of the critical point depends on the sign of the eigenvalue. To prove stability/instability of the critical point we use similar arguments presented in Case–I.

Sub case–II(a) Suppose $\lambda < 0$.

In this case, we show that $(0, 0)$ is asymptotically stable. We need the following inequalities to prove the stability

$$te^{\lambda t} \leq \frac{1}{e|\lambda|}, \qquad \|T\| = \sqrt{2 + t^2} \leq (t + 2), \ t \geq 0.$$

Hence from (6.17), for $t > 0$ we obtain (see Sub case–I(a) for details)

$$
\begin{aligned}
\|Y(t)\| &\leq e^{\lambda t} \|P\| \|T\| \|C\| \\
&\leq (2 + t)e^{\lambda t} \|P\| \|P^{-1} Y(0)\| \\
&\leq \left(2 + \frac{1}{e|\lambda|}\right) \|P\| \|P^{-1}\| \|Y(0)\|.
\end{aligned}
\tag{6.21}
$$

Let $\epsilon > 0$ be arbitrary. On choosing

$$\delta = \frac{e|\lambda|\epsilon}{(2e|\lambda| + 1)\|P\| \|P^{-1}\|},$$

and using (6.21), we get

$$\|Y(t; Y(0))\| < \epsilon, \ t \geq 0, \text{ whenever } \|Y(0)\| < \delta.$$

This proves that $(0, 0)$ is a stable critical point.

Again, from (6.19) it is straightforward to conclude that $(0, 0)$ is an asymptotically stable critical point. Moreover we have

$$\|Y(t; Y(0))\| \to 0 \text{ as } t \to \infty, \ Y(0) \in \mathbb{R}^2.$$

In this sub case $(0, 0)$ is called an *asymptotically stable node*.

Sub case–II(b). Suppose $\lambda > 0$.

In this sub case, we show that $(0,0)$ is unstable using the same arguments as in Sub case I(b). For, we fix $\epsilon = 1$ and let $\delta > 0$ be arbitrary. Let $k > 0$ be such that $k\|\mathbf{v_1}\| < \delta$. On choosing $Y(0) = k\mathbf{v_1}$ and using (6.20), we obtain

$$\|Y(t)\| = e^{\lambda t}\|Y(0)\| \to \infty \text{ as } t \to \infty.$$

Hence for any $\delta > 0$, we find $Y(0) \in \mathbb{R}^2$ and $\tilde{t} > 0$ such that

$$\|Y(0)\| < \delta \text{ and } \|Y(\tilde{t}; Y(0))\| > 1.$$

Therefore $(0,0)$ is an unstable critical point. Again, using the triangle inequality in (6.18) we get

$$\|Y(t)\| \geq e^{\lambda t}\left|t|c_2|\|\mathbf{v_1}\| - \|c_1\mathbf{v_1} + c_2\mathbf{v_2}\|\right|, \ t \geq 0.$$

Thus, it immediately follows that

$$\|Y(t; Y(0))\| \to \infty \text{ as } t \to \infty, \ Y(0) \in \mathbb{R}^2 \backslash \{(0,0)\}.$$

In this sub case also, we say that $(0,0)$ is an *unstable node*.

Case–III. Suppose A has complex eigenvalues. Let $\lambda_1 = \alpha + i\beta$, $\lambda_2 = \alpha - i\beta$, $\beta \neq 0$ be the eigenvalues of A. Then there exists a nonsingular matrix P with real entries such that

$$A = P\begin{pmatrix} \alpha & \beta \\ -\beta & \alpha \end{pmatrix}P^{-1}.$$

This is an elementary result in matrix theory and is left as an exercise. Using Theorem 5.3.8 and Example 5.3.7, we compute

$$\exp\left(\begin{pmatrix} \alpha & \beta \\ -\beta & \alpha \end{pmatrix}t\right) = \exp\left(\begin{pmatrix} \alpha t & 0 \\ 0 & \alpha t \end{pmatrix} + \begin{pmatrix} 0 & \beta t \\ -\beta t & 0 \end{pmatrix}\right)$$

$$= \exp\left(\begin{pmatrix} \alpha t & 0 \\ 0 & \alpha t \end{pmatrix}\right)\exp\left(\begin{pmatrix} 0 & \beta t \\ -\beta t & 0 \end{pmatrix}\right)$$

$$= e^{\alpha t}R(t),$$

where $R(t) = \begin{pmatrix} \cos(\beta t) & \sin(\beta t) \\ -\sin(\beta t) & \cos(\beta t) \end{pmatrix}$. Therefore the solution to (6.8) in this case is

$$Y(t) = e^{\alpha t}PR(t)P^{-1}Y(0), \ t \geq 0. \tag{6.22}$$

This case has three important sub cases. They are:

(a) The eigenvalues of A are purely imaginary, i.e., $\alpha = 0$.
(b) The real part of the eigenvalues of A is negative., i.e., $\alpha < 0$.
(c) The real part of the eigenvalues of A is positive., i.e., $\alpha > 0$.

Sub case–III(a). We assume that $\alpha = 0$.
Then the solution to (6.8) is

$$Y(t) = PR(t)P^{-1}Y(0), \ t \geq 0. \tag{6.23}$$

In this sub case, we show that $(0,0)$ is a stable critical point. To that end, we first note that for every $X = (x_1, x_2) \in \mathbb{R}^2$, we have

$$\|R(t)X\|^2 = (x_1 \cos(\beta t) + x_2 \sin(\beta t))^2 + (x_2 \cos(\beta t) - x_1 \sin(\beta t))^2 = \|X\|^2.$$

Hence for $t \geq 0$, we estimate

$$\begin{aligned} \|Y(t)\| &\leq \|P\|\|R(t)P^{-1}Y(0)\| \\ &= \|P\|\|P^{-1}Y(0)\| \\ &\leq \|P\|\|P^{-1}\|\|Y(0)\|. \end{aligned} \quad (6.24)$$

To prove stability, for given $\epsilon > 0$, we choose

$$\delta = \frac{\epsilon}{\|P\|\|P^{-1}\|}$$

so that

$$\|Y(t; Y(0))\| < \epsilon, \ t \geq 0, \text{ whenever } \|Y(0)\| < \delta.$$

Thus the critical point $(0,0)$ is stable. On the other hand, since $R(t + \frac{2\pi}{\beta}) = R(t)$, $t \geq 0$, we have

$$Y(t + \tfrac{2\pi}{\beta}) = Y(t), \ t \geq 0.$$

In view of (6.23), one can easily conclude that the trajectories in this sub case are either circles or ellipses (the proof is left to the reader). The trajectories of (6.8) when the eigenvalues of A purely imaginary are shown in Figure 6.5. The critical point $(0,0)$ in this sub case is called a *center*.

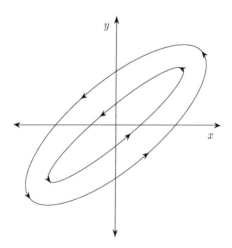

FIGURE 6.5: The trajectories when the eigenvalues of A are purely imaginary.

Sub case–III(b). Let the real part α of the eigenvalues be negative.

In this case, we show that the critical point $(0,0)$ is asymptotically stable. The proof of stability is the same as the one in sub case III(a). In view of (6.22). for $t \geq 0$ we get

$$
\begin{aligned}
\|Y(t)\| & \leq e^{\alpha t}\|P\|\|R(t)P^{-1}Y(0)\| \\
& = e^{\alpha t}\|P\|\|P^{-1}Y(0)\| \\
& \leq e^{\alpha t}\|P\|\|P^{-1}\|\|Y(0)\| \\
& \leq \|P\|\|P^{-1}\|\|Y(0)\|.
\end{aligned}
\tag{6.25}
$$

Let $\epsilon > 0$ be arbitrary. On choosing

$$
\delta = \frac{\epsilon}{\|P\|\|P^{-1}\|},
$$

we obtain

$$
\|Y(t; Y(0))\| < \epsilon, \ t \geq 0, \text{ whenever } \|Y(0)\| < \delta.
$$

which proves the stability of $(0,0)$. Asymptotic stability of $(0,0)$ is an immediate consequence of (6.25). Furthermore, from (6.25), it follows that

$$
\|Y(t; Y(0))\| \to 0 \text{ as } t \to \infty, \ Y(0) \in \mathbb{R}^2.
$$

Moreover observe that the trajectories of (6.8) in this sub case are spirals around the critical point $(0,0)$ and they are given in Figures 6.6(a)–6.6(b). In this sub case, the critical point $(0,0)$ is called a *stable spiral.*

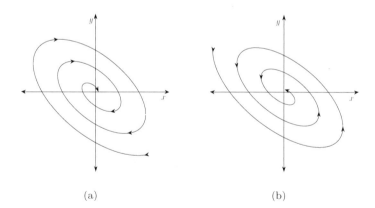

(a)   (b)

FIGURE 6.6: The trajectories when the eigenvalues of A are complex numbers with negative real part. The trajectories spiral around the origin while converging to it; (a) A trajectory spiraling in the clockwise direction; (b) A trajectory spiraling in the anti-clockwise direction.

Sub case–III(c). Assume that $\alpha > 0$.
In this sub case, we show that $(0,0)$ is unstable using the argument presented in the earlier cases. Let $\epsilon = 1$, and $\delta > 0$ be arbitrary. We define a sequence

(t_n) where $t_n = 2n\pi/\beta$, $n \in \mathbb{N}$, so that $R(t_n)$ becomes the 2×2 identity matrix. Then in view of (6.22) we have

$$Y(t_n; Y(0)) = e^{\alpha t_n} Y(0), \ n \in \mathbb{N}, \ Y(0) \in \mathbb{R}^2.$$

For each $Y(0)$ with $0 < \|Y(0)\| < \delta$, there exists a sufficiently large N such that

$$\|Y(t_N; Y(0))\| = e^{\alpha t_N} \|Y(0)\| > 1.$$

This shows that $(0,0)$ is unstable. Furthermore it is not difficult to show that for every $Y(0) \neq (0,0)$

$$\|Y(t; Y(0))\| \to \infty, \text{ as } t \to \infty,$$

and the details are left to the reader. As in Sub case–III(b), in this sub case also the trajectories are spirals around the origin. The only difference is $Y(t)$ moves away from the origin as t increases in this sub case, whereas $Y(t)$ moves toward the origin in Sub case–III(b). The trajectories of (6.8) in this sub case are given in Figures 6.7(a)–6.7(b).

In this sub case, the critical point $(0,0)$ is called an *unstable spiral.*

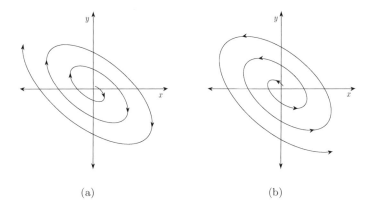

(a)   (b)

FIGURE 6.7: The trajectories when the eigenvalues of A are complex numbers with positive real part. Notice that they are moving away from the origin; (a) A trajectory spiraling in the clockwise direction; (b) A trajectory spiraling in the anti-clockwise direction.

We summarize our previous discussion in the following theorem.

Theorem 6.2.1. *Assume that A is a 2×2 invertible matrix. Then the following hold true for the critical point $(0,0)$ of $Y'(t) = AY(t)$:*
(a) If both the eigenvalues of A are negative real numbers, then $(0,0)$ is an asymptotically stable (asymptotically stable node).
(b) If one of the eigenvalues of A is a positive real number, then $(0,0)$ is unstable.

(c) If the eigenvalues of A are purely imaginary, then $(0,0)$ is stable (center).
(d) If the real part of the complex eigenvalues of A is negative, then (0.0) is asymptotically stable (stable spiral).
(e) If the real part of the complex eigenvalues of A is positive, then (0.0) is unstable (unstable spiral). □

Though the statement of Theorem 6.2.1 seems lengthy. the essence of the theorem is the following: The critical point $(0,0)$ of $Y'(t) = AY(t)$ is unstable if and only if there is at least one eigenvalue of A which is positive or has positive real part. In Theorem 6.2.1, we have characterized the long time behavior of the solution to $Y' = AY$, in terms of the signs of (the real part of) the eigenvalues of A. We now write the eigenvalues of A in terms of its trace $Tr(A)$ and determinant $\det(A)$ (it is possible because A is a 2×2 matrix). To that end, we can write the characteristic equation of a 2×2 matrix A as

$$\det(A - \lambda I) = \lambda^2 - Tr(A)\lambda + \det(A) = 0.$$

The eigenvalues of A say, λ_1 and λ_2 are given by

$$\lambda_1 = \frac{Tr(A) - \sqrt{(Tr(A))^2 - 4\det(A)}}{2}, \quad \lambda_2 = \frac{Tr(A) + \sqrt{(Tr(A))^2 - 4\det(A)}}{2}.$$

In the next lemma, we provide relations between the signs of $Tr(A)$ and $\det(A)$ and the signs of the eigenvalues.

Lemma 6.2.2. *Let λ_1 and λ_2 be the eigenvalues of a 2×2 matrix A. Then the following hold true.*

1. *If $Tr(A) < 0$, $\det(A) > 0$, and $4\det(A) < (Tr(A))^2$, then $\lambda_1, \lambda_2 < 0$.*

2. *If $Tr(A) > 0$. $\det(A) > 0$. and $4\det(A) < (Tr(A))^2$. then $\lambda_1. \lambda_2 > 0$.*

3. *If $\det(A) < 0$. then $\lambda_1. \lambda_2$ have opposite signs.*

4. *If $Tr(A) < 0$. and $4\det(A) = (Tr(A))^2$. then $\lambda_1 = \lambda_2 < 0$.*

5. *If $Tr(A) > 0$, and $4\det(A) = (Tr(A))^2$, then $\lambda_1 = \lambda_2 > 0$.*

6. *If $Tr(A) = 0$, and $\det(A) > 0$, then λ_1 and λ_2 purely imaginary.*

7. *If $Tr(A) < 0$, $\det(A) > 0$, and $4\det(A) > (Tr(A))^2$. then $\lambda_1, \lambda_2 \in \mathbb{C}$ and the real part of λ_1 and λ_2 is negative.*

8. *If $Tr(A) > 0$, $\det(A) > 0$, and $4\det(A) > (Tr(A))^2$, then $\lambda_1, \lambda_2 \in \mathbb{C}$ the real part of λ_1 and λ_2 is positive.*

These are precisely the eight situations which were discussed so far in this section. We present a schematic picture which might help the readers to remember the above conclusions and the results in Theorem 6.2.1.

In order to do that, consider a 2-D plane in which the horizontal and vertical axes are labeled as the trace (p) and det (q) axes, respectively. We first plot the curve $q = \frac{1}{4}p^2$, and locate (p_0, q_0) in the pq-plane, where p_0 and q_0 are the trace and the determinant of the matrix A. Depending on the location of (p_0, q_0), we conclude the nature of the critical point $(0, 0)$ as shown in Figure 6.8.

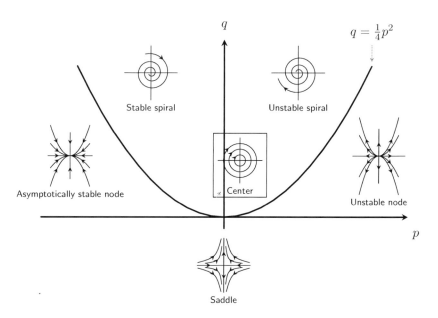

FIGURE 6.8: The pq-plane with different regions showing the nature of the critical point.

From Figure 6.8, it is clear that $(0, 0)$ is stable if and only if (p_0, q_0) is in the 2nd quadrant $(p < 0, q > 0)$ or on the positive q axis.

In the next example, we investigate the nature of the critical points of linear systems using the theory developed so far.

Example 6.2.3. Find the nature of the critical points for the systems $Y' = AY$, where A is given as follows

(i) $\begin{pmatrix} 2 & -4 \\ 3 & -4 \end{pmatrix}$, (ii) $\begin{pmatrix} 2 & -3 \\ 3 & -4 \end{pmatrix}$, (iii) $\begin{pmatrix} 5 & -3 \\ 3 & -4 \end{pmatrix}$, (iv) $\begin{pmatrix} 7 & 5 \\ 3 & 4 \end{pmatrix}$.

Solution: We first note that $(0, 0)$ is the critical point for all the problems given in this example.

(i) On computing the trace and the determinant of A we get

$$p = Tr(A) = -2, \ q = \det(A) = 4 \ \text{and} \ q > \tfrac{1}{4}p^2.$$

From Lemma 6.2.2, we get that the eigenvalues of A are complex numbers with a negative real part. Using Theorem 6.2.1, we conclude that $(0,0)$ is a stable spiral. One can arrive at the same conclusion by computing the eigenvalues of A ($\lambda_\pm = -1 \pm i\sqrt{3}$ in this case) directly.

(ii) As in (i) we compute the trace and the determinant of A to obtain

$$p = Tr(A) = -2, \ q = \det(A) = 1 \text{ and } q = \tfrac{1}{4}p^2.$$

Using Lemma 6.2.2, we obtain that the eigenvalues of A are equal and negative. In view of Theorem 6.2.1, we get that $(0,0)$ is a stable node. One can conclude the same by computing the eigenvalues of A (-1, and -1 in this case) directly.

(iii) We have

$$p = 1, \ q = -11,$$

in this case. Again, from Lemma 6.2.2 and Theorem 6.2.1, we get that the critical point is unstable and a saddle point.

(iv) For this A we obtain

$$p = 11, \ q = 13 \text{ and } q < \tfrac{1}{4}p^2.$$

Using the same argument, we conclude that $(0,0)$ is an unstable node. □

Example 6.2.4. Find the nature of the critical point of the system

$$\begin{cases} x' = -3x + 2y - 7, \\ y' = x - y + 3. \end{cases}$$

Solution: The critical point of the given system is the solution of the system of algebraic equations given by

$$-3\hat{x} + 2\hat{y} - 7 = 0, \ \hat{x} - \hat{y} + 3 = 0.$$

Clearly $\hat{x} = -1$ and $\hat{y} = 2$ is a unique solution to this system of equations. In other words $(-1, 2)$ is a critical point of the given system. In view of Remark 6.1.6. we introduce new functions $\tilde{x}(t) := x(t) + 1$ and $\tilde{y}(t) = y(t) - 2$. $t \geq 0$. Then the given system of ODEs reduces to

$$\begin{cases} \tilde{x}' = -3\tilde{x} + 2\tilde{y}, \\ \tilde{y}' = \tilde{x} - \tilde{y}, \end{cases} \tag{6.26}$$

for which $(0,0)$ is the critical point. If we write (6.26) in the standard form $Y' = AY$, then $A = \begin{pmatrix} -3 & 2 \\ 1 & -1 \end{pmatrix}$. As in the previous example, we compute the trace and the determinant of A to get

$$p = -4, \ q = 1$$

and thus $q < \tfrac{1}{4}p^2$. Hence $(0,0)$ is a stable node for (6.26). Therefore, the critical point $(-1, 2)$ of the given system is a stable node. □

Example 6.2.5. Discuss the nature of the critical point of $\begin{cases} x' = 2x + ay, \\ y' = x + 3y, \end{cases}$

for different values of a.

Solution: If $a = 6$, then there is no isolated critical point. Hence we assume that $a \neq 6$. Then for the given system $(0,0)$ is the critical point and

$$p = 5, \; q = 6 - a.$$

Moreover, the condition $q - \frac{1}{4}p^2 > 0$ is the same as $a < -\frac{1}{4}$. Using Lemma 6.2.2, Theorem 6.2.1 the following conclusions can be drawn.

(i) If $a > 6$, then the critical point is a saddle (which is indeed unstable).

(ii) If $a < 6$, then the critical point is unstable.

(iii) If $a < -\frac{1}{4}$, then the critical point is an unstable spiral.

(iv) If $-\frac{1}{4} \leq a < 6$, then the critical point is an unstable node.   □

Example 6.2.6. Discuss the nature of the critical point of $\begin{cases} x' = ax - 5y, \\ y' = 2x - y, \end{cases}$

for different values of a.

Solution: If $a = 10$, then there is no isolated critical point. Therefore we assume that $a \neq 10$ so that $(0,0)$ is a unique critical point of the given system. It is straightforward to get

$$p = a - 1, \; q = -a + 10,$$

and $q - \frac{1}{4}p^2 > 0$ if and only if

$$a^2 + 2a - 39 < 0, \text{ or } -1 - \sqrt{40} < a < -1 + \sqrt{40}.$$

Using Lemma 6.2.2, Theorem 6.2.1, the following conclusions can be drawn.

(i) If $a > 10$, then $q < 0$ and hence $(0,0)$ is a saddle point (which is unstable).

(ii) If $a < 10$, then the critical point is stable if and only if $p \leq 0$, i.e., $a \leq 1$.

(iii) If $a \leq -1 - \sqrt{40}$, then $(0,0)$ is a stable node.

(iv) If $-1 - \sqrt{40} < a < 1$, then the critical point is a stable spiral.

(v) If $a = 1$, then $(0,0)$ is stable and a center.

(vi) If $1 < a < -1 + \sqrt{40}$, then the critical point is an unstable spiral.

(vii) If $-1 + \sqrt{40} \leq a < 10$, then $(0,0)$ is an unstable node.   □

The discussion relating the eigenvalues of A and the long time behavior of the solutions to the corresponding system of linear ODEs is limited to the autonomous systems. We conclude this section by presenting an example of a non-autonomous system of linear equations for which the critical point is unstable, but the eigenvalues of the corresponding matrix are negative. In other words, the following example shows that Theorem 6.2.1 cannot be extended to the non-autonomous case.

Example 6.2.7. Consider the following system of equations with variable coefficients

$$\begin{cases} x'(t) = -2x(t) + a(t)y(t), \; t > 0, \\ y'(t) = -y(t), \; t > 0. \end{cases} \tag{6.27}$$

The matrix associated with the given system is

$$\begin{pmatrix} -2 & a(t) \\ 0 & -1 \end{pmatrix}$$

and its eigenvalues are -1 and -2 for all $t \geq 0$. We now find a function $a : [0, \infty) \to \mathbb{R}$ such that the critical point $(0,0)$ of (6.27) is unstable. Since the system under consideration is decoupled. from the second equation we find that

$$y(t) = y(0)e^{-t}, \ t \geq 0.$$

On substituting y in the first equation of (6.27), we get $x'(t) = -2x(t) + y(0)a(t)e^{-t}$, whose solution is

$$x(t) = x(0)e^{-2t} + y(0)e^{-2t} \int_0^t a(s)e^s ds, \ t \geq 0.$$

We now choose $a(t) = e^{3t}, \ t \geq 0$. With this choice of a, we show that $(0,0)$ is unstable. For, as usual we take $\epsilon = 1$ and let $\delta > 0$ be arbitrary. By taking $(x(0), y(0)) = (0, \frac{\delta}{2})$, we get

$$\|(x(t), y(t))\| > |x(t)| = \tfrac{\delta}{8}(e^{2t} - e^{-2t}) \to \infty \text{ as } t \to \infty.$$

Thus there exists $\tilde{t}$ such that $\|(x(\tilde{t}), y(\tilde{t}))\| \geq 1$. Hence the critical point $(0,0)$ is unstable. $\square$

6.3 Lyapunov energy function

In this section, we present a very useful tool to prove the stability of the critical point for an autonomous system of equations (both linear and non-linear). Moreover, we introduce a function called a Lyapunov energy function for the system whose existence is sufficient for a critical point to be stable.

Definition 6.3.1. Let $\Omega \subseteq \mathbb{R}^2$ be an open set containing the origin $(0,0)$. We assume that $h : \Omega \to \mathbb{R}$ is a C^1-function. The function h is said to be positive definite in Ω if $h(0,0) = 0$ and $h(x,y) > 0$, $(x,y) \in \Omega \backslash \{(0,0)\}$. The function h is said to be positive semi-definite if $h(0,0) = 0$ and $h(x,y) \geq 0$, $(x,y) \in \Omega \backslash \{(0,0)\}$. The function h is said to be negative (semi) definite if $-h$ is positive (semi) definite.

Example 6.3.2. The functions $f_1(x,y) = ax^2 + by^2$, $a, b > 0$ and $f_2(x,y) = e^{x^2+y^2} - 1$, are positive definite functions on $\mathbb{R}^2$. On the other hand, $f_3(x,y) = \sin^2 x$, and $f_4(x,y) = x^2(y-2)^2$ are positive semi-definite functions on $\mathbb{R}^2$. $\square$

Lemma 6.3.3. *The function $H : \mathbb{R}^2 \to \mathbb{R}$ defined by $H(x,y) = ax^2 + 2hxy + by^2$, is positive definite if and only if $a > 0$ and $ab > h^2$.*

Proof. We begin with the assumption that H is positive definite. Then $a = H(1,0) > 0$, and

$$H(x_0, 1) = ax_0^2 + 2hx_0 + b > 0, \quad x_0 \in \mathbb{R}.$$

Hence the discriminant of $H(x_0, 1)$ is negative, i.e., $4(h^2 - ab) < 0$. Conversely, let $a > 0$ and $ab > h^2$. Since $ab > h^2$, the quadratic equation $H(\mu, 1) = 0$ has no real roots. Furthermore, since $a > 0$, we have $H(\mu, 1) > 0$, $\mu \in \mathbb{R}$. Let $(x_0, y_0) \neq (0, 0)$. Suppose $y_0 \neq 0$, then it follows that $H(x_0, y_0) = y_0^2 H(\frac{x_0}{y_0}, 1) > 0$. On the other hand, if $y_0 = 0$ then $H(x_0, y_0) = ax_0^2 > 0$. ☐

Consider the nonlinear autonomous system of equations, for $t > 0$,

$$\begin{cases} x'(t) = g_1(x(t), y(t)), \\ y'(t) = g_2(x(t), y(t)). \end{cases} \tag{6.28}$$

We now define a Lyapunov energy function for (6.28).

Definition 6.3.4. Let $\Omega \subseteq \mathbb{R}^2$ be an open set containing $(0, 0)$. A C^1-function $E : \Omega \to \mathbb{R}$, $(u, v) \mapsto E(u, v)$, is said to be a Lyapunov energy function for system of ODEs (6.28), if E is positive definite and $g_1 \frac{\partial E}{\partial u} + g_2 \frac{\partial E}{\partial v}$ is negative semi-definite.

Example 6.3.5. For the system

$$\begin{cases} x'(t) = -y(t), \\ y'(t) = x(t), \end{cases} \tag{6.29}$$

the function $E : \mathbb{R}^2 \to \mathbb{R}$ defined by $E(u, v) = u^2 + v^2$, is a Lyapunov energy function. For, we first compare the given system with (6.28) to arrive at $g_1(u, v) = -v$, $g_2(u, v) = u$. Then notice that E is a positive definite function and $g_1 \frac{\partial E}{\partial u} + g_2 \frac{\partial E}{\partial v} \equiv 0$ which is a negative definite function. Hence E is a Lyapunov energy function for the given system ☐

Remark 6.3.6. A Lyapunov energy function for system (6.28) need not be unique. For, it is easy to verify that if E is a Lyapunov energy function for (6.28), then E^2 is also a Lyapunov energy function for the same system.

Remark 6.3.7. Suppose E is a Lyapunov energy function, and $(x(t), y(t))$ is a solution to (6.28). Then $\mathcal{E}(t) := E(x(t), y(t))$ is a non-increasing function of t. For, on differentiating $\mathcal{E}$ and using the definition of E, we get

$$\begin{aligned} \frac{d\mathcal{E}}{dt}(t) &= x'(t) \frac{\partial E}{\partial u}(x(t), y(t)) + y'(t) \frac{\partial E}{\partial v}(x(t), y(t)) \\ &= g_1(x(t), y(t)) \frac{\partial E}{\partial u}(x(t), y(t)) + g_2(x(t), y(t)) \frac{\partial E}{\partial v}(x(t), y(t)) \\ &\leq 0, \end{aligned}$$

for all $t > 0$. In other words, E along the trajectories, i.e., $\mathcal{E}$ is non-increasing (see Figures 6.9(a)–6.9(b)).

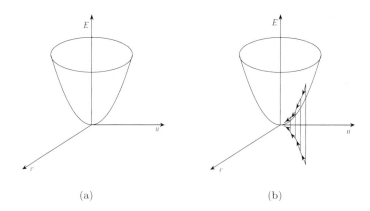

(a)   (b)

FIGURE 6.9: (a) A schematic representation of a surface given by a Lyapunov energy function; (b) The function E is decreasing along the trajectories of (6.28). Note that we have taken the trajectory in the uv-plane.

We now prove the main theorem of this section.

Theorem 6.3.8 (Lyapunov). *Assume that $(0,0)$ is a critical point of (6.28). If there exists a Lyapunov energy function E for (6.28), then $(0,0)$ is a stable critical point. Moreover, if $g_1 \frac{\partial E}{\partial u} + g_2 \frac{\partial E}{\partial v}$ is negative definite, then $(0,0)$ is asymptotically stable.*

Proof. The proof is divided into several steps.
Step 1. We first prove the stability of the critical point.
The main result which will be used in this step is Remark 6.3.7. Let $\epsilon > 0$ be given. Without loss of generality, we assume that

$$\{(u, v) : \|(u, v)\| \le \epsilon\} \subset \Omega.$$

Let $S_\epsilon = \{(u, v) : \|(u, v)\| = \epsilon\}$. Since S_ϵ is closed, bounded, and E is positive definite, there exists $m > 0$ such that

$$m = \min_{(u,v) \in S_\epsilon} E(u, v).$$

On the other hand, since $E(0,0) = 0$, and E is continuous, we can choose $\delta \in \mathbb{R}$ such that $0 < \delta < \epsilon$, and

$$E(u, v) < \tfrac{m}{2}, \text{ whenever } \|(u, v)\| < \delta.$$

Let $Y(t; Y_0)$ denote the solution to (6.28) with $(x(0), y(0)) = Y_0$ where $\|Y_0\| < \delta$. In view of Remark 6.3.7, we notice that

$$E\big(Y(t; Y_0)\big) \le E(Y_0) < \frac{m}{2}, \quad t \ge 0. \tag{6.30}$$

To prove the stability of $(0,0)$, we shall establish that

$$\|Y(t; Y_0)\| < \epsilon, \ t \geq 0. \tag{6.31}$$

On the contrary we suppose (6.31) fails. Then the curve $Y(t; Y_0)$ must cross the circle S_ϵ. Owing to the continuity of the function $t \mapsto \|Y(t; Y_0)\|$, there exists $t_0 > 0$ such that $Y(t_0; Y_0) \in S_\epsilon$. Thus from the definition of m, we obtain

$$E\big(Y(t_0; Y_0)\big) \geq m,$$

which is a contradiction to (6.30). Hence (6.31) holds true.

Step 2. We now prove asymptotic stability of $(0,0)$ when $g_1 \frac{\partial E}{\partial u} + g_2 \frac{\partial E}{\partial v}$ is negative definite.

Let us fix $\epsilon > 0$. For this ϵ, let δ and Y_0 be as in Step 1 such that (6.31) hold true. Let $\mathcal{E}(t) := E(Y(t; Y_0))$, $t \geq 0$. We first observe that since $Y(t; Y_0)$ is a bounded function of t, so is $\mathcal{E}$. Next, in view of Remark 6.3.7, $\mathcal{E}(t)$ decreases as t increases. Therefore there exists $L \geq 0$ such that $\mathcal{E}(t) \to L$ as $t \to \infty$.

Claim 1. We wish to prove that $L = 0$.

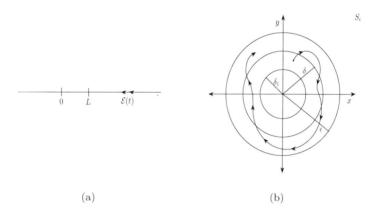

(a)   (b)

FIGURE 6.10: (a) The function E is decreasing to L; (b) Every trajectory of (6.28) is inside the annulus region $\mathcal{A}$.

On the contrary, we assume that $L > 0$. Since $E(0,0) = 0$ and E is continuous, we can choose $0 < \delta_1 < \delta$ such that $E(u, v) < L$, whenever $\|(u, v)\| < \delta_1$. Hence we have

$$Y(t; Y_0) \in \mathcal{A}, \ t \geq 0,$$

where $\mathcal{A} := \{(u, v) \in \mathbb{R}^2 : \delta_1 \leq \|(u, v)\| \leq \epsilon\}$ (see Figures 6.10(a)–6.10(b)). Since $\mathcal{A}$ is closed and bounded, there exists $k > 0$ such that

$$\max_{(u,v) \in \mathcal{A}} \left\{ g_1(u, v) \frac{\partial E}{\partial u}(u, v) + g_2(u, v) \frac{\partial E}{\partial v}(u, v) \right\} = -k.$$

We now notice that for $t \geq 0$,

$$
\begin{aligned}
\mathcal{E}(t) &= \mathcal{E}(0) + \int_0^t \frac{d}{d\tau}\mathcal{E}(\tau)d\tau \\
&= E(Y_0) + \int_0^t \Big\{ g_1(Y(\tau;Y_0))\frac{\partial E}{\partial u}(Y(\tau;Y_0)) \\
&\quad + g_2(Y(\tau;Y_0))\frac{\partial E}{\partial v}(Y(\tau;Y_0)) \Big\}d\tau \\
&\leq E(Y_0) - kt. \qquad\qquad (6.32)
\end{aligned}
$$

In particular, for $t = \frac{2E(Y_0)}{k}$, it follows that $\mathcal{E}(t) < 0$, which is a contradiction. We now analyze the reason for this contradiction. The contradiction is because $k > 0$, which is due to the existence of $\delta_1 > 0$. Finally, δ_1 is positive due to the assumption that $L > 0$. Therefore we have $L = 0$ and this proves Claim 1.

Claim 2. We prove that $Y(t;Y_0) \to (0,0)$ as $t \to \infty$.
Suppose not, since $Y(t;Y_0)$ is a bounded function of t, there exist a sequence (t_n) and a point $(\alpha,\beta) \neq (0,0)$, such that $Y(t_n;Y_0) \to (\alpha,\beta)$ as $t_n \to \infty$. Since E is continuous, we have

$$
E(Y(t_n;Y_0)) \to E(\alpha,\beta) \text{ as } t \to \infty.
$$

In view of Claim 1, we have $E(Y(t_n;Y_0)) \to 0$ and $E(\alpha,\beta) = 0$. This is a contradiction to the positive definiteness of E.
This completes the proof of the theorem. $\qquad\qquad\square$

We now discuss another type of functions analogous to the Lyapunov energy function whose existence assures us that the critical point is unstable. The details are given in the following result.

Proposition 6.3.9. *Let $(0,0)$ be a critical point of (6.28). Let $\Omega \subseteq \mathbb{R}^2$ be an open set containing $(0,0)$. Suppose there exists a C^1-function $E : \Omega \to \mathbb{R}$ such that E is a positive definite function and $g_1\frac{\partial E}{\partial u} + g_2\frac{\partial E}{\partial v}$ is positive definite, then $(0,0)$ is unstable.*

Proof. We use the technique presented in Step 2 of Theorem 6.3.8 to prove this result.
We suppose that $(0,0)$ is stable. Let $\epsilon > 0$ be such that

$$
\{(u,v) : \|(u,v)\| \leq \epsilon\} \subset \Omega.
$$

Then there exists $0 < \delta < \epsilon$, such that $\|Y(t;Y_0)\| < \epsilon$, $t \geq 0$ whenever $\|Y_0\| < \delta$. We now fix $Y_0 \neq 0$, with $\|Y_0\| < \delta$. Let

$$
\mathcal{E}(t) := E(Y(t;Y_0)), \ t \geq 0.
$$

By the same argument used in Remark 6.3.7, we obtain that $\mathcal{E}$ increases with t. In particular, we get $\mathcal{E}(t) > \mathcal{E}(0) = E(Y_0)$, $t \geq 0$. Since $E(0,0) = 0$ and E is continuous, there exists $0 < r < \delta$ such that $E(u,v) < E(Y_0)/2$ whenever

$\|(u, v)\| < r$. Hence it follows that

$$r < \|Y(t; Y_0)\| < \epsilon, \ t \geq 0.$$

Since the trajectory $Y(t; Y_0)$, $t \geq 0$, is bounded, so is $\mathcal{E}(t)$. On the other hand, for $t \geq 0$, we have

$$
\begin{aligned}
\mathcal{E}(t) &= \mathcal{E}(0) + \int_0^t \frac{d}{d\tau} \mathcal{E}(\tau) d\tau \\
&= E(Y_0) + \int_0^t \{ g_1(Y(\tau; Y_0)) \frac{\partial E}{\partial u}(Y(\tau; Y_0)) \\
&\quad + g_2(Y(\tau; Y_0)) \frac{\partial E}{\partial v}(Y(\tau; Y_0)) \} d\tau \\
&\geq E(Y_0) + kt,
\end{aligned}
\tag{6.33}
$$

where $k = \min\limits_{r \leq \|(u,v)\| \leq \epsilon} \{ g_1(u, v) \frac{\partial E}{\partial u}(u, v) + g_2 \frac{\partial E}{\partial v} \}$. In view of (6.33) we get that $\mathcal{E}$ is unbounded which is a contradiction. Hence $(0, 0)$ is an unstable critical point. □

Though the existence of a Lyapunov energy function guarantees the stability of the critical point, there is no definite algorithm to find it. In some cases, where (6.28) represents the equations of motion of particle(s), we can take the total energy of the system as E. This is the reason for referring E to be a Lyapunov energy function. In the next few examples, we introduce the method of undetermined coefficients-powers to construct the Lyapunov energy function explicitly. The same method can be employed to find E such that the hypotheses of Proposition 6.3.9 is satisfied.

Example 6.3.10. Investigate the nature of the critical points of

$$
\begin{cases}
x' = -4xy, \\
y' = x^2 - 5y^3.
\end{cases}
$$

Solution. We first observe that $(0, 0)$ is the only critical point of the given system. Let $g_1(u, v) = -4uv$ and $g_2(u, v) = u^2 - 5v^3$. In order to find a Lyapunov energy function, we begin with the ansatz

$$E(u, v) = au^{2m} + bv^{2n}, \ (u, v) \in \mathbb{R}^2,$$

where $m, n \in \mathbb{N}$ and $a, b > 0$. Clearly E is positive definite. We would like to find a, b, m, n such that $g_1 \frac{\partial E}{\partial u} + g_2 \frac{\partial E}{\partial v}$ is either negative (semi) definite or positive definite.
To this end, we begin with computing

$$g_1 \frac{\partial E}{\partial u} + g_2 \frac{\partial E}{\partial v} = -8mau^{2m}v + 2nbu^2 v^{2n-1} - 10nbv^{2n+2}, \ (u, v) \in \mathbb{R}^2.$$

Out of the three terms on the right hand side of the above expression, the first and the second terms change their signs due to the presence of the odd powers of u or v whereas the third one is non-positive. We equate the powers of u (and v) in the first and the second terms to get $m = n = 1$. With this choice of m and n, we obtain

$$g_1 \frac{\partial E}{\partial u} + g_2 \frac{\partial E}{\partial v} = -8au^2v + 2bu^2v - 10bv^4, \ (u,v) \in \mathbb{R}^2.$$

Then we choose $a = 1$ and $b = 4$, to get

$$g_1 \frac{\partial E}{\partial u} + g_2 \frac{\partial E}{\partial v} = -40v^4, \ (u,v) \in \mathbb{R}^2,$$

which is a negative semi-definite function. Therefore $E(u,v) = u^2 + 4v^2$, is a Lyapunov energy function for the given system of equations. In view of Theorem 6.3.8, we get that the critical point $(0,0)$ is stable. □

Using the Lyapunov energy function constructed in the previous example, we could not conclude that the critical point $(0,0)$ is asymptotically stable as $g_1 \frac{\partial E}{\partial u} + g_2 \frac{\partial E}{\partial v}$ is a positive semi-definite function. The Lyapunov energy function that we are going to construct in the next example gives that the critical point is asymptotically stable.

Example 6.3.11. Investigate the behavior of the critical points of the system

$$\begin{cases} x' = -y - x^3, \\ y' = -y^3 + x^5. \end{cases}$$

Solution. We first note that $(0,0)$ is the only critical point of the given system Let $g_1(u,v) = -v - u^3$ and $g_2(u,v) = -v^3 + u^5$. As in the previous example, we begin with the ansatz

$$E(u,v) = au^{2m} + bv^{2n}, \ (u,v) \in \mathbb{R}^2,$$

where $m, n \in \mathbb{N}$ and $a, b > 0$. Clearly E is positive definite. An easy computation yields, for $(u,v) \in \mathbb{R}^2$,

$$g_1 \frac{\partial E}{\partial u} + g_2 \frac{\partial E}{\partial v} = -2mau^{2m-1}v - 2mau^{2m+2} - 2nbv^{2n+2} + 2nbu^5v^{2n-1}.$$

We notice that the first and the last terms on the right hand side of the above expression change their signs due to presence of odd powers of u and v. Therefore we equate the powers of u (and v) in those terms to get $2m - 1 = 5$ and $2n - 1 = 1$. Hence we choose $m = 3$ and $n = 1$ to obtain

$$g_1 \frac{\partial E}{\partial u} + g_2 \frac{\partial E}{\partial v} = -6au^5v - 6au^8 - 2bv^4 + 2bu^5v, \ (u,v) \in \mathbb{R}^2.$$

Finally, on choosing $a = 1$ and $b = 3$, we get

$$g_1 \frac{\partial E}{\partial u} + g_2 \frac{\partial E}{\partial v} = -6u^8 - 6v^4, \ (u,v) \in \mathbb{R}^2,$$

which is a negative definite function. Therefore $E(u,v) = u^6 + 3v^2$, is a Lyapunov energy function for the given system of equations. In view of Theorem 6.3.8, we get that the critical point $(0,0)$ is asymptotically stable. □

Example 6.3.12. Investigate the behavior of the critical points of the system

$$\begin{cases} x' = -2xy + x^3, \\ y' = x^2 + y^3. \end{cases}$$

Solution. In this example also, $(0,0)$ is the only critical point of the given system. As usual, we begin with a positive definite function $E(u,v) = au^{2m} + bv^{2n}$, $(u,v) \in \mathbb{R}^2$, with $a,b > 0$, $m,n \in \mathbb{N}$. An easy computation gives

$$g_1 \frac{\partial E}{\partial u} + g_2 \frac{\partial E}{\partial v} = -4amu^{2m}v + 2amu^{2m+2} + 2bnu^2v^{2n-1} + 2bnv^{2n+2}.$$

As the first and the third terms on the right hand side of the above expression change their signs, we set $m = 1$ and $n = 1$ to get

$$g_1 \frac{\partial E}{\partial u} + g_2 \frac{\partial E}{\partial v} = -4au^2v + 2au^4 + 2bu^2v + 2bv^4, \quad (u,v) \in \mathbb{R}^2.$$

We now choose $a = 1$ and $b = 2$ to obtain

$$g_1 \frac{\partial E}{\partial u} + g_2 \frac{\partial E}{\partial v} = 2u^4 + 4v^4, \quad (u,v) \in \mathbb{R}^2,$$

which is a positive definite function. In view of Proposition 6.3.9, it follows that $(0,0)$ is unstable. □

Example 6.3.13. Investigate the behavior of the critical points of the system

$$\begin{cases} x' = 2y, \\ y' = -x - y - x^3. \end{cases}$$

Solution. Here also $(0,0)$ is the only critical point of the given system. If we begin with the ansatz $E(u,v) = au^{2m} + bv^{2n}$, $(u,v) \in \mathbb{R}^2$, with $a,b > 0$, $m,n \in \mathbb{N}$, then we get

$$g_1 \frac{\partial E}{\partial u} + g_2 \frac{\partial E}{\partial v} = 4amu^{2m-1}v - 2bnuv^{2n-1} - 2bnv^{2n} - 2bnu^3v^{2n-1}.$$

On the right hand side of the above expression there are three terms which change their signs, namely the first, the second and the fourth. For no value of m, the powers of u in these three terms are the same. Hence $E = au^{2m} + bv^{2n}$ cannot be a Lyapunov energy function for the given system.

Thus we add another term to the above ansatz to get a new trail function

$$E_1(u,v) = au^{2l} + bu^{2m} + cv^{2n}, \quad (u,v) \in \mathbb{R}^2, \ a,b,c > 0, \ l,m,n \in \mathbb{N}.$$

Clearly E_1 is a positive definite function. Again we compute

$$g_1 \frac{\partial E_1}{\partial u} + g_2 \frac{\partial E_1}{\partial v} = 4alu^{2l-1}v + 4bmu^{2m-1}v - 2cnuv^{2n-1}$$
$$-2cnv^{2n} - 2cnu^3v^{2n-1}.$$

We now choose $l = n = 1$, and $m = 2$ to get

$$g_1 \frac{\partial E_1}{\partial u} + g_2 \frac{\partial E_1}{\partial v} = 4auv + 8bu^3v - 2cuv - 2cv^2 - 2cu^3v.$$

We take $a = 2, b = 1$, and $c = 4$ so that

$$g_1 \frac{\partial E_1}{\partial u} + g_2 \frac{\partial E_1}{\partial v} = -8v^2,$$

which is a negative semi-definite function. Therefore a Lyapunov energy function for the given system is $E_1(u,v) = 2u^2 + u^4 + 4v^2$ and the critical point $(0,0)$ is stable. $\hfill\square$

6.4 Perturbed linear systems

In this section, we use the Lyapunov energy function method developed in the previous section to study the nature of critical points for special type of nonlinear systems. In particular, we discuss the situation where the linear system is perturbed by adding a nonlinear term. We show that if the nonlinear term is 'small' in magnitude, then the nature of the critical point of the fully nonlinear system is the same as that of the corresponding linear system in many cases.

Let $A = \begin{pmatrix} a_{11} & a_{12} \\ a_{21} & a_{22} \end{pmatrix} \in M_2(\mathbb{R})$. Assume that $f, g : \mathbb{R}^2 \to \mathbb{R}$ are nonlinear functions which are continuous and $f(0,0) = g(0,0) = 0$. Consider the nonlinear system of equations

$$\begin{pmatrix} x'(t) \\ y'(t) \end{pmatrix} = A \begin{pmatrix} x(t) \\ y(t) \end{pmatrix} + \begin{pmatrix} f(x(t), y(t)) \\ g(x(t), y(t)) \end{pmatrix}, \quad t > 0, \tag{6.34}$$

and the associated linear system

$$\begin{pmatrix} x'(t) \\ y'(t) \end{pmatrix} = A \begin{pmatrix} x(t) \\ y(t) \end{pmatrix}, \quad t > 0. \tag{6.35}$$

We now quantify what we mean by f and g being 'small' in the following definition.

Definition 6.4.1. The critical point $(0,0)$ of (6.34) is said to be a simple critical point if

$$\lim_{(u,v)\to(0,0)} \frac{|f(u,v)|}{\sqrt{u^2+v^2}} = \lim_{(u,v)\to(0,0)} \frac{|g(u,v)|}{\sqrt{u^2+v^2}} = 0. \tag{6.36}$$

Remark 6.4.2. The main technique which is used to prove that $(0,0)$ is a simple critical point of (6.34) is to convert the Cartesian coordinates (u,v) to the polar coordinates (r, θ) by setting $u = r\cos\theta$, $v = r\sin\theta$, or $r =$

$\sqrt{u^2 + v^2}$, $\theta = \tan^{-1}(\frac{v}{u})$. If $\tilde{f}(r) := \max\limits_{0 \leq \theta \leq 2\pi} |f(r \cos \theta, r \sin \theta)|$, and $\tilde{g}(r) :=$ $\max\limits_{0 \leq \theta \leq 2\pi} |g(r \cos \theta, r \sin \theta)|$, $r \ll 1$, then the condition

$$\lim_{r \to 0^+} \frac{\tilde{f}(r)}{r} = \lim_{r \to 0^+} \frac{\tilde{g}(r)}{r} = 0,$$

implies (6.36). The advantage in this procedure is that instead of working with the limits in 2-D, it is enough to work with the limits in 1-D which are relatively simpler to deal with.

Example 6.4.3. Show that $(0,0)$ is a simple critical point of (6.34) if

(*i*) $f(u, v) = u^2 + v^2$, $g(u, v) = v \sin u$,

(*ii*) $f(u, v) = u - \sin u$, $g(u, v) = 1 - \cos(u + v)$,

(*iii*) $f(u, v) = v - v e^{uv}$, $g(u, v) = u^{4/3} - 4v^3$.

Solution. We use Remark 6.4.2 to prove that $(0,0)$ is a simple critical point. (*i*) To that end, we compute

$$\tilde{f}(r) = \max_{0 \leq \theta \leq 2\pi} |f(r \cos \theta, r \sin \theta)| = \max_{0 \leq \theta \leq 2\pi} |r^2 \cos^2 \theta + r^2 \sin^2 \theta| = r^2.$$

Therefore we obtain

$$\lim_{r \to 0} \frac{\tilde{f}(r)}{r} = \lim_{r \to 0} r = 0.$$

On the other hand, since $|\sin \gamma| \leq |\gamma|$, $\gamma \in \mathbb{R}$, it follows that

$$|g(r \cos \theta, r \sin \theta)| \leq r |\sin(r \cos \theta)| \leq r^2 |\cos \theta| \leq r^2.$$

Hence we have $\tilde{g}(r) \leq r^2$, $r > 0$. Thus we obtain

$$0 \leq \lim_{r \to 0} \frac{\tilde{g}(r)}{r} \leq \lim_{r \to 0} \frac{r^2}{r} = 0.$$

In view of Remark 6.4.2, we conclude that $(0,0)$ is a simple critical point.

(*ii*) We consider

$$\lim_{(u,v) \to (0,0)} \frac{|u - \sin u|}{\sqrt{u^2 + v^2}} \leq \lim_{(u,v) \to (0,0)} \frac{|u - \sin u|}{|u|} = \lim_{u \to 0} \left| 1 - \frac{\sin u}{u} \right| = 0.$$

On the other hand, we estimate

$$|g(u, v)| = 1 - \cos(u + v) = 2 \sin^2(\tfrac{u+v}{2}) \leq \tfrac{1}{2} |u + v|^2.$$

Hence we have

$$\tilde{g}(r) \leq \frac{r^2}{2} \max_{0 \leq \theta \leq 2\pi} |\cos \theta + \sin \theta|^2 = r^2 \max_{0 \leq \theta \leq 2\pi} \left| \sin(\theta + \frac{\pi}{4}) \right|^2 = r^2.$$

As in (*i*), one can show that $\lim\limits_{r \to \infty} \frac{\tilde{g}(r)}{r} = 0$, which readily implies that $(0,0)$ is a simple critical point.

(*iii*) Using the methods from elementary calculus, we readily obtain that

$$
\begin{aligned}
\tilde{f}(r) &= \max_{0 \leq \theta \leq 2\pi} \left| (1 - e^{\frac{r^2}{2}\sin(2\theta)})r\sin\theta \right| \\
&\leq r \max_{0 \leq \theta \leq 2\pi} \left| e^{\frac{r^2}{2}\sin(2\theta)} - 1 \right| \\
&\leq r(e^{\frac{r^2}{2}} - 1)
\end{aligned}
$$

and thus

$$
0 \leq \lim_{r \to 0} \frac{|\tilde{f}(r)|}{r} \leq \lim_{r \to 0} e^{\frac{r^2}{2}} - 1 = 0.
$$

We turn our attention toward the other function g. Using the triangle inequality we obtain

$$
\tilde{g}(r) = \max_{\theta} \left| r^{\frac{4}{3}}\cos^{\frac{4}{3}}\theta - 4r^3\sin^3\theta \right| \leq r^{\frac{4}{3}} + 4r^3.
$$

Therefore we estimate

$$
0 \leq \lim_{r \to 0} \frac{\tilde{g}(r)}{r} \leq \lim_{r \to 0} \frac{r^{\frac{4}{3}} + 4r^3}{r} = \lim_{r \to 0}(r^{\frac{1}{3}} + 4r^2) = 0.
$$

Hence $(0,0)$ is a simple critical point. $\qquad\square$

In the next theorem, we show that if $(0,0)$ is a simple critical point of (6.34), then except in the case when A has purely imaginary eigenvalues, it is a stable critical point of system (6.34) whenever it is so for the associated linear system (6.35). Before we state the next theorem, we recall from Section 6.2 that $(0,0)$ is the stable critical point of linear system (6.35) if $Tr(A) < 0$ and $\det(A) > 0$.

Theorem 6.4.4. *Assume that* $(0,0)$ *is a simple critical point of* (6.34). *Assume that* $p := a_{11} + a_{22} < 0$ *and* $q := a_{11}a_{22} - a_{12}a_{21} > 0$. *Then* $(0,0)$ *is an asymptotically stable critical point of* (6.34).

Proof. In order to show that $(0,0)$ is asymptotically stable, we construct a Lyapunov energy function for (6.34) explicitly. For, we begin with the ansatz

$$
E(u,v) = \frac{1}{2}(C_1 u^2 + 2C_2 uv + C_3 v^2), \quad (u,v) \in \mathbb{R}^2. \tag{6.37}
$$

Our objective is to find C_1, C_2, and C_3 such that E satisfies the hypotheses of Theorem 6.3.8.

Step 1. In this step, we find C_1, C_2, and C_3 such that

$$
(a_{11}u + a_{12}v)\frac{\partial E}{\partial u} + (a_{21}u + a_{22}v)\frac{\partial E}{\partial v} = -(u^2 + v^2). \tag{6.38}
$$

A straightforward calculation yields

$$
\begin{cases}
(a_{11}u + a_{12}v)\frac{\partial E}{\partial u} + (a_{21}u + a_{22}v)\frac{\partial E}{\partial v} = (a_{11}C_1 + a_{21}C_2)u^2 \\
+(a_{12}C_1 + (a_{11} + a_{22})C_2 + a_{21}C_3)uv + (a_{12}C_2 + a_{22}C_3)v^2.
\end{cases} \tag{6.39}
$$

On comparing the coefficients of u^2, uv, and v^2 in (6.38)–(6.39), we notice that (C_1, C_2, C_3) is a solution of

$$a_{11}C_1 + a_{21}C_2 = -1,$$
$$a_{12}C_1 + (a_{11} + a_{22})C_2 + a_{21}C_3 = 0,$$
$$a_{12}C_2 + a_{22}C_3 = -1.$$

Therefore we have

$$C_1 = \frac{a_{21}^2 + a_{22}^2 + q}{-pq}, \quad C_2 = \frac{a_{11}a_{21} + a_{12}a_{22}}{pq}, \quad C_3 = \frac{a_{11}^2 + a_{12}^2 + q}{-pq}, \qquad (6.40)$$

where $p = a_{11} + a_{22}$ and $q = a_{11}a_{22} - a_{12}a_{21}$.

Step 2. In this step, we show that E given in (6.37) together with (6.40) is a positive definite function.

In view of Lemma 6.3.3, it is enough to show that $C_1 > 0$ and $C_2^2 < C_1 C_3$. We first notice that $C_1 > 0$ because $p < 0$ and $q > 0$. We then compute

$$
\begin{aligned}
C_2^2 - C_1 C_3 &= \frac{1}{p^2 q^2}\Big((a_{11}a_{21} + a_{12}a_{22})^2 - (a_{21}^2 + a_{22}^2 + q)(a_{11}^2 + a_{12}^2 + q)\Big) \\
&= \frac{1}{p^2 q^2}\Big(2a_{11}a_{12}a_{21}a_{22} - a_{11}^2 a_{22}^2 - a_{21}^2 a_{12}^2 \\
&\qquad -q(a_{11}^2 + a_{12}^2 + a_{21}^2 + a_{22}^2) - q^2\Big) \\
&= \frac{-1}{p^2 q^2}\Big(2q^2 + q(a_{11}^2 + a_{12}^2 + a_{21}^2 + a_{22}^2)\Big) \\
&< 0.
\end{aligned}
$$

Therefore E is positive definite. Hence E is a Lyapunov energy function for linear system (6.35).

Step 3. In this step, we show that E is a Lyapunov energy function for (6.34). Firstly, we let

$$g_1(u, v) = a_{11}u + a_{12}v + f(u, v), \quad g_1(u, v) = a_{21}u + a_{22}v + g(u, v), \quad (u, v) \in \mathbb{R}^2.$$

Using the result proved in Step 1, we find that

$$
\begin{aligned}
g_1 \frac{\partial E}{\partial u} + g_2 \frac{\partial E}{\partial v} &= -u^2 - v^2 + (C_1 u + C_2 v)f(u, v) + (C_2 u + C_3 v)g(u, v) \\
&\leq -(u^2 + v^2) + C(|u| + |v|)(|f(u, v)| + |g(u, v)|), \quad (6.41)
\end{aligned}
$$

where $C = \max\{C_1, |C_2|, C_3\}$.

Since $(0, 0)$ is a simple critical point, there exists $\delta > 0$ such that

$$\frac{|f(u, v)|}{\sqrt{u^2 + v^2}} < \frac{1}{8C}, \quad \frac{|g(u, v)|}{\sqrt{u^2 + v^2}} < \frac{1}{8C}, \qquad (6.42)$$

whenever $\|(u,v)\| < \delta$. From (6.41)–(6.42), it follows that

$$g_1\frac{\partial E}{\partial u}(u,v) + g_2\frac{\partial E}{\partial v}(u,v) \leq -(u^2+v^2) + \frac{C(|u|+|v|)}{4C}\sqrt{u^2+v^2}$$
$$\leq -(u^2+v^2) + \frac{2\sqrt{u^2+v^2}}{4}\sqrt{u^2+v^2}$$
$$= -\frac{u^2+v^2}{2},$$

whenever $\|(u,v)\| < \delta$. Hence E satisfies hypotheses of Theorem 6.3.8 with $\Omega = \{(u,v) : \|(u,v)\| < \delta\}$. Therefore $(0,0)$ is an asymptotically stable critical point of (6.34). □

We now turn our attention toward the situation when $p = a_{11} + a_{22} > 0$ and $q = a_{11}a_{22} - a_{21}a_{12} > 0$. Recall from Section 6.2 that $(0,0)$ an unstable node for (6.35) provided $p > 0$, and $q > 0$. In this case also, we show that if $(0,0)$ is a simple critical point of (6.34), then the nature of $(0,0)$ is the same for both (6.34) and (6.35). This is given in the following theorem.

Theorem 6.4.5. *Assume that $(0,0)$ is a simple critical point of (6.34). Assume that $p := a_{11} + a_{22} > 0$ and $q := a_{11}a_{22} - a_{12}a_{21} > 0$. Then $(0,0)$ is an unstable critical point of (6.34).*

Proof. In order to prove this result, we show the existence of a function $\tilde{E}$ which satisfies the hypotheses of Proposition 6.3.9. In order to construct such a function, we follow the procedure described in Steps 1-2 of Theorem 6.4.4. We first begin with

$$\tilde{E}(u,v) = \frac{1}{2}(\tilde{C}_1 u^2 + 2\tilde{C}_2 uv + \tilde{C}_3 v^2), \ (u,v) \in \mathbb{R}^2.$$

We wish to find $\tilde{C}_1, \tilde{C}_2$, and $\tilde{C}_3$ such that

$$(a_{11}u + a_{12}v)\frac{\partial \tilde{E}}{\partial u} + (a_{21}u + a_{22}v)\frac{\partial \tilde{E}}{\partial v} = (u^2+v^2).$$

A straightforward calculation yields $(\tilde{C}_1, \tilde{C}_2, \tilde{C}_3) = (-C_1, -C_2, -C_3)$ where C_1, C_2, and C_3 are given in (6.40). Since $p, q > 0$ it follows that $\tilde{C}_1 > 0$. Moreover, from Step 2 of Theorem 6.4.4, we obtain

$$\tilde{C}_2^2 - \tilde{C}_1\tilde{C}_3 = C_2^2 - C_1 C_3 < 0.$$

Hence $\tilde{E}$ is a positive definite function.
Let g_1, g_2 be as in Step 3 of Theorem 6.4.4 and $\tilde{C} = \max\{\tilde{C}_1, |\tilde{C}_2|, \tilde{C}_3\}$. Then we have

$$g_1\frac{\partial E}{\partial u} + g_2\frac{\partial E}{\partial v} \geq (u^2+v^2) - \tilde{C}(|u|+|v|)(|f(u,v)| + |g(u,v)|).$$

If $\delta > 0$ be such that

$$\frac{|f(u,v)|}{\sqrt{u^2 + v^2}} < \frac{1}{8\tilde{C}}, \quad \frac{|g(u,v)|}{\sqrt{u^2 + v^2}} < \frac{1}{8\tilde{C}},$$

whenever $\|(u,v)\| < \delta$. Then we obtain

$$g_1 \frac{\partial E}{\partial u}(u,v) + g_2 \frac{\partial E}{\partial v}(u,v) \geq (u^2 + v^2) - (|u| + |v|)\frac{\sqrt{u^2 + v^2}}{4} \geq \frac{u^2 + v^2}{2}.$$

Hence $\tilde{E}$ satisfies the hypotheses of Proposition 6.3.9 which readily implies that $(0,0)$ is an unstable critical point of (6.34). $\qquad\square$

Another interesting case which is not covered in Theorems 6.4.4–6.4.5 is $p = 0$. Our proofs of Theorems 6.4.4–6.4.5 cannot be extended to this case for the obvious reason that p is in the denominators of C_is and $\tilde{C}_i$s. In fact, in the following example we present two systems having:

(i) $(0,0)$ as a simple critical point,
(ii) the same associated linear system which satisfies $a_{11} + a_{22} = 0$,
(iii) a different behavior of the critical point.

Example 6.4.6. Discuss the nature of the critical point $(0,0)$ of the systems

$$(i) \begin{cases} x' = y - x^3, \\ y' = -3x - 5y^5, \end{cases} \qquad (ii) \begin{cases} x' = y + x^3, \\ y' = -3x + 5y^5. \end{cases}$$

Solution. We begin with estimating

$$0 \leq \lim_{(u,v)\to(0,0)} \frac{|u^3|}{\sqrt{u^2 + v^2}} \leq \lim_{(u,v)\to(0,0)} \frac{|u^3|}{\sqrt{u^2}} = \lim_{u\to 0} u^2 = 0.$$

Again, using the same argument we obtain that

$$0 \leq \lim_{(u,v)\to(0,0)} \frac{5|v^5|}{\sqrt{u^2 + v^2}} = \lim_{(u,v)\to(0,0)} \frac{5|v^5|}{\sqrt{v^2}} = \lim_{v\to 0} 5v^4 = 0.$$

Therefore $(0,0)$ is a simple critical point for the given systems.

Claim 1. The critical point $(0,0)$ is asymptotically stable for system (i). Using the method introduced in Section 6.3, we construct a Lyapunov energy function to show the asymptotic stability of $(0,0)$. For, we begin with the ansatz

$$E(u,v) = au^{2m} + bv^{2n}, \quad (u,v) \in \mathbb{R}^2, \ a,b > 0, \ m,n \in \mathbb{N}.$$

Let $g_1(u,v) = v - u^3$ and $g_2(u,v) = -3u - 5v^5$. An easy computation gives

$$g_1 \frac{\partial E}{\partial u} + g_2 \frac{\partial E}{\partial v} = 2amu^{2m-1}v - 2amu^{2m+2} - 6bnuv^{2n-1} - 10bnv^{2n+4}.$$

We now choose $a = 3, b = 1$, and $m = n = 1$, so that

$$g_1(u,v)\frac{\partial E}{\partial u}(u,v) + g_2(u,v)\frac{\partial E}{\partial v}(u,v) = -6u^4 - 10v^6, \quad (u,v) \in \mathbb{R}^2,$$

which is a negative definite function. Hence $E(u,v) = 3u^2 + v^2$, is a Lyapunov energy function for the system given in (i) and $(0,0)$ is an asymptotically stable critical point.

Claim 2. The critical point $(0,0)$ is unstable for system (ii).
We construct a function satisfying the hypotheses of Proposition 6.3.9 to prove this claim. In fact we use the same strategy and calculations used in the proof of Claim 1. For, we consider the positive definite function

$$E(u,v) = 3u^2 + v^2, \quad (u,v) \in \mathbb{R}^2.$$

On setting $g_1(u,v) = v + u^3$ and $g_2(u,v) = -3u + 5v^5$, we get

$$g_1 \frac{\partial E}{\partial u}(u,v) + g_2 \frac{\partial E}{\partial v}(u,v) = 6u^4 + 10v^6, \quad (u,v) \in \mathbb{R}^2,$$

which is a positive definite function. In view of Proposition 6.3.9, we conclude that $(0,0)$ is an unstable critical point of the system given in (ii). □

Example 6.4.7. Discuss the nature of the critical point $(0,0)$ of the system

$$\begin{cases} x' = y - x^3 y^2, \\ y' = -4x - 9x^2 y^3. \end{cases}$$

Solution. We first notice that $(0,0)$ is a simple critical point and the associated linear system for the given system is

$$\begin{pmatrix} x' \\ y' \end{pmatrix} = \begin{pmatrix} 0 & 1 \\ -4 & 0 \end{pmatrix} \begin{pmatrix} x \\ y \end{pmatrix}.$$

Since the trace of the matrix is zero, Theorems 6.4.4–6.4.5 are not applicable. Hence we use the direct method which is discussed in Section 6.3.
Claim. The critical point $(0,0)$ of the given system is asymptotically stable.
We prove this claim by constructing a Lyapunov energy function. As usual, we begin with the ansatz

$$E(u,v) = au^{2m} + bv^{2n}, \quad (u,v) \in \mathbb{R}^2, \ a,b > 0, \ m,n \in \mathbb{N}.$$

and compute

$$g_1 \frac{\partial E}{\partial u}(u,v) + g_2 \frac{\partial E}{\partial v}(u,v) = 2amu^{2m-1}v - 2amu^{2m+2}v^2$$
$$-8bnuv^{2n-1} - 18bnu^2 v^{2n+2}.$$

We now choose $a = 4, b = 1$, and $m = n = 1$ so that

$$g_1 \frac{\partial E}{\partial u}(u,v) + g_2 \frac{\partial E}{\partial v}(u,v) = -8u^4 v^2 - 18u^2 v^4, \quad (u,v) \in \mathbb{R}^2,$$

which is a positive semi-definite function. Hence $(0,0)$ is a stable critical point.
□

In the case when $a_{11}a_{22} - a_{12}a_{21} < 0$ also, the behavior of the simple critical point $(0,0)$ for (6.34) is the same as that of the associated linear system (6.35). This is given in the next result, whose proof is out of the scope of this book.

Theorem 6.4.8. *If $a_{11}a_{22} - a_{12}a_{21} < 0$, and $(0,0)$ is a simple critical point of nonlinear system (6.34), then $(0,0)$ is an unstable critical point of that system.*

Proof. A proof of this result can be found in [11]. □

We now provide some examples which use Theorems 6.4.4–6.4.5 and 6.4.8.

Example 6.4.9. Discuss the nature of the critical point $(0,0)$ of the system

$$\begin{cases} x' = -2x - 3y + 7y\sin(2x), \\ y' = 2x - 5y. \end{cases}$$

Solution. Let $f(u,v) = 7v\sin(2u)$, and $g(u,v) \equiv 0$. Then it follows that

$$0 \le \lim_{(u,v) \to (0,0)} \frac{7|v\sin(2u)|}{\sqrt{u^2 + v^2}} \le \lim_{(u,v) \to (0,0)} \frac{14|u||v|}{\sqrt{u^2 + v^2}} \le \lim_{(u,v) \to (0,0)} 7\sqrt{u^2 + v^2} = 0.$$

Since there is no nonlinear term in the second equation of the given system, we conclude that $(0,0)$ is a simple critical point.

The associated linear system for the given system is

$$\begin{pmatrix} x' \\ y' \end{pmatrix} = \begin{pmatrix} -2 & -3 \\ 2 & -5 \end{pmatrix} \begin{pmatrix} x \\ y \end{pmatrix}.$$

For this linear system, we find that $p = -7$ and $q = 16$. Hence $(0,0)$ is a stable critical point for the linear system. In view of Theorem 6.4.4, we conclude that $(0,0)$ is a stable critical point of the given system. □

Example 6.4.10. Discuss the nature of the critical point $(0,0)$ of the system

$$\begin{cases} x' = x - 7x^{3/5}y^2, \\ y' = 2y - x^2 y^{2/3}. \end{cases}$$

Solution. Let $f(u,v) = -7u^{3/5}v^2$ and $g(u,v) = -u^2 v^{2/3}$. By converting f and g into the polar coordinates, we obtain

$$\tilde{f}(r) = \max_{\theta} |f(r\cos\theta, r\sin\theta)| \le 7r^{3/5}r^2,$$

and

$$\tilde{g}(r) = \max_{\theta} |g(r\cos\theta, r\sin\theta)| \le r^{2/3}r^2.$$

Now it is straightforward to show that

$$0 \le \lim_{r \to 0} \frac{\tilde{f}(r)}{r} \le \lim_{r \to 0} 7r^{8/5} = 0,$$

and

$$0 \le \lim_{r \to 0} \frac{\tilde{g}(r)}{r} \le \lim_{r \to 0} r^{5/3} = 0.$$

Hence $(0,0)$ is a simple critical point of the given system. The associated linear system for the given system is

$$\begin{pmatrix} x' \\ y' \end{pmatrix} = \begin{pmatrix} 1 & 0 \\ 0 & 2 \end{pmatrix} \begin{pmatrix} x \\ y \end{pmatrix}.$$

Here $p = 3$, $q = 2$ and hence $(0,0)$ is an unstable critical point of the linear system. In view of Theorem 6.4.5, we conclude that $(0,0)$ is an unstable critical point of the given system. □

Example 6.4.11. Discuss the nature of the critical point $(0,0)$ of the system

$$\begin{cases} x' = 3x - 2y - 4\sin y, \\ y' = x + y - 5(y - x)\cos(x + 2y). \end{cases}$$

Solution. We should not be tempted to write the corresponding linear system as

$$\begin{pmatrix} x' \\ y' \end{pmatrix} = \begin{pmatrix} 3 & -2 \\ 1 & 1 \end{pmatrix} \begin{pmatrix} x \\ y \end{pmatrix},$$

and conclude that $(0,0)$ is unstable (since $p = 4$, $q = 5$). It is because that the nonlinear terms $f(u,v) = -4\sin v$, $g(u,v) = -5(v - u)\cos(u + 2v)$, do not satisfy condition (6.36).

Therefore our objective is to write the given system as the sum of 'linear part' and 'nonlinear part' such that the nonlinear part satisfies (6.36) so that Theorems 6.4.4–6.4.5 and 6.4.8 are applicable. It is enough to write the nonlinear terms f and g as the sum of linear and nonlinear terms. In order to do that we begin with the following heuristic argument using the Taylor series. We write

$$f(u,v) = -4\sin v = -4v + \frac{4v^3}{3!} - \frac{4v^5}{5!} + \cdots = -4v + f_1(v),$$

and

$$\begin{aligned} g(u,v) &= -5(v - u)\cos(u + 2v) \\ &= -5(v - u)\left(1 - \frac{(u + 2v)^2}{2!} + \cdots\right) \\ &= -5(v - u) + g_2(u,v). \end{aligned}$$

Hence f and g are written as
$$-4\sin v = -4v + (4v - \sin v),$$

and
$$-5(v - u)\cos(u + 2v) = -5(v - u) + 5(v - u)\big(1 - \cos(u + 2v)\big).$$

Therefore we write the given system as

$$\begin{cases} x' = 3x - 6y + (4y - \sin y), \\ y' = 6x - 4y + 5(y - x)\big(1 - \cos(x + 2y)\big). \end{cases} \tag{6.43}$$

Let $f_1(v) = 4v - 4\sin v$, and $g_1(u,v) = 5(v - u)\big(1 - \cos(u + 2v)\big)$. To prove $(0,0)$ is a simple critical point of (6.43) we estimate

$$0 \le \lim_{(u,v)\to(0,0)} \frac{4|v - \sin v|}{\sqrt{u^2 + v^2}} \le \lim_{(u,v)\to(0,0)} \frac{4|v - \sin v|}{|v|} = 4\lim_{v\to 0}\left|1 - \frac{\sin v}{v}\right| = 0.$$

Similarly, we estimate

$$
\begin{aligned}
|g_1(u,v)| &\leq 5(|u|+|v|)(1-\cos(u+2v)) \\
&\leq 10(|u|+|v|)\sin^2(\frac{u+2v}{2}) \\
&\leq \frac{5}{2}(|u|+|v|)\,|u+2v|^2 \\
&\leq \frac{5}{2}(|u|+2|v|)^3.
\end{aligned}
$$

By converting to the polar coordinates, we get

$$
\tilde{g}_1(r) = \max_\theta |g_1(r\cos\theta, r\sin\theta)| \leq \frac{5}{2}(r+2r)^3 = \frac{135}{2}r^3.
$$

Therefore we have

$$
0 \leq \lim_{r\to 0} \frac{\tilde{g}_1(r)}{r} \leq \lim_{r\to 0} \frac{135}{2}r^2 = 0.
$$

Hence $(0,0)$ is a simple critical point of (6.43). The associated linear system for (6.43) is

$$
\begin{pmatrix} x' \\ y' \end{pmatrix} = \begin{pmatrix} 3 & -6 \\ 6 & -4 \end{pmatrix} \begin{pmatrix} x \\ y \end{pmatrix},
$$

for which $(0,0)$ is a stable critical point ($p = -1$, $q = 24$). In view of Theorem 6.4.4, $(0,0)$ is a stable critical point of (6.43). Thus $(0,0)$ is a stable critical point of the given system.   □

Example 6.4.12. Discuss the nature of the critical points of the system

$$
\begin{cases} x' = -x + xy, \\ y' = -4y + x^2 - 5y^2. \end{cases}
$$

Solution. To find the critical points (u,v) of the given system of ODEs, we need to solve the system of nonlinear equations

$$
\begin{cases} \hat{x}(-1+\hat{y}) = 0, \\ -4\hat{y} + \hat{x}^2 - 5\hat{y}^2 = 0. \end{cases}
$$

It is straightforward to solve this system to find that the critical points are precisely $(0,0), (0, -\frac{4}{5}), (3,1), (-3,1)$. We investigate the behavior of each critical point separately. We begin with the critical point $(0,0)$. It is easy to show that $(0,0)$ is indeed a simple critical point. Moreover, the associated linear system is

$$
\begin{pmatrix} x' \\ y' \end{pmatrix} = \begin{pmatrix} -1 & 0 \\ 0 & -4 \end{pmatrix} \begin{pmatrix} x \\ y \end{pmatrix}.
$$

Since $p = -5$ and $q = 4$, we conclude that $(0,0)$ is a stable critical point of the linear system. Thus $(0,0)$ is a stable critical point of the given system of ODE.

To study the nature of the critical point $(3, 1)$, we set $\tilde{x} = x - 3$ and $\tilde{y} = y - 1$. On substituting $x = \tilde{x} + 3$ and $y = \tilde{y} + 1$ in the given system. we get

$$\begin{cases} \tilde{x}' = 3\tilde{y} + \tilde{x}\tilde{y}, \\ \tilde{y}' = 6\tilde{x} - 14\tilde{y} + \tilde{x}^2 - 5\tilde{y}^2. \end{cases} \tag{6.44}$$

Therefore (0.0) is a simple critical point of (6.44) (which is equivalent to saying that $(3, 1)$ is a simple critical point of the given system). The associated linear system for (6.44) is

$$\begin{pmatrix} \tilde{x}' \\ \tilde{y}' \end{pmatrix} = \begin{pmatrix} 0 & 3 \\ 6 & -14 \end{pmatrix} \begin{pmatrix} \tilde{x} \\ \tilde{y} \end{pmatrix}.$$

Since $p = -14$ and $q = -18$, we conclude that (0.0) is an unstable critical point of the (6.44). Thus $(3, 1)$ is an unstable critical point of the given system of ODE.

To study the nature of the critical point $(-3, 1)$, we introduce new variables $\tilde{x} = x + 3$ and $\tilde{y} = y - 1$. On substituting $x = \tilde{x} - 3$ and $y = \tilde{y} + 1$ in the given system, we obtain

$$\begin{cases} \tilde{x}' = -3\tilde{y} + \tilde{x}\tilde{y}, \\ \tilde{y}' = -6\tilde{x} - 14\tilde{y} + \tilde{x}^2 - 5\tilde{y}^2. \end{cases} \tag{6.45}$$

Hence $(0, 0)$ is a simple critical point of (6.45) and the associated linear system for (6.45) is

$$\begin{pmatrix} \tilde{x}' \\ \tilde{y}' \end{pmatrix} = \begin{pmatrix} 0 & -3 \\ -6 & -14 \end{pmatrix} \begin{pmatrix} \tilde{x} \\ \tilde{y} \end{pmatrix}.$$

Since $p = -14$ and $q = -18$, we conclude that (0.0) is an unstable critical point of the (6.45). Hence (-3.1) is an unstable critical point of the given system of ODE.

Finally, to understand the behavior of $(0, -\frac{4}{5})$. as before we set $\tilde{y} = y + \frac{4}{5}$. Then the given system becomes

$$\begin{cases} x' = -\frac{9}{5}x + x\tilde{y}. \\ \tilde{y}' = 4\tilde{y} + x^2 - 5\tilde{y}^2. \end{cases} \tag{6.46}$$

Thus $(0, 0)$ is a simple critical point of (6.46). Using the same arguments presented so far for the other critical points, we conclude that $(0, -\frac{4}{5})$ is an unstable critical point of the given system of ODE. $\qquad\square$

6.5   Periodic solutions

In this section, we provide some necessary conditions and sufficient conditions for the existence of periodic solutions/closed paths. Moreover, we discuss the relation between the behavior of bounded solutions and closed paths of the system of first order ODE.

Consider the 2×2 nonlinear autonomous system of ODEs, for $t > 0$,

$$\begin{cases} x'(t) = g_1(x(t), y(t)), \\ y'(t) = g_2(x(t), y(t)). \end{cases} \tag{6.47}$$

We begin with the definition of a periodic solution (6.47).

Definition 6.5.1. A solution $(x(t), y(t))$, $t \geq 0$ to (6.47) is said to be periodic if there exists $T > 0$ such that

$$(x(t + T), y(t + T)) = (x(t), y(t)), \ t \geq 0,$$

and

$$T_0 = \inf\{T : (x(t + T), y(t + T)) = (x(t), y(t)), \ t \geq 0\}$$

is called the period of the solution $(x(t), y(t))$.

It is easy to observe that if T_0 is the period of $(x(t), y(t))$, then

$$(x(t + T_0), y(t + T_0)) = (x(t), y(t)), \ t \geq 0. \tag{6.48}$$

Moreover, $T_0 = 0$ if and only if $(x(t), y(t))$ is a constant function of t. The details are left to the reader as an easy exercise. Henceforth we assume that $T_0 > 0$.

Definition 6.5.2. A solution $(x(t), y(t))$, $t \geq 0$ to (6.47) is said to be a *closed path* of (6.47) if $(x(t), y(t))$, $t \geq 0$ forms a closed curve[3] in the phase plane.

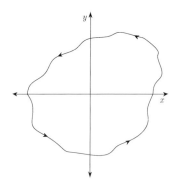

FIGURE 6.11: An example of a closed path in the phase plane.

A typical closed path is shown in Figure 6.11. The following lemma establishes the equivalence of periodic solutions and closed paths in the phase plane for (6.47).

---

[3]A curve $\gamma : [a, b] \to \mathbb{R}^2$ with $\gamma(a) = \gamma(b)$ is called a closed curve in $\mathbb{R}^2$.

Lemma 6.5.3. *If $(x(t), y(t))$, $t \geq 0$ is a periodic solution to (6.47), then it is a closed path. Conversely, if $(x(t), y(t))$, $t \geq 0$ is a closed path of (6.47), then it is a periodic solution.*

Proof. From the definitions of the periodic solution and the closed path of (6.47), it is straightforward to notice that every periodic solution is a closed path.

Conversely, we assume that $(x(t), y(t))$, $t \geq 0$, represents a closed path C in the phase plane. By setting

$$m := \min_{(x,y) \in C} \sqrt{g_1^2(x, y) + g_2^2(x, y)}, \quad v(t) := \sqrt{\left(x'(t)\right)^2 + \left(y'(t)\right)^2}, \ t \geq 0,$$

we obtain

$$v(t) \geq m > 0, \ t \geq 0. \tag{6.49}$$

Moreover the number

$$\int_0^\tau v(s)ds$$

is equal to the length of the portion of the curve C between the points $A = (x(0), y(0))$ and $B = (x(\tau), y(\tau))$ measured along the increasing direction of t. Now from (6.49), it follows that

$$\int_0^\tau v(s)ds \geq m\tau, \ \tau > 0.$$

Since C has finite length, say L, there exists $T_1 > 0$ such that

$$\int_0^{T_1} v(s)ds = L.$$

Therefore we get $(x(T_1), y(T_1)) = (x(0), y(0))$. This readily implies that $(x(t), y(t))$, $t > 0$ is periodic (see Exercise 6.12). □

From Section 6.2, we make the following observations regarding the trajectories of $Y' = AY$, where A is a constant matrix:

(i) Every bounded trajectory is either a closed path or it converges to $(0, 0)$.

(ii) If there exists a closed path then every trajectory is a closed path.

Whereas in the nonlinear case neither (i) nor (ii) hold true. In the next couple of examples, we demonstrate this feature by computing the closed paths explicitly.

Example 6.5.4. Discuss the behavior of the solutions to

$$\begin{cases} x' = -y + x(1 - 3x^2 - 3y^2), \\ y' = x + y(1 - 3x^2 - 3y^2). \end{cases}$$

Solution. We write the given system in polar coordinates. For, we set

$$x(t) = r(t)\cos(\theta(t)), \ y(t) = r(t)\sin(\theta(t)),$$

or

$$r^2(t) = x^2(t) + y^2(t), \tag{6.50}$$

$$\theta(t) = \tan^{-1}\left(\frac{y(t)}{x(t)}\right). \tag{6.51}$$

On differentiating (6.50) with respect to t and using the given ODEs, we find that

$$
\begin{aligned}
rr' &= xx' + yy' \\
&= -xy + x^2(1 - 3x^2 - 3y^2) + xy + y^2(1 - 3x^2 - 3y^2) \\
&= r^2(1 - 3r^2).
\end{aligned}
$$

Therefore we have

$$r' = r(1 - 3r^2). \tag{6.52}$$

Similarly on differentiating (6.51) with respect to t, we obtain

$$
\begin{aligned}
\theta' &= \frac{1}{1 + \frac{y^2}{x^2}}\left(\frac{xy' - yx'}{x^2}\right) \\
&= \frac{1}{r^2}(xy' - yx') \\
&= \frac{1}{r^2}\left(x^2 + xy(1 - 3x^2 - 3y^2) + y^2 - xy(1 - 3x^2 - 3y^2)\right) \\
&= 1.
\end{aligned}
\tag{6.53}
$$

We see that the given system reduces to (6.52)–(6.53) which is a decoupled system. The solution to (6.53) is

$$\theta(t) = t + \theta(0), \ t \geq 0. \tag{6.54}$$

Therefore every non-constant trajectory spirals around the origin in the anti-clockwise direction. Using the method of separation of variables, we solve (6.52) to get

$$r(t) = \frac{r(0)}{\sqrt{(1 - 3r^2(0))e^{-2t} + 3r^2(0)}}, \ t \geq 0, \tag{6.55}$$

provided $r(0) \neq \frac{1}{\sqrt{3}}$. On the other hand, if $r(0) = \frac{1}{\sqrt{3}}$ then we obtain

$$r(t) = \frac{1}{\sqrt{3}}, \ t > 0. \tag{6.56}$$

From (6.55), we find that

$$r(t) \to \frac{1}{\sqrt{3}} \text{ as } t \to \infty,$$

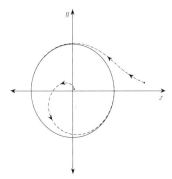

FIGURE 6.12: All the trajectories starting inside the circle C with center at the origin, radius $\frac{1}{\sqrt{3}}$, not only spiral around the origin but also move arbitrary close to every point of C as t increases. Similarly, all the trajectories starting outside the circle also have this property. Finally, the circle is also a trajectory of the given ODE.

for every choice of $r(0) > 0$. Hence the system has a unique periodic solution (closed path) given by (6.56) and every other solution approaches the periodic solution as $t \to \infty$. A schematic representation of the solution to the given system is presented in Figure 6.12. □

The closed path $r = \frac{1}{\sqrt{3}}$ that is obtained in Example 6.5.4 has two more special properties. The first one is that there is no other closed path 'near' this closed path. In general, any closed path having this property is called as an isolated closed path (analogous to the notion of isolated point). The second property is that every neighboring non-periodic trajectory asymptotically approaches $r = \frac{1}{\sqrt{3}}$. In general, any periodic solution to (6.47) satisfying these two properties is called a limit cycle. For various types of limit cycles the reader can refer to [11]. We present another example where we find a limit cycle.

Example 6.5.5. Discuss the behavior of the solutions to
$$\begin{cases} x' = 8x - 5y - 2x(x^2 + y^2), \\ y' = 5x + 8y - 2y(x^2 + y^2). \end{cases}$$

Solution. As in the previous example, we set
$$x(t) = r(t)\cos(\theta(t)), \; y(t) = r(t)\sin(\theta(t)),$$
or
$$r^2(t) = x^2(t) + y^2(t), \tag{6.57}$$
$$\theta(t) = \tan^{-1}\left(\frac{y(t)}{x(t)}\right). \tag{6.58}$$

On differentiating (6.57)–(6.58) with respect to t, we obtain

$$r' = \frac{xx' + yy'}{r} = 2r(4 - r^2), \tag{6.59}$$

$$\theta' = \frac{1}{r^2}(xy' - yx') = 5. \tag{6.60}$$

Using the method of separation of variables, we find that if $r(0) \neq 2$, then

$$r(t) = \frac{2r(0)}{\sqrt{\left(4 - r^2(0)\right)e^{-16t} + r^2(0)}}, \quad t \geq 0, \tag{6.61}$$

and if $r(0) = 2$, then

$$r(t) = 2, \ t \geq 0. \tag{6.62}$$

Moreover, we have

$$\theta(t) = 5t + \theta_0, \ t > 0. \tag{6.63}$$

As in the previous example, every solution to the given system is a curve which spirals around the origin in the anti-clockwise direction. Finally, we conclude that

$$\lim_{t \to \infty} r(t) = 2,$$

for every choice of $r(0) > 0$. Hence the system has a unique periodic solution (closed path) given by (6.62) and every other solution approaches the periodic solution as $t \to \infty$. Thus $r = 2$ is a limit cycle of the system. $\quad\square$

In fact, for a general nonlinear 2×2 system, it is often a difficult task to find a closed path explicitly though we know *a priori* that it exists. Therefore for practical purposes, we often find a bounded region where the closed path lies. We now state some results which provide the necessary conditions for the existence of a closed path in a region. The first one is due to Bendixson.

Theorem 6.5.6 (Bendixson's negative criterion). *Assume that* $\dfrac{\partial g_1}{\partial x} + \dfrac{\partial g_2}{\partial y}$ *takes only positive (negative) values, possibly except at a finite number of points in a simply connected domain R. Then (6.47) does not have a closed path in R.*

Proof. Without loss of generality, assume that

$$\frac{\partial g_1}{\partial x}(x, y) + \frac{\partial g_2}{\partial y}(x, y) > 0, \ (x, y) \in R. \tag{6.64}$$

Let $C = \{(x(t), y(t)) : t \geq 0\}$ be a closed path of (6.47) which lies in R and T

be the period of the closed path. Let S be the interior of the closed curve C. Then in view of Green's theorem, we have

$$
\begin{aligned}
\iint_S \left(\frac{\partial g_1}{\partial x} + \frac{\partial g_2}{\partial y} \right) dx dy &= \oint_C g_1 dy - g_2 dx \\
&= \int_0^T \left[g_1\big(x(t), y(t)\big) y'(t) - g_2\big(x(t), y(t)\big) x'(t) \right] dt \\
&= \int_0^T \left[g_1\big(x(t), y(t)\big) g_2\big(x(t), y(t)\big) \right. \\
&\qquad \left. - g_1\big(x(t), y(t)\big) g_2\big(x(t), y(t)\big) \right] dt \\
&= 0,
\end{aligned}
$$

which is a contradiction to (6.64). Hence there is no closed path in R. This completes the proof of the theorem. $\square$

Example 6.5.7. Show that the system $\begin{cases} x' = -x^5 + 3y^2 \sin x, \\ y' = -y^7 - y^3 \cos x, \end{cases}$

has no closed paths in $\mathbb{R}^2$.

Solution. Let $g_1(x, y) = -x^5 + 3y^2 \sin x$ and $g_2(x, y) = -y^3 \cos(x) - y^7$. An easy computation yields

$$
\frac{\partial g_1}{\partial x} + \frac{\partial g_2}{\partial y} = -5x^4 + 3y^2 \cos(x) - 3y^2 \cos(x) - 7y^6 < 0, \ (x, y) \neq (0, 0).
$$

In view of the Bendixson theorem, we conclude that the given system does not admit a closed path. $\square$

Using Green's theorem, we prove another necessary condition for the existence of closed paths.

Theorem 6.5.8 (Curl theorem). *Assume that*

$$
\frac{\partial g_1}{\partial y} - \frac{\partial g_2}{\partial x} = 0, \tag{6.65}
$$

possibly except at a finite number of points in a simply connected domain R. Then (6.47) does not have a closed path in R.

Proof. We suppose $C = \{(x(t), y(t)) : t \geq 0\}$ is a closed path of (6.47) which lies entirely in R and T is the period of the closed path. Let S be the interior of the closed curve C. We first notice that g_1 and g_2 simultaneously do not vanish at the points on C. Next, using (6.65) and Green's theorem, we get

$$
\begin{aligned}
0 &= \iint_S \left(\frac{\partial g_1}{\partial y} - \frac{\partial g_2}{\partial x} \right) dx dy \\
&= - \oint_C g_1 dx + g_2 dy \\
&= - \int_0^T \left[g_1^2(x(t), y(t)) + g_2^2(x(t), y(t)) \right] dt \\
&< 0,
\end{aligned}
$$

which is a contradiction. Hence there is no closed path in R. □

Another necessary condition for the existence of a closed path that we present in the next theorem is due to Poincaré. The proof of this theorem is out of the scope of this book. For a proof see for instance [9, 28, 32]

Theorem 6.5.9 (Poincaré). *Every closed path of system* (6.47) *necessarily surrounds at least one of the critical points of the system.* □

We now state the main theorem of this section, which is known as the Poincaré–Bendixson theorem. The theorem essentially talks about the trichotomy of non-trivial bounded solutions[4] to (6.47). Every non-trivial bounded solution is one of the following: (i) a trajectory converging to a critical point, (ii) a closed path, (iii) a trajectory which spirals toward a closed path.

Theorem 6.5.10 (Poincaré–Bendixson). *Let R be a closed and bounded region in the phase plane which does not contain any critical points of* (6.47). *If $C = \{(x(t), y(t)) : t > 0\}$ is a trajectory of* (6.47) *such that $C \subset R$, then either C is a closed path or C spirals toward a closed path.* □

In Figure 6.13, we depict a trajectory which is confined to the annulus region R. Moreover, we have shown a closed path whose existence is guaranteed by the Poincaré–Bendixson theorem.

Remark 6.5.11. Suppose $R \subset \mathbb{R}^2$ satisfies the hypothesis of Theorem 6.5.10. Then R cannot be simply connected. For, if R is simply connected, then from Theorem 6.5.10 R has a closed path, say, $\tilde{C}$. Due to Theorem 6.5.9 the interior of $\tilde{C}$ which is a subset of R has a critical point, which is a contradiction. Thus R is not simply connected.

In each of Examples 6.5.4–6.5.5, we verify that there is a region R in the phase plane satisfying the hypotheses of the Poincaré–Bendixson theorem to conclude that closed paths exist for the systems in those examples without

---

[4]The constant solutions given by the critical points of system (6.47) are taken as trivial bounded solutions to the system.

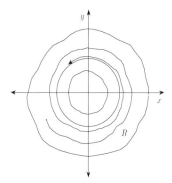

FIGURE 6.13: If there is a trajectory in the annulus region R, then it is either a closed path or encircles a closed path.

explicitly computing them. In order to do this, we need to find a closed and bounded region R without having any critical point and a non-trivial solution to the system of ODEs which lies in R for all $t > 0$.

Example 6.5.4 (Revisited) Use the Poincaré–Bendixson theorem to show that there exists a periodic solution to

$$\begin{cases} x' = -y + x(1 - 3x^2 - 3y^2), \\ y' = x + y(1 - 3x^2 - 3y^2). \end{cases}$$

Solution. We recall that the given system in polar coordinates is

$$\begin{cases} r' = r(1 - 3r^2), \\ \theta' = 1. \end{cases} \tag{6.66}$$

Observe that

$$r(1 - 3r^2) > 0, \ r \in (0, \tfrac{1}{\sqrt{3}}), \text{ and } r(1 - 3r^2) < 0, \ r \in (\tfrac{1}{\sqrt{3}}, \infty).$$

Using the theory developed in Chapter 2 (see Section 2.5), we conclude that if $r(0) = \tfrac{1}{2}$ then $r(t)$ is an increasing function of t and $r(t) \to \tfrac{1}{\sqrt{3}}$ as $t \to \infty$. Let

$$R = \{(x, y) \in \mathbb{R}^2 : \tfrac{1}{100} \le x^2 + y^2 \le 1\}.$$

Then the trajectory of (6.66) with $r(0) = \tfrac{1}{2}$, $\theta(0) = 0$ lies in R for $t \ge 0$. Hence the hypotheses of the Poincaré–Bendixson theorem is satisfied. Therefore there exists a periodic solution to the given system in R. □

Example 6.5.5 (Revisited) Use the Poincaré–Bendixson theorem to show that there exists a periodic solution to

$$\begin{cases} x' = 8x - 5y - 2x(x^2 + y^2), \\ y' = 5x + 8y - 2y(x^2 + y^2). \end{cases}$$

Solution. We recall that the given system in polar coordinates is

$$\begin{cases} r' = 2r(4 - r^2), \\ \theta' = 5. \end{cases} \tag{6.67}$$

It is easy to observe that

$$2r(4 - r^2) > 0, \ r \in (0, 2), \text{ and } 2r(4 - r^2) < 0, \ r \in (2, \infty).$$

As in the previous example, we notice that if $r(0) = 1$ then r is an increasing function and $r(t) \to 2$ as $t \to \infty$. Let

$$R = \{(x, y) \in \mathbb{R}^2 : \tfrac{1}{2} \le x^2 + y^2 \le 9\}.$$

Then the trajectory of (6.67) with $r(0) = 1$, $\theta(0) = 0$ lies in R for $t \ge 0$. Hence the hypotheses of the Poincaré–Bendixson theorem is satisfied and subsequently, there exists a closed path to the given system in R. ☐

Example 6.5.12. Use the Poincaré–Bendixson theorem to show that there exists a periodic solution to

$$\begin{cases} x' = 5x - 7y - x(5x^2 + 3y^2), \\ y' = 7x + 3y - y(5x^2 + 3y^2). \end{cases}$$

Solution. As before, we write the given system in polar coordinates to get

$$\begin{aligned} r' &= r(1 - r^2)(5\cos^2(\theta) + 3\sin^2\theta) \\ &= r(1 - r^2)(4 + \cos(2\theta)), \end{aligned}$$

and

$$\theta' = 7 - \sin(2\theta).$$

For $0 < r \le 1$, it follows that

$$3r(1 - r^2) \le r' \le 5r(1 - r^2).$$

Using the comparison theorems (see Section 2.4 and Examples 2.5.16–2.5.17 in Chapter 2), for the choice of $r(0) = \tfrac{3}{4}$, $\theta(0) = 0$, we obtain that

$$\tfrac{1}{2} \le r(t) \le 1, \ t \ge 0.$$

Let $R = \{(x, y) \in \mathbb{R}^2 : \tfrac{1}{5} \le x^2 + y^2 \le 4\}$. Then the trajectory starting at $r(0) = \tfrac{3}{4}$, $\theta(0) = 0$ lies in R. Therefore from the Poincaré–Bendixson theorem, there exists a closed path to the given system in the annulus region R. ☐

Example 6.5.13. Use the Poincaré–Bendixson theorem to show that there exists a periodic solution to

$$\begin{cases} x' = 2x - 7y - x(3x^2 + 4y^2), \\ y' = 7x + 2y - y(3x^2 + 4y^2). \end{cases}$$

Solution. The given system in polar coordinates is

$$r' = r\left(2 - r^2(3\cos^2\theta + 4\sin^2\theta)\right)$$
$$= \frac{r}{2}\left(4 - r^2[7 - \cos(2\theta)]\right),$$

and

$$\theta' = 7.$$

It is straightforward to notice that

$$2r(1 - 2r^2) \leq r' \leq r(2 - 3r^2).$$

Using the comparison theorem, for the choice of $\theta(0) = 0$, $r(0) = \sqrt{\frac{3}{5}}$ (in general for any $\frac{1}{\sqrt{2}} \leq r(0) \leq \sqrt{\frac{2}{3}}$ and $\theta \in \mathbb{R}$), we obtain that

$$\frac{1}{\sqrt{2}} \leq r(t) \leq \sqrt{\frac{2}{3}}, \ t \geq 0.$$

Therefore from the Poincaré–Bendixson theorem, there exists a periodic solution to the given system in the annulus region $\{(x, y) \in \mathbb{R}^2 : \frac{1}{2} \leq x^2 + y^2 \leq \frac{2}{3}\}$. □

Exercise 6.1. Let $g_1, g_2 : \mathbb{R}^2 \to \mathbb{R}$ be C^1-functions. Consider the autonomous system of ODEs, for $t > 0$

$$\begin{cases} x'(t) = g_1\big(x(t), y(t)\big), \\ y'(t) = g_2\big(x(t), y(t)\big). \end{cases} \tag{6.68}$$

Let $(\hat{x}, \hat{y}) \in \mathbb{R}^2$ be an isolated critical point of (6.68). Then show that $(\hat{x}, \hat{y})$ is stable if and only if for given $\epsilon > 0$ there exists $\delta > 0$ such that if for every $t_1 > 0$ with $\|(x(t_1), y(t_1)) - (\hat{x}, \hat{y})\| < \delta$, then $\|(x(t), y(t)) - (\hat{x}, \hat{y})\| < \epsilon, t > t_1$.

Exercise 6.2. Let $A, B \in M_2(\mathbb{R})$ be such that $AB = BA$. Assume that $(0, 0)$ is an asymptotically stable critical point of the systems $Z' = AZ$ and $U' = BU$. Then show that $(0, 0)$ is an asymptotically stable critical point of $Y' = (A + B)Y$.

Exercise 6.3. Let $A \in M_2(\mathbb{R})$ be such that it has at least one eigenvalue with a positive real part. Prove that for a given $(a_1, a_2) \in \mathbb{R}^2$ and for every $\epsilon > 0$ there exists a solution Y to $Y' = AY$ with $\|Y(0) - (a_1, a_2)\| < \epsilon$ and $\|Y(t)\| \to \infty$ as $t \to \infty$.

Exercise 6.4. Prove Exercise 6.3 when $A \in M_n(\mathbb{R})$.

Exercise 6.5. Assume that $A : [0, \infty) \to M_n(\mathbb{R})$ is continuous and Φ is a fundamental matrix of $Y'(t) = A(t)Y(t)$. Moreover assume that $\lim_{t\to\infty} \Phi(t)$ exists in $\mathbb{R}^{n \times n}$. Then prove the following statements.

(i) If Y is a solution to $Y' = AY$ then $\lim_{t\to\infty} Y(t)$ exists.

(ii) For a given $\tilde{Y} \in \mathbb{R}^n$ there exists a solution Y to $Y' = AY$ such that $\lim_{t\to\infty} Y(t) = \tilde{Y}$.

Exercise 6.6. Discuss the stability of the critical point $(0,0)$ of the following systems by constructing Lyapunov (type) functions for them explicitly:

(i) $\begin{cases} x' = 4x^2 y - x^3 y^4, \\ y' = -3x^3 - x^2 y^3. \end{cases}$

(ii) $\begin{cases} x' = -5x^3 + xy^3 - y^5, \\ y' = -x^2 + xy^2. \end{cases}$

(iii) $\begin{cases} x' = -y + xy + x^3, \\ y' = y + 5x^3 - 5x^4. \end{cases}$

(iv) $\begin{cases} x' = y^2, \\ y' = -xy - y^3 - 3x^3 y. \end{cases}$

Exercise 6.7. Suppose $h : \mathbb{R}^2 \to \mathbb{R}$ is a C^1-function, $\frac{\partial h}{\partial y}$ and $y\frac{\partial h}{\partial y}$ are bounded. Consider the system

$$\begin{cases} x' = y - xh(x, y), \\ y' = -x - yh(x, y), \end{cases}$$

and prove the following results:

(i) If $h(u, v) \geq 0$, $(u, v) \in \mathbb{R}^2$ then $(0,0)$ is stable.

(ii) If $h(u, v) > 0$, $(u, v) \in \mathbb{R}^2\backslash\{(0,0)\}$, then $(0,0)$ is asymptotically stable.

(iii) If $h(u, v) < 0$, $(u, v) \in \mathbb{R}^2\backslash\{(0,0)\}$, then $(0,0)$ is unstable.

Hint: Take the Lyapunov energy function to be $E(u, v) = u^2 + v^2$, $(u, v) \in \mathbb{R}^2$.

Exercise 6.8. Consider the equation representing the motion of an undamped simple pendulum $y''(t) + w^2 \sin(y(t)) = 0$, $\frac{-\pi}{2} \leq y \leq \frac{\pi}{2}$.

(i) Write this equation as a 2×2 system of first order ODEs by introducing $x(t) = y'(t)$, $t > 0$.

(ii) Show that the critical point $(0,0)$ of the 2×2 system is stable by considering the Lyapunov energy function $E = \frac{1}{2}u^2 + w^2(1 - \cos v)$ (sum of the kinetic energy and the potential energy).

Exercise 6.9. Consider the equation $y''(t) + y'(t) + \sin(y(t)) = 0$.

(i) Write this equation as a 2×2 system of first order ODEs by introducing $x(t) = y'(t)$, $t > 0$.

(*ii*) Show that the critical point $(0,0)$ of the 2×2 system is stable by considering the Lyapunov energy function $E = \frac{1}{2}u^2 + (1 - \cos v)$ (sum of the kinetic energy and the potential energy).

(*iii*) Show that $(0,0)$ is asymptotically stable by considering the Lyapunov energy function $E_1(u, v) = u^2 + (u + v)^2 + 4(1 - \cos v)$.

Exercise 6.10. Discuss the stability of all the critical points of the following systems:

(*i*)
$$
\begin{cases}
x' = x - y, \\
y' = -15x + 2y - 4\sin y.
\end{cases}
$$

(*ii*)
$$
\begin{cases}
x' = -6x + 6y, \\
y' = -4x + xy + 3.
\end{cases}
$$

(*iii*)
$$
\begin{cases}
x' = x^2 - y^2, \\
y' = 3y + x^2 + 2.
\end{cases}
$$

(*iv*)
$$
\begin{cases}
x' = x + y + x^2, \\
y' = x - 20y + x^2.
\end{cases}
$$

(*v*)
$$
\begin{cases}
x' = x + y + (2x + y)\cos(3x + 5y), \\
y' = 2x + y + 4\sin(x - 2y).
\end{cases}
$$

(For (*v*) discuss the stability of only $(0,0)$.)

Exercise 6.11 (Perron). Let the (real part of the) eigenvalues of a 2×2 matrix A be negative. Let f, g be continuous functions and $(0,0)$ be a simple critical point of the system

$$
\begin{pmatrix} x'(t) \\ y'(t) \end{pmatrix} = A \begin{pmatrix} x(t) \\ y(t) \end{pmatrix} + \begin{pmatrix} f(x(t), y(t)) \\ g(x(t), y(t)) \end{pmatrix}.
$$

Show that there exists $\delta > 0$ such that if $\|(x(0), y(0))\| < \delta$ then $(x(t), y(t))$ exists for every $t > 0$. Moreover show that $(0,0)$ is asymptotically stable.

Hint: Show that the solution $(x(t), y(t))$ to the given system satisfies

$$
\begin{pmatrix} x(t) \\ y(t) \end{pmatrix} = e^{At} \begin{pmatrix} x(0) \\ y(0) \end{pmatrix} + \int_0^t e^{(t-s)A} \begin{pmatrix} f(x(s), y(s)) \\ g(x(s), y(s)) \end{pmatrix} ds.
$$

Take the norm and apply the Gronwall lemma to complete the proof.

Exercise 6.12. Let g_1, g_2 be Lipschitz functions. Let $(x(t), y(t))$, $t \geq 0$ be the solution to

$$
\begin{cases}
x' = g_1(x, y), \\
y' = g_2(x, y), \\
(x(0), y(0)) = (x_0, y_0).
\end{cases}
$$

If there exists T such that $\big(x(T), y(T)\big) = (x_0, y_0)$, then show that the solution is periodic.

Exercise 6.13. Show that the system
$$
\begin{cases}
x' = y + x(1 + 5y)(x^2 + y^2 + 1), \\
y' = -x + (y - 5x^2)(x^2 + y^2 + 1),
\end{cases}
$$
does not admit a non-constant periodic solution.

Hint: Write this system in the polar coordinates (r, θ) and show that $r' > 0$ whenever $r > 0$.

Exercise 6.14. Show that every non-constant solution to the following systems is periodic:

(i)
$$
\begin{cases}
x' = y + y(x^2 + y^2 + 2), \\
y' = -2x - 2x(x^2 + y^2 + 2).
\end{cases}
$$

(ii)
$$
\begin{cases}
x' = 3ye^{x^2+2}, \\
y' = -5xe^{x^2+2}.
\end{cases}
$$

Hint: Consider $\frac{x'}{y'}$ and show that any solution $(x(t), y(t)), t > 0$ to the given system is a closed curve in the phase plane.

Exercise 6.15. Show that the following systems do not admit any non-constant periodic solutions:

(i)
$$
\begin{cases}
x' = xy^2 + 9x^3y^4, \\
y' = 2x^2y^3 + y\sin^2 x.
\end{cases}
$$

(ii)
$$
\begin{cases}
x' = 3xe^{2y} - 4x^2y^3, \\
y' = 2xy^4.
\end{cases}
$$

(iii)
$$
\begin{cases}
x' = 2x + y + y^2(x^3 + y^2 + 2), \\
y' = x - 2y + x^2(y^3 + x^2 + 2).
\end{cases}
$$

Hint: Use the Bendixson theorem.

Exercise 6.16. Show that the following systems do not admit any non-constant periodic solutions:

(i)
$$
\begin{cases}
x' = e^x y^2 - e^y \cos(x + y), \\
y' = 2ye^x - \sqrt{2}e^y \sin(x + y + \dfrac{\pi}{4}).
\end{cases}
$$

(ii)
$$
\begin{cases}
x' = 2xy^3 + y^4 e^{x+y}, \\
y' = 3x^2y^2 + e^{x+y}(y^4 + 4y^3).
\end{cases}
$$

Hint: Use Theorem 6.5.8.

Exercise 6.17 (Dulac negative criterion). Let $g_1, g_2 \, : \, \mathbb{R}^2 \to \mathbb{R}$ be C^1-functions. Consider the system of ODEs, for $t > 0$

$$\begin{cases} x'(t) = g_1\big(x(t), y(t)\big), \\ y'(t) = g_2\big(x(t), y(t)\big). \end{cases}$$

Assume that there exists a C^1-function ψ such that $\dfrac{\partial(\psi g_1)}{\partial x} + \dfrac{\partial(\psi g_2)}{\partial y}$ takes either only positive values or only negative values, possibly except at a finite number of points in a simply connected domain R. Then show that the given system of ODEs does not have a closed path in R.

Exercise 6.18. Discuss the existence of periodic solutions to the following system:

$$\begin{cases} x' = y, \\ y' = a_1 x + a_2 y + a_3 x^2 + a_4 y^2, \ a_i \neq 0, \ 1 \leq i \leq 4. \end{cases}$$

Hint: Use Exercise 6.17 with $\psi = e^{-2a_4 x}$.

Exercise 6.19. Let $g_1, g_2 : \mathbb{R}^2 \to \mathbb{R}$ be C^1-functions. Consider the system of ODEs, for $t > 0$

$$\begin{cases} x'(t) = g_1\big(x(t), y(t)\big), \\ y'(t) = g_2\big(x(t), y(t)\big). \end{cases}$$

Assume that there exists a C^1-function ψ such that it takes only positive values, except possibly at a finite number of points in a simply connected domain R. If $\dfrac{\partial(\psi g_1)}{\partial y} = \dfrac{\partial(\psi g_2)}{\partial x}$ in R, then show that the given system does not have a closed path in R.

Exercise 6.20. Verify the Poincaré–Bendixson theorem for the following systems:

(i)
$$\begin{cases} x' = -x + 2y + x(5 - x^2 - y^2), \\ y' = -2x - y + y(5 - x^2 - y^2). \end{cases}$$

(ii)
$$\begin{cases} x' = 4x - 5y - x(x^2 + y^2)^2, \\ y' = 5x + 4y - y(x^2 + y^2)^2. \end{cases}$$

(iii)
$$\begin{cases} x' = x + y - 2x(x^2 + y^2), \\ y' = -x + y - 2y(x^2 + y^2). \end{cases}$$

Exercise 6.21. Use the Poincaré–Bendixson theorem to prove that there exist closed paths to the following systems:

(i)
$$\begin{cases} x' = 6x - y - 2x(x^2 + y^2), \\ y' = x + 2y - 2y(x^2 + y^2). \end{cases}$$

(ii) $\begin{cases} x' = x + y - x^3, \\ y' = -x + y - y^3. \end{cases}$

(iii) $\begin{cases} x' = 2x - y - x(4x^2 + y^2), \\ y' = x + 2y - y(4x^2 + y^2). \end{cases}$

Chapter 7

Series solutions

7.1 Introduction

In the treatise on 'fluxional equations,' Newton mentioned that all differential equations admit power series solutions. This statement influenced many mathematicians of eighteenth century including Euler. In the quest of finding the power series solutions to various ODEs, Euler introduced 'cylindrical functions' which were studied extensively by Bessel (we now call them as Bessel's functions). Later on, the reasons for the success as well as limitations of this method are identified. We discuss some of them in this chapter.

The main objective of this chapter is to introduce the methods to find the general solution to the second order linear homogeneous ODEs when the coefficients have a special property, viz., analyticity. Furthermore, we study the properties of the solutions to two special ODEs, namely the Legendre equation and the Bessel equation. Consider the homogeneous linear ODE with variable coefficients

$$y''(x) + a_1(x)y'(x) + a_2(x)y(x) = 0, \ x \in J, \tag{7.1}$$

where J is an open interval. We begin our discussion by showing that if the coefficients in (7.1) are infinitely differentiable then so is its solution.

Lemma 7.1.1. *Suppose a_1 and a_2 are infinitely differentiable functions. Then there exist two linearly independent solutions to (7.1) which are infinitely differentiable.*

Proof. By the existence and uniqueness theorem (see Theorem 3.2.3), there exist two linearly independent solutions, say ϕ_1 and ϕ_2 which are twice differentiable. On substituting $y = \phi_1$ in (7.1), we get the identity

$$\phi_1'' = -a_1\phi_1' - a_2\phi_1.$$

Since ϕ_1 is twice differentiable, a_1 and a_2 are infinitely differentiable, therefore the right hand side of the above identity is differentiable. Thus ϕ_1'' is differentiable, i.e., ϕ_1 is thrice differentiable. Proceeding in the same way, one can prove by induction that for a given $n \in \mathbb{N}$, the function ϕ_1 is differentiable n-times. Hence ϕ_1 is infinitely differentiable. Similarly one can prove that ϕ_2 is also infinitely differentiable. This completes the proof. □

One can easily extend this result to the non-homogeneous case provided the non-homogeneous term is infinitely differentiable. It is natural to ask whether the solution to (7.1) is analytic if a_1 and a_2 are analytic. We have an affirmative answer for this question. Before we get into the details (see Section 7.2), we recall the definition of an analytic function and some of its properties which will be useful in this chapter.

Definition 7.1.2. Let J be an open interval. A function $f : J \to \mathbb{R}$ is said to be an analytic function if for every $x_0 \in J$, there exist $\delta > 0$ and a sequence (c_n) such that

$$f(x) = \sum_{n=0}^{\infty} c_n(x - x_0)^n, \ x \in (x_0 - \delta, x_0 + \delta) \subseteq J.$$

Example 7.1.3. All polynomials, $e^{\alpha x}$, $\sin(\alpha x)$, $\cos(\alpha x)$, $\alpha \in \mathbb{R}$, their finite products and linear combinations are analytic in $\mathbb{R}$.   □

Theorem 7.1.4. *Assume that* $J = (x_0 - \delta, x_0 + \delta)$, *and* $f : J \to \mathbb{R}$ *is an analytic function given by* $f(x) = \sum_{n=0}^{\infty} c_n(x - x_0)^n$, $x \in J$. *Then the following hold true.*

(i) $f \equiv 0$ *if and only if* $c_n = 0$, $n \in \mathbb{N}$.

(ii) The function f *is differentiable and* f' *is also an analytic function. Moreover* f' *is given by* $f'(x) = \sum_{n=1}^{\infty} nc_n(x - x_0)^{n-1}$, $x \in J$.

(iii) The function f *is infinitely differentiable.*

(iv) If $0 < \delta_1 < \delta$ *then the series* $\sum_{n=0}^{\infty} c_n(x - x_0)^n$ *converges absolutely and uniformly in* $(x_0 - \delta_1, x_0 + \delta_1)$.

7.2 Existence of analytic solutions

In this section, we show that if the coefficients a_1 and a_2 in (7.1) are analytic, then the solutions to (7.1) are also analytic. In particular, if a_1 and a_2 are constants then it is clear that any solution of (7.1) is analytic (see Chapter 3). If a_1 and a_2 are analytic in $(x_0 - \delta, x_0 + \delta)$ for some $\delta > 0$ then we say that x_0 is an *ordinary point* for (7.1).

We now state the main theorem of this section.

Theorem 7.2.1. *Suppose* $J = (x_0 - \delta, x_0 + \delta)$, a_1 *and* a_2 *are analytic functions given by*

$$a_1(x) = \sum_{n=0}^{\infty} p_n(x - x_0)^n \text{ and } a_2(x) = \sum_{n=0}^{\infty} q_n(x - x_0)^n, \ x \in J.$$

Then there exists a unique solution to the homogeneous linear ODE with analytic coefficients

$$\begin{cases} y''(x) + a_1(x)y'(x) + a_2(x)y(x) = 0, \\ y(x_0) = \alpha_0, \ y'(x_0) = \alpha_1, \end{cases} \tag{7.2}$$

which is analytic in J.

Proof. Without loss of generality we assume that $x_0 = 0$. By the existence-uniqueness result there exists a unique solution to (7.2). We need to show that the solution is analytic.

Step 1. In this step we derive a necessary and sufficient condition on the sequence (c_n) such that $\phi(x) = \sum_{n=0}^{\infty} c_n x^n$, $|x| < \delta$ is a solution to (7.2) provided ϕ is analytic.

Using Theorem 7.1.4, we get

$$\phi'(x) = \sum_{n=0}^{\infty} (n + 1)c_{n+1}x^n,$$

and

$$\phi''(x) = \sum_{n=0}^{\infty} (n + 2)(n + 1)c_{n+2}x^n.$$

On substituting ϕ' and ϕ'' in (7.2), we obtain

$$\sum_{n=0}^{\infty}(n+2)(n+1)c_{n+2}x^n + \left(\sum_{n=0}^{\infty} p_n x^n\right)\left(\sum_{n=0}^{\infty}(n+1)c_{n+1}x^n\right) + \left(\sum_{n=0}^{\infty} q_n x^n\right)\left(\sum_{n=0}^{\infty} c_n x^n\right) = 0.$$

After taking the product of series, we arrive at

$$\sum_{n=0}^{\infty}\left((n + 2)(n + 1)c_{n+2} + \sum_{k=0}^{n}(k + 1)c_{k+1}p_{n-k} + \sum_{k=0}^{n} c_k q_{n-k}\right)x^n = 0.$$

Thus we have

$$(n + 2)(n + 1)c_{n+2} = -\sum_{k=0}^{n}(k + 1)c_{k+1}p_{n-k} - \sum_{k=0}^{n} c_k q_{n-k}, \ n \geq 0. \tag{7.3}$$

Hence if an analytic function $\phi(x) = \sum_{n=0}^{\infty} c_n x^n$, $|x| < \delta$ is a solution to (7.2), then $c_0 = \alpha_0$, $c_1 = \alpha_1$ and c_n for $n \geq 2$ is given by (7.3). Conversely, if $\sum_{n=0}^{\infty} c_n x^n$ converges in $|x| < \delta$, $c_0 = \alpha_0$, $c_1 = \alpha_1$, and c_n for $n \geq 2$ is given by (7.3), then $\phi(x) = \sum_{n=0}^{\infty} c_n x^n$, $|x| < \delta$ is the solution to (7.2).

Step 2. If $c_0 = \alpha_0$, $c_1 = \alpha_1$, c_n for $n \geq 2$ is given by (7.3), then $\sum\limits_{n=0}^{\infty} c_n x^n$ converges for $x \in J$.

Let $\delta_1 \in (0, \delta)$. We first prove that $\sum\limits_{n=0}^{\infty} c_n x^n$ converges for $|x| < \delta_1$. Since both series $\sum\limits_{n=0}^{\infty} p_n x^n$ and $\sum\limits_{n=0}^{\infty} q_n x^n$ converge absolutely and uniformly on $|x| \leq \delta_1$, we can choose $M > 0$ such that

$$|p_n| \delta_1^n \leq M, \ |q_n| \delta_1^n \leq M, \text{ for } n \geq 0.$$

In view of (7.3), we have

$$\delta_1^n (n+2)(n+1)|c_{n+2}| \leq M \sum_{k=0}^{n} \big((k+1)|c_{k+1}| + |c_k|\big) \delta_1^k.$$

We define a new sequence (C_n) of nonnegative numbers as $C_0 = |c_0|$, $C_1 = |c_1|$ and

$$\delta_1^n (n+2)(n+1) C_{n+2} = M \sum_{k=0}^{n} \big((k+1)C_{k+1} + C_k\big) \delta_1^k + T_n, \ n \geq 0, \quad (7.4)$$

where $T_n \geq 0$, will be chosen later.

It is very easy to note that $|c_n| \leq C_n$, for $n \geq 0$. On the other hand, by replacing n with $(n-1)$ in (7.4), we get

$$\delta_1^{n-1} n(n+1) C_{n+1} = M \sum_{k=0}^{n-1} \big((k+1)C_{k+1} + C_k\big) \delta_1^k + T_{n-1}. \quad (7.5)$$

On subtracting (7.5) from (7.4), we obtain

$$(n+1)\big(\delta_1(n+2)C_{n+2} - nC_{n+1}\big) = M\delta_1\big((n+1)C_{n+1} + C_n\big) + \frac{T_n - T_{n-1}}{\delta_1^{n-1}}. (7.6)$$

We note that (7.6) is a relation between C_{n+2}, C_{n+1}, and C_n. We choose T_n such that it reduces to a relation between C_{n+2} and C_{n+1}. To this end, we choose $T_n = M\delta_1^{n+1} C_{n+1}$ so that (7.6) becomes

$$(n+1)\big(\delta_1(n+2)C_{n+2} - nC_{n+1}\big) = M\delta_1(n+1+\delta_1)C_{n+1}.$$

After rearrangement of terms, we find that

$$\frac{C_{n+2}}{C_{n+1}} = \frac{n(n+1) + M\delta_1(n+1+\delta_1)}{\delta_1(n+1)(n+2)} \longrightarrow \frac{1}{\delta_1} \text{ as } n \to \infty.$$

Then we have

$$\lim_{n\to\infty} \frac{C_{n+1}|x|^{n+1}}{C_n |x|^n} = |x| \lim_{n\to\infty} \frac{C_{n+1}}{C_n} = \frac{|x|}{\delta_1}.$$

By the ratio test, the series $\sum\limits_{n=0}^{\infty} C_n |x|^n$ converges when $|x| < \delta_1$.

Since $|c_n| \leq C_n$, $n \geq 0$, owing to the comparison test, the series $\sum_{n=0}^{\infty} c_n x^n$ converges for $|x| < \delta_1$. Moreover, since δ_1 is arbitrary, $\sum_{n=0}^{\infty} c_n x^n$, converges when $x \in J$.

This completes the proof of Step 2 and hence the theorem. □

Example 7.2.2. Using the power series method, solve $y''(x) + y(x) = 0$, $x \in \mathbb{R}$.

Solution: Let $\phi(x) = \sum_{n=0}^{\infty} c_n x^n$, $x \in \mathbb{R}$ be the solution. Then we compute

$$\phi'(x) = \sum_{n=0}^{\infty} (n+1)c_{n+1} x^n,$$

$$\phi''(x) = \sum_{n=0}^{\infty} (n+2)(n+1)c_{n+2} x^n.$$

On substituting ϕ and ϕ'' in the given equation, we get

$$\sum_{n=0}^{\infty} \left((n+2)(n+1)c_{n+2} + c_n \right) x^n = 0, \ x \in \mathbb{R}.$$

Thus we have

$$c_{n+2} = -\frac{c_n}{(n+2)(n+1)}, \ n \geq 0. \tag{7.7}$$

By putting $n = 2k$, $k \geq 0$, in (7.7) we obtain

$$c_{2k+2} = -\frac{c_{2k}}{(2k+2)(2k+1)} = \cdots = \frac{(-1)^{k+1}c_0}{(2k+2)!}.$$

Similarly, by putting $n = 2k + 1$, $k \geq 0$, in (7.7) we arrive at

$$c_{2k+3} = \frac{(-1)^{k+1}c_1}{(2k+3)!}.$$

Hence

$$\phi(x) = c_0 \sum_{k=0}^{\infty} \frac{(-1)^k x^{2k}}{(2k)!} + c_1 \sum_{k=0}^{\infty} \frac{(-1)^k x^{2k+1}}{(2k+1)!} = c_0 \cos x + c_1 \sin x,$$

is a solution to the given equation. □

It is not always possible to sum up the series to obtain the solution in a closed form. In that case we stop after writing the solution as a power series.

Example 7.2.3. Solve $(1 + x^2)y'' + 2xy' - 2y = 0$, $x \in \mathbb{R}$.

Solution. As in the previous example, we begin with $\phi(x) = \sum_{n=0}^{\infty} c_n x^n$, $x \in \mathbb{R}$, and compute

$$\phi'(x) = \sum_{n=0}^{\infty} (n+1)c_{n+1} x^n,$$

$$\phi''(x) = \sum_{n=0}^{\infty} (n+2)(n+1)c_{n+2} x^n.$$

On substituting ϕ, ϕ', and ϕ'' in the given equation, we find that

$$(1 + x^2) \sum_{n=0}^{\infty} (n+2)(n+1)c_{n+2}x^n + 2 \sum_{n=0}^{\infty} (n+1)c_{n+1}x^{n+1} - 2 \sum_{n=0}^{\infty} c_n x^n = 0, \ x \in \mathbb{R}.$$

On equating the coefficient of x^n to zero, we obtain

$$(n+2)(n+1)c_{n+2} + \big(n(n-1) + 2n - 2\big)c_n = 0, \ n \in \mathbb{N}.$$

After rearranging the terms, we have

$$c_{n+2} = -\frac{(n-1)c_n}{n+1}, \ n \geq 0. \tag{7.8}$$

On putting $n = 1$ in (7.8), we find $c_3 = 0$. Since $c_3 = 0$, we get $c_{2k+1} = 0$, $k \geq 1$.
On the other hand, for $n = 2k$, $k \geq 1$, it follows that

$$c_{2k+2} = -\frac{(2k-1)c_{2k}}{2k+1} = \frac{(2k-3)c_{2k-2}}{2k+1} = \cdots = \frac{(-1)^k c_2}{2k+1}.$$

Hence

$$\phi(x) = c_1 x + c_0 \sum_{k=0}^{\infty} \frac{(-1)^{k-1} x^{2k}}{2k-1}, \ x \in \mathbb{R},$$

is the general solution to the given problem. $\square$

7.3   The Legendre equation

In this section, we study the properties of polynomial solutions to the Legendre equation. We call these polynomials (after normalizing) as the Legendre polynomials. The Legendre polynomials are widely used in many branches of mathematics. For instance, they are used in the Gauss quadrature formulae to find the numerical values of integrals over $[-1, 1]$. Furthermore, they appear more often in the study of some important partial differential equations in the curvilinear coordinate systems.

Assume that $\alpha \in \mathbb{R}$. The equation

$$(1 - x^2)y'' - 2xy' + \alpha(\alpha + 1)y = 0, \ x \in (-1, 1), \tag{7.9}$$

is called the Legendre equation. This equation can be written in the standard form as

$$y'' - \frac{2x}{1 - x^2}y' + \frac{\alpha(\alpha + 1)}{1 - x^2}y = 0.$$

Note that, $(1-x^2)^{-1}$ is analytic in $(-1,1)$ because it is the sum of the geometric series $\sum_{n=0}^{\infty} x^{2n}$. Since the coefficients of y' and y in (7.9) are analytic in $(-1,1)$, due to Theorem 7.2.1, we can seek the power series solution to the Legendre equation.

Assume that $\phi(x) = \sum_{n=0}^{\infty} c_n x^n$, $x \in (-1,1)$, is a solution to (7.9). On substituting ϕ, ϕ', ϕ'' in (7.9), we get

$$\sum_{n=0}^{\infty} \left((n+2)(n+1)c_{n+2}(1-x^2) - 2(n+1)c_{n+1}x + \alpha(\alpha+1)c_n\right)x^n = 0,$$

or

$$\sum_{n=0}^{\infty} \left((n+2)(n+1)c_{n+2} - n(n-1)c_n - 2nc_n + \alpha(\alpha+1)c_n\right)x^n = 0. \quad (7.10)$$

The function ϕ is a solution of (7.9) if and only if the coefficients of the powers of x^n in (7.10) are equal to zero for every $n \in \mathbb{N} \cup \{0\}$. Therefore we have

$$c_{n+2} = \frac{\left(n(n-1)+2n-\alpha(\alpha+1)\right)c_n}{(n+2)(n+1)} = \frac{(n-\alpha)(n+\alpha+1)c_n}{(n+2)(n+1)}. \quad (7.11)$$

Hence, for $n=2k$, $k \geq 1$, we obtain

$$c_{2k} = \frac{(2k-\alpha-2)(2k+\alpha-1)}{2k(2k-1)}c_{2k-2}$$

$$= \frac{(2k-\alpha-2)(2k-\alpha-4)(2k+\alpha-1)(2k+\alpha-3)}{2k(2k-1)(2k-2)(2k-4)}c_{2k-4}$$

$$\vdots$$

$$= \frac{\left(\prod_{m=1}^{k}(2(k-m)-\alpha)\right)\left(\prod_{m=1}^{k}(2(k-m)+\alpha+1)\right)c_0}{(2k)!}.$$

Similarly, for $n=2k+1$, $k \geq 1$, we get

$$c_{2m+1} = \frac{\left(\prod_{m=1}^{k}(2(k-m)-\alpha+1)\right)\left(\prod_{m=1}^{k}(2(k-m)+\alpha+2)\right)c_1}{(2k+1)!}.$$

Therefore the general solution to the Legendre equation is

$$\phi(x) = c_0\phi_0(x) + c_1\phi_1(x), \quad x \in (-1,1), \quad (7.12)$$

where

$$\phi_0(x) = 1 + \sum_{k=1}^{\infty} \frac{\left(\prod_{m=1}^{k}(2(k-m)-\alpha)\right)\left(\prod_{m=1}^{k}(2(k-m)+\alpha+1)\right)x^{2k}}{(2k)!},$$

TABLE 7.1:   The polynomial solutions
(ϕ_0 and ϕ_1) to the Legendre equation for
$\alpha = 0, 1, 2, 3, 4, 5$

α	ϕ_0	α	ϕ_1
0	1	1	x
2	$1 - 3x^2$	3	$x - \dfrac{5}{3}x^3$
4	$1 - 10x^2 + \dfrac{35}{3}x^4$	5	$x - \dfrac{14}{3}x^3 + \dfrac{21}{5}x^5$

$$\phi_1(x) = x + \sum_{k=1}^{\infty} \frac{\left(\prod_{m=1}^{k} \big(2(k-m) - \alpha + 1\big) \right)\left(\prod_{m=1}^{k} \big(2(k-m) + \alpha + 2\big) \right) x^{2k+1}}{(2k+1)!}.$$

The following remarks are in order from the general solution to the Legendre equation.

Remark 7.3.1. Suppose α is a nonnegative even integer, i.e., $\alpha = 2l$, $l \in \mathbb{N} \cup \{0\}$. Then, in view of (7.11), one has $c_{2l+2} = c_{2l+4} = \cdots = 0$. Therefore, ϕ_0 is a polynomial of degree $2l$ and it has only even powers of x. Similarly, if α is a positive odd number (say, $2l + 1$, $l \geq 0$), then ϕ_1 is a polynomial (of degree $2l + 1$) solution of the Legendre equation and it has only odd powers of x. Moreover, if α is even (resp. odd), then ϕ_1 (resp. ϕ_0) is not a polynomial.

In fact, we are interested in studying the polynomial solutions of the Legendre equation. Therefore now onward, unless specified otherwise, we assume $\alpha = n$, a nonnegative integer. We now define the Legendre polynomials.

Definition 7.3.2. A polynomial solution P_n, $n \in \mathbb{N} \cup \{0\}$, to the equation

$$(1 - x^2)y'' - 2xy' + n(n+1)y = 0, \tag{7.13}$$

is said to be the n-th Legendre polynomial if $P_n(1) = 1$.

We have already seen that for every $n \in \mathbb{N} \cup \{0\}$, there exists a polynomial solution of degree n, say Q_n, to (7.13). If $Q_n(1) \neq 0$, then $\frac{Q_n(x)}{Q_n(1)}$ is a Legendre polynomial. The difficulty here is that, it is not *a priori* clear whether for each $n \in \mathbb{N}$ there is any polynomial solution to (7.13) which does not vanish at $x = 1$. Before addressing this issue, we observe that none of the polynomial solutions to (7.13) given in Table 7.1 vanish at $x = 1$. This motivates us to prove the following existence and uniqueness result for P_n.

Theorem 7.3.3. *For a given $n \in \mathbb{N} \cup \{0\}$, there exists a unique Legendre polynomial P_n of degree n.*

Proof. In order to prove the existence part, it is enough to find a polynomial solution Q_n to (7.13) such that $Q_n(1) \neq 0$. To that end, we set

$$h(x) = (x^2 - 1)^n$$

and observe that

$$h'(x) = 2nx(x^2 - 1)^{n-1}.$$

Hence we have

$$(x^2 - 1)h'(x) = 2nxh(x). \tag{7.14}$$

On differentiating (7.14) $(n+1)$ times, we get

$$(x^2 - 1)h^{(n+2)} + 2(n+1)xh^{(n+1)} + n(n+1)h^{(n)} = 2nxh^{(n+1)} + 2n(n+1)h^{(n)}.$$

Rearrangement of terms gives

$$(x^2 - 1)h^{(n+2)} + 2xh^{(n+1)} - n(n+1)h^{(n)} = 0.$$

This readily implies that $h^{(n)}$ is a solution to (7.13). Moreover $h^{(n)}$ is a polynomial of degree n.

Claim. $h^{(n)}(1) = n!\,2^n$.

For, consider

$$
\begin{aligned}
h^{(n)}(1) &= \frac{d^n}{dx^n}\left((x+1)^n(x-1)^n\right)\Big|_{x=1} \\
&= (x+1)^n \frac{d^n}{dx^n}(x-1)^n\Big|_{x=1} + n\frac{d}{dx}(x+1)^n \frac{d^{n-1}}{dx^{n-1}}(x-1)^n\Big|_{x=1} \\
&\quad + \cdots + \left(\frac{d^n}{dx^n}(x+1)^n\right)(x-1)^n\Big|_{x=1} \\
&= (1+1)^n\, n! + 0 + 0 \cdots + 0 \\
&= n!\,2^n.
\end{aligned}
$$

Thus we can take $Q_n(x) = h(x)$ and a Legendre polynomial of degree n is $\frac{Q_n(x)}{Q_n(1)}$. In other words, we have

$$P_n(x) = \frac{1}{n!2^n}\frac{d^n}{dx^n}(x^2 - 1)^n. \tag{7.15}$$

We now turn our attention toward uniqueness. Let P_n and $\tilde{P}_n$ be two Legendre polynomials of degree n. In view of (7.12), there exist c_0, c_1, c_2, and c_3 such that

$$P_n(x) = c_0\phi_0(x) + c_1\phi_1(x), \quad x \in (-1,1), \tag{7.16}$$
$$\tilde{P}_n(x) = c_2\phi_0(x) + c_3\phi_1(x), \quad x \in (-1,1). \tag{7.17}$$

Without loss of generality assume that n is even. In view of Remark (7.3.1), we find that ϕ_0 is a polynomial and ϕ_1 is not a polynomial. Since $c_1\phi_1$ is not a polynomial for $c_1 \neq 0$, from (7.16) we conclude that $c_1 = 0$. Similarly we have $c_3 = 0$. On the other hand, since $P_n(1) = 1$, we find that $\phi_0(1) \neq 0$. On subtracting (7.16) from (7.17) and putting $x = 1$, we get

$$0 = \tilde{P}_n(1) - P_n(1) = (c_2 - c_0)\phi_0(1).$$

As $\phi_0(1) \neq 0$, we obtain $c_0 = c_2$. This proves $P_n = \tilde{P}_n$ and hence there exists a unique Legendre polynomial of degree n. ☐

Formula (7.15) is called Rodrigue's formula. We list here the first few Legendre polynomials that are computed using Rodrigue's formula. Notice that these polynomials match with $\frac{\phi_i}{\phi_i(1)}$, $i = 0, 1$ are computed from Table 7.1.

$$P_0(x) = 1, \ P_1(x) = x, \ P_2(x) = \frac{3}{2}x^2 - \frac{1}{2}, \ P_3(x) = \frac{5}{2}x^3 - \frac{3}{2}x,$$

$$P_4(x) = \frac{35}{8}x^4 - \frac{15}{4}x^2 + \frac{3}{8}, \ P_5(x) = \frac{63}{8}x^5 - \frac{35}{4}x^3 + \frac{15}{8}x.$$

7.3.1 Applications of Rodrigue's formula

In this subsection, we present some interesting applications of the Rodrigue formula. As the first application, we prove a result regarding the symmetry of the graph of the polynomial P_n.

Lemma 7.3.4. *The polynomial P_n is an even function[1] whenever n is even. Moreover, P_n is an odd function whenever n is odd.*

Proof. Since $(x^2 - 1)^n$ is an even function, we have that the derivative of $(x^2 - 1)^n$ is odd and the double derivative of $(x^2 - 1)^n$ is even. On continuing in this way we get that the n-th derivative of $(x^2 - 1)^n$ is even (resp. odd) if n is even (resp. odd). Therefore from Rodrigue's formula P_n is even (resp. odd) if n is even (resp. odd). ☐

Lemma 7.3.5. *If $f \in C^n([-1, 1])$, $n \in \mathbb{N}$, then*

$$\int_{-1}^{1} f(x)P_n(x)dx = \frac{(-1)^n}{n!2^n} \int_{-1}^{1} f^{(n)}(x)(x^2 - 1)^n dx. \tag{7.18}$$

Proof. Let $h(x) = (x^2 - 1)^n$, $x \in [-1, 1]$.

Claim. $h^{(k)}(\pm 1) = 0$, whenever $0 \leq k \leq n - 1$.

---

[1] A function $f : [-1, 1] \to \mathbb{R}$ is said to be an even function if $f(-x) = f(x)$, $x \in [-1, 1]$. We say that f is an odd function if $f(-x) = -f(x)$, $x \in [-1, 1]$.

Using the Leibniz formula we obtain

$$
\begin{aligned}
\frac{d^k}{dx^k} h(x) &= \frac{d^k}{dx^k}\left((x+1)^n (x-1)^n\right) \\
&= \sum_{r=0}^{k} \binom{k}{r} \frac{d^r}{dx^r}\left((x+1)^n\right) \frac{d^{k-r}}{dx^{k-r}}\left((x-1)^n\right) \\
&= \sum_{r=0}^{k} \binom{k}{r} \frac{n!}{(n-r)!}(x+1)^{n-r} \frac{n!}{(n-k+r)!}(x-1)^{n-k+r}.
\end{aligned}
$$

On substituting $x = \pm 1$, we get

$$
h^{(k)}(\pm 1) = 0, \ 0 \le k \le n-1.
$$

We now consider

$$
\begin{aligned}
\int_{-1}^{1} f(x) P_n(x) dx &= \frac{1}{n! 2^n} \int_{-1}^{1} f(x) h^{(n)}(x) dx \\
&= \frac{-1}{n! 2^n} \left[\int_{-1}^{1} f'(x) h^{(n-1)}(x) dx - f(x) h^{(n-1)}(x) \Big|_{x=-1}^{x=1} \right] \\
&= \frac{1}{n! 2^n} \left[\int_{-1}^{1} f''(x) h^{(n-2)}(x) dx - f'(x) h^{(n-2)}(x) \Big|_{x=-1}^{x=1} \right] \\
&\quad \vdots \\
&= \frac{(-1)^n}{n! 2^n} \left[\int_{-1}^{1} f^{(n)}(x) h(x) dx - f^{(n-1)}(x) h(x) \Big|_{x=-1}^{x=1} \right].
\end{aligned}
$$

Since $h^{(k)}(\pm 1) = 0, \ 0 \le k \le n-1$, all the boundary terms in the above expressions vanish. This establishes (7.18). $\qquad\square$

As an immediate consequence of Lemma 7.3.5, we obtain the following result.

Theorem 7.3.6. *For $m, n \in \mathbb{N} \cup \{0\}$, one has*

(i) the orthogonal property given by, $\displaystyle\int_{-1}^{1} P_m(x) P_n(x) dx = 0$, *if $m \neq n$.*

(ii) $\displaystyle\int_{-1}^{1} P_n^2(x) dx = \frac{2}{2n+1}.$

Proof. If $m \neq n$, without loss of generality we assume that $m < n$. Then the n-th derivative of P_m is identically zero. We put $f = P_m$ in (7.18) to get

$$
\int_{-1}^{1} P_m(x) P_n(x) dx = \frac{(-1)^n}{n! 2^n} \int_{-1}^{1} \left(\frac{d^n}{dx^n} P_m(x) \right) (x^2 - 1)^n dx = 0.
$$

This proves (i).

In order to prove (ii), we consider

$$
\frac{d^n}{dx^n} P_n(x) = \frac{1}{n! 2^n} \frac{d^{2n}}{dx^{2n}} (x^2 - 1)^n = \frac{(2n)!}{n! 2^n}. \tag{7.19}
$$

By putting $f = P_n$ in (7.18), we obtain

$$\int_{-1}^{1} P_n^2(x)dx = \frac{(-1)^n}{n!2^n} \int_{-1}^{1} \left(\frac{d^n}{dx^n} P_n(x)\right)(x^2 - 1)^n dx$$

$$= \frac{(2n)!}{(n!)^2 2^{2n}} \int_{-1}^{1} (1 - x^2)^n dx. \qquad (7.20)$$

To complete the proof we prove the following claim.

Claim. $\displaystyle\int_{-1}^{1} (1 - x^2)^n dx = \frac{2^{2n+1}(n!)^2}{(2n + 1)!}.$

We begin with the observation

$$\int_{-1}^{1} (1 - x^2)^n dx = 2 \int_{0}^{1} (1 - x^2)^n dx = 2 \int_{0}^{\pi/2} \cos^{2n+1}(t)dt =: 2A_n.$$

We evaluate A_n to prove the claim. Using the integration by parts formula, we find

$$A_n = \int_{0}^{\pi/2} (\cos^{2n} t)(\sin t)' \, dt$$

$$= 2n \int_{0}^{\pi/2} (\cos^{2n-1} t) \sin^2 t \, dt$$

$$= 2n \int_{0}^{\pi/2} (\cos^{2n-1} t - \cos^{2n+1} t) dt$$

$$= 2nA_{n-1} - 2nA_n.$$

Therefore A_n's satisfy

$$A_n = \frac{2n}{2n + 1} A_{n-1}, \ n \in \mathbb{N}$$

and $A_0 = 1$. From this iterative formula we see that

$$A_n = \frac{2n}{2n + 1} A_{n-1} = \frac{2n(2n - 2)}{(2n + 1)(2n - 1)} A_{n-2} = \cdots = \frac{2n(2n - 2)\cdots 2}{(2n + 1)(2n - 1)\cdots 3}.$$

Hence

$$A_n = \frac{(n!2^n)^2}{(2n + 1)!}$$

and the claim is proved. Now Theorem 7.3.6(*ii*) follows immediately from (7.20). $\qquad \square$

There is an interesting application of Theorem 7.3.6. We first denote

$$\Pi_N := \{P/P \text{ is a polynomial with real coefficients, } \deg(P) \leq N\}.$$

Clearly Π_N is an $(N + 1)$ dimensional vector space over $\mathbb{R}$. Since the elements in the set $S = \{P_0, \dots, P_N\} \subset \Pi_N$ are mutually orthogonal, it follows that S is a linearly independent (why?). Since the dimension of Π_N and the cardinality of S are the same, we conclude that S forms a basis for Π_N. Thus for every $P \in \Pi_N$, there exist $c_0, \ldots, c_N$ such that

$$P(x) = c_0 P_0(x) + \cdots + c_N P_N(x). \tag{7.21}$$

To find c_n, $0 \le n \le N$, we multiply (7.21) with P_n, integrate over $[-1, 1]$ and use Theorem 7.3.6 (i) to get

$$\int_{-1}^{1} P(x) P_n(x) dx = c_n \int_{-1}^{1} P_n^2(x) dx = \frac{2c_n}{2n+1}.$$

Hence the coefficients in (7.21) are given by

$$c_n = \frac{2n+1}{2} \int_{-1}^{1} P(x) P_n(x) dx, \ 0 \le n \le N. \tag{7.22}$$

The reader is advised to write the polynomials x, x^2, x^3 in terms of Legendre polynomials.

7.4 Linear ODEs with regular singular points

In this section, we extend the power series method to the situation where the coefficients a_1 and a_2 in (7.1) need not be analytic. Consider the equation

$$(x - x_0)^2 y''(x) + (x - x_0) b_1(x) y'(x) + b_0(x) y(x) = 0, \ x \in J, \tag{7.23}$$

where J is an open interval containing x_0 and b_0, b_1 are analytic in J. As the coefficient of y'' in (7.23) indeed vanishes in J (namely at x_0), the usual existence and uniqueness theory is not applicable for this equation.

The points at which the coefficient of the highest derivative term in a linear ODE with variable coefficients vanishes are called the *singular points* of that ODE. In particular, x_0 is a singular point of (7.23).

A special case: If $b_0(x_0) = b_0'(x_0) = 0 = b_1(x_0)$, then there exist two analytic functions a_0 and a_1 such that

$$b_0(x) = (x - x_0)^2 a_0(x), \ b_1(x) = (x - x_0) a_1(x)$$

in an interval containing J. Hence (7.23) becomes

$$y''(x) + a_1(x) y'(x) + a_0(x) y(x) = 0, \ x \in J,$$

which can be solved using the method described in Section 7.2. In fact, $x = x_0$ is not a singular point in this case. We present another situation in which the series solution can be obtained. Define

$$a_1(x) := \frac{b_1(x)}{(x - x_0)}, \ a_2(x) := \frac{b_0(x)}{(x - x_0)^2}. \tag{7.24}$$

Then (7.23) reduces to

$$y''(x) + a_1(x)y'(x) + a_2(x)y(x) = 0. \qquad (7.25)$$

Definition 7.4.1. We say that the point $x = x_0$ is a regular singular point of (7.25) if
(*i*) at least one of a_1, a_2 is not analytic in J and
(*ii*) both $(x - x_0)a_1$ and $(x - x_0)^2 a_2$ are analytic in J.
If (*i*) or (*ii*) does not hold, then we say that $x = x_0$ is an irregular singular point

Example 7.4.2. For the equation

$$x^2 y'' + xe^x y' + \sin(x)y = 0, \ x \in (-1, 1)$$

$x = 0$ is a regular singular point. For, we observe that here $a_1(x) = \dfrac{e^x}{x}$ and $a_2(x) = \dfrac{\sin x}{x^2}$. Since xa_1 and $x^2 a_0$ are analytic, $x = 0$ is a regular singular point.

On the other hand, it is straightforward to observe that $x = 0$ is not a regular singular point of

$$x^3 y'' + xe^x y' + \sin(x)y = 0, \ x \in (-1, 1).$$

Here both $xa_1(x) = \dfrac{e^x}{x}$ and $x^2 a_2(x) = \dfrac{\sin x}{x}$ are not analytic in $(-1, 1)$. □

Definition 7.4.3. A solution to (7.25) which is of the form

$$\phi(x) = |x - x_0|^m \sum_{n=0}^{\infty} c_n (x - x_0)^n, \ x \in J \backslash \{x_0\},$$

where $m \in \mathbb{R}$, is said to be a Frobenius solution.

We now state the main theorem of this section which deals with the existence of a Frobenius solution.

Theorem 7.4.4. *Let $x = x_0$ be a regular singular point of (7.25). Furthermore, let $(x - x_0)a_1$ and $(x - x_0)^2 a_0$ be given by*

$$(x - x_0)a_1(x) = \sum_{n=0}^{\infty} p_n(x - x_0)^n, \ (x - x_0)^2 a_2(x) = \sum_{n=0}^{\infty} q_n(x - x_0)^n,$$

in $J := (x_0 - \delta, x_0 + \delta)$. Assume that the indicial polynomial which is defined by

$$f(m) := m(m - 1) + mp_0 + q_0$$

has distinct real roots m_1 and m_2 with $|m_1 - m_2| \notin \mathbb{N} \cup \{0\}$. Then (7.25) has two linearly independent Frobenius solutions of the form

$$\psi_1(x) = |x - x_0|^{m_1} \sum_{n=0}^{\infty} c_n(x - x_0)^n, \ \psi_2(x) = |x - x_0|^{m_2} \sum_{n=0}^{\infty} \tilde{c}_n(x - x_0)^n, \ x \in J \backslash \{x_0\},$$

$$(7.26)$$

*where c_n's (resp. $\tilde{c}_n$'s) are computed by substituting ψ_1 (resp. ψ_2) in (7.25).
Moreover, both $\sum_{n=0}^{\infty} c_n(x - x_0)^n$, $\sum_{n=0}^{\infty} \tilde{c}_n(x - x_0)^n$ are convergent in J.*

Proof. The proof of this theorem closely follows that of Theorem 7.2.1. Without loss of generality, we assume that $x_0 = 0$. Consider the formal expression

$$\phi(x) = |x|^m \sum_{n=0}^{\infty} c_n x^n, \ 0 < |x| < \delta. \tag{7.27}$$

Throughout the proof, we assume that $0 < x < \delta$. All the arguments we present here can be easily extended to the case $-\delta < x < 0$.

Step 1. In this step, we derive a necessary and sufficient condition on m. c_n. $\forall n \geq 0$ so that ϕ given in (7.27) is a solution to (7.25) provided the power series in ϕ converges.

On differentiating we get

$$\phi'(x) = \sum_{n=0}^{\infty} (n + m)c_n x^{n+m-1},$$

$$\phi''(x) = \sum_{n=0}^{\infty} (n + m)(n + m - 1)c_n x^{n+m-2}.$$

As before we compute

$$
\begin{aligned}
a_1(x)\phi'(x) &= \frac{1}{x}\left(\sum_{n=0}^{\infty} p_n x^n\right)\left(\sum_{n=0}^{\infty}(n + m)c_n x^{n+m-1}\right) \\
&= \sum_{n=0}^{\infty}\sum_{k=0}^{n} p_{n-k}(k + m)c_k x^{n+m-2},
\end{aligned}
$$

and

$$
\begin{aligned}
a_2(x)\phi(x) &= \frac{1}{x^2}\left(\sum_{n=0}^{\infty} q_n x^n\right)\left(\sum_{n=0}^{\infty} c_n x^{n+m}\right) \\
&= \sum_{n=0}^{\infty}\sum_{k=0}^{n} q_{n-k}c_k x^{n+m-2}.
\end{aligned}
$$

On substituting ϕ'', $a_1\phi'$, and $a_0\phi$ in (7.25), we obtain

$$
\begin{aligned}
0 = \sum_{n=0}^{\infty}\Bigg(&\big[(n + m)(n + m - 1) + (m + n)p_0 + q_0\big]c_n \\
&+ \sum_{k=0}^{n-1}\big((k + m)p_{n-k} + q_{n-k}\big)c_k\Bigg)x^{n+m-2}. \tag{7.28}
\end{aligned}
$$

We now equate the coefficient of x^{m-2} in (7.28) to zero to realize

$$\big(m(m-1)+mp_0+q_0\big)c_0 = 0.$$

The equation

$$m(m-1)+mp_0+q_0 = 0 \qquad (7.29)$$

is called the indicial equation of (7.25). (Recall that $f(m) = m(m-1)+mp_0+q_0$ is called the indicial polynomial.)

By equating the coefficient of x^{m-1} to zero in (7.28) one gets

$$\big[(m+1)m+(m+1)p_0+q_0\big]c_1 + (mp_1+q_1)c_0 = 0.$$

This can be rewritten as

$$f(m+1)c_1 = -(mp_1+q_1)c_0. \qquad (7.30)$$

We proceed in the same way (by equating the coefficient of x^{n+m-2} to zero) to find that

$$f(m+n)c_n = -\sum_{k=0}^{n-1}\big((m+k)p_{n-k}+q_{n-k}\big)c_k,\ n \geq 1. \qquad (7.31)$$

Therefore if ϕ is a solution to (7.25), then m must be a root of the indicial equation (7.29), c_0 is arbitrary, and c_n's are given by (7.31) for $n \geq 1$.

Conversely, assume that m is a solution of $f(m) = 0$, c_0 is arbitrary and c_n, $n \geq 1$, is given by (7.31). Moreover if $\sum_{n=0}^{\infty} c_n x^n$ converges in $0 < x < \delta$, then ϕ given in (7.27) satisfies

$$\phi'' + a_1\phi' + a_0\phi = c_0 f(m)x^{m-2} = 0,\ 0 < x < \delta. \qquad (7.32)$$

Without loss of generality, we assume that the roots of the indicial polynomial satisfy $m_1 > m_2$. If we fix $m = m_1$, then in view of (7.31), we have

$$f(m_1+n)c_n = -\sum_{k=0}^{n-1}\big((m_1+k)p_{n-k}+q_{n-k}\big)c_k,\ n \geq 1. \qquad (7.33)$$

We observe that $f(m_1+n)$ never vanishes for any $n \in \mathbb{N}$. Hence for a fixed c_0, the sequence (c_n) is uniquely determined by (7.33).

Step 2. In this step we prove that $\sum_{n=0}^{\infty} c_n x^n$ indeed converges in $|x| < \delta$, where c_n's are given in (7.33).

We first fix $\delta_1 \in (0,\delta)$ and choose $M > 0$ such that,

$$|p_n|\delta_1^n \leq M,\ |q_n|\delta_1^n \leq M,\ n \in \mathbb{N} \cup \{0\}.$$

Since m_1 and m_2 are the roots of (7.29), we have $1 - p_0 = m_1 + m_2$. Now a straightforward computation shows that

$$\begin{aligned}
f(n+m_1) &= (n+m_1)^2 - (n+m_1) + (n+m_1)p_0 + q_0 \\
&= n(n+2m_1-1+p_0) \\
&= n(n+m_1-m_2) \\
&\geq n(n-|m_1-m_2|). \qquad (7.34)
\end{aligned}$$

Combining (7.33) and (7.34), we get

$$n(n - |m_1 - m_2|)\delta_1^n|c_n| \leq M \sum_{k=0}^{n-1}(|m_1| + k + 1)\delta_1^k|c_k|. \qquad (7.35)$$

We define a new sequence C_n by

$$C_n = |c_n|, \ 0 \leq n \leq m_1 - m_2,$$

and

$$n(n - |m_1 - m_2|)\delta_1^n C_n = M \sum_{k=0}^{n-1}(|m_1| + k + 1)\delta_1^k C_k, \ n > m_1 - m_2. \qquad (7.36)$$

From the definition of (C_n) and (7.35), it is clear that $|c_n| \leq C_n$ for $n \in \mathbb{N}\cup\{0\}$. We now prove that $\sum_{n=0}^{\infty} C_n x^n$ converges for $|x| < \delta_1$. For, if we let $n > |m_1 - m_2| + 1$, and replace n with $(n - 1)$ in (7.36), then we find that

$$(n - 1)(n - 1 - |m_1 - m_2|)\delta_1^{n-1}C_{n-1} = M \sum_{k=0}^{n-2}(|m_1| + k + 1)\delta_1^k C_k. \qquad (7.37)$$

On subtracting (7.37) from (7.36) and rearranging the terms, we get

$$n(n - |m_1 - m_2|)\delta_1 C_n = \big[M(|m_1| + n) + (n - 1)(n - 1 - |m_1 - m_2|)\big]C_{n-1}.$$

Therefore one has

$$\lim_{n\to\infty}\frac{C_n}{C_{n-1}} = \lim_{n\to\infty}\frac{M(|m_1| + n) + (n - 1)(n - 1 - |m_1 - m_2|)}{n(n - |m_1 - m_2|)\delta_1} = \frac{1}{\delta_1}.$$

As before, we use the ratio test to discuss the convergence of the series $\sum_{k=0}^{\infty} C_n|x|^n$. Consider,

$$\lim_{n\to\infty}\frac{C_{n+1}|x|^{n+1}}{C_n|x|^n} = |x| \lim_{n\to\infty}\frac{C_{n+1}}{C_n} = \frac{|x|}{\delta_1}.$$

From the ratio test, the series $\sum_{n=0}^{\infty} C_n|x|^n$ converges when $|x| < \delta_1$.

Since $|c_n| \leq C_n$, $n \geq 0$, owing to the comparison test, the series $\sum_{n=0}^{\infty} c_n x^n$ converges whenever $|x| < \delta_1$. As δ_1 is arbitrary, we have the convergence of the series $\sum_{n=0}^{\infty} c_n x^n$, $x \in J$.

Step 3. To find the other solution, we start with $m = m_2$ in (7.31) to get a sequence $(\tilde{c}_n)$ given by

$$f(m_2 + n)\tilde{c}_n = -\sum_{k=0}^{n-1}\big((m_2 + k)p_{n-k} + q_{n-k}\big)\tilde{c}_k, \qquad (7.38)$$

and $\tilde{c}_0$ is arbitrary. We note that $f(m_2 + n)$ never vanishes for any $n \in \mathbb{N}$, because $m_1 - m_2 \notin \mathbb{N} \cup \{0\}$. Hence $\tilde{c}_n$ is well defined once we fix $\tilde{c}_0$. The convergence of $\sum\limits_{n=0}^{\infty} \tilde{c}_n x^n$ can be established along the same lines as Step 2. $\square$

There are a couple of remarks which are immediate from the proof of Theorem 7.4.4.

Remark 7.4.5. In the case when $m_1 = m_2$, in view of (7.33) and (7.38) one has $c_n = \tilde{c}_n$, $n \geq 1$. In other words, there exists a unique Frobenius solution in this case. The form of the second solution is discussed at the end of this section.

Remark 7.4.6. In the case when $(m_1 - m_2) \in \mathbb{N}$, one Frobenius solution $\psi_1(x) = |x - x_0|^{m_1} \sum\limits_{n=0}^{\infty} c_n(x - x_0)^n$ exists where c_0 is arbitrary and c_n for $n \geq 1$ is given by (7.33). We now investigate the existence of the second Frobenius solution independent of ψ_1. We first observe that the left hand side of (7.38) vanishes at $n = m_1 - m_2$. If the right hand side (RHS) of (7.38) also vanishes, then we take $\tilde{c}_{m_1-m_2}$ as an arbitrary constant and proceed further to find the second Frobenius solution as $\psi_2(x) = |x - x_0|^{m_2} \sum\limits_{n=0}^{\infty} \tilde{c}_n(x - x_0)^n$. Suppose the RHS does not vanish for $n = m_1 - m_2$, then there exists no second Frobenius solution. The form of the second solution is given at the end of this section.

Remark 7.4.7. If the roots of the indicial equation are complex, then (c_n) is a sequence of complex numbers. The tools in complex analysis that are required to deal with the convergence of the series $\sum\limits_{n=0}^{\infty} c_n(x - x_0)^n$ are out of the scope of this book.

We present an example where we find Frobenius solutions using the technique in Theorem 7.4.4.

Example 7.4.8. Solve $2x^2 y'' + xy' - (x + 1)y = 0$.
Solution. We write the given equation in the standard form, i.e.,

$$y'' + \frac{1}{2x} y' - \frac{(x + 1)}{2x^2} y = 0.$$

Here $a_1(x) = \frac{1}{2x}$, and $a_2(x) = \frac{x+1}{2x^2}$. Clearly, $x = 0$ is a regular singular point. Let

$$\phi(x) = x^m \sum_{k=0}^{\infty} c_n x^n.$$

As in the proof of Theorem 7.4.4, we compute ϕ', ϕ'', substitute them in the given equation and cancel x^{m-2} to find that (see (7.28))

$$\sum_{n=0}^{\infty} \left(\left[(n+m)(n+m-1) + \frac{(m+n)}{2} - \frac{1}{2} \right] c_n - \frac{1}{2} c_{n-1} \right) x^n = 0, \quad (7.39)$$

where $c_{-1} = 0$. The indicial equation is obtained by equating the constant term in the above series to zero, i.e.,

$$c_0 \left(m(m-1) + \tfrac{1}{2} - \tfrac{1}{2} \right) = 0,$$

or

$$c_0 (2m^2 - m - 1) = 0.$$

Hence the roots of the indicial equation are $m_1 = 1$ and $m_2 = -\frac{1}{2}$. Therefore from Theorem 7.4.4, there exist two Frobenius solutions. To find the first Frobenius solution, we substitute $m = 1$ in (7.39) and equate coefficient of x^n, to zero for all $n \in \mathbb{N}$. On equating the coefficient of x to zero, we have

$$c_1 \left(2 + 1 - \frac{1}{2} \right) - \frac{c_0}{2} = 0, \text{ or } c_1 = \frac{c_0}{5}.$$

Similarly, if we equate the coefficient of x^2 to zero, then we find that

$$c_2 \left(6 + \tfrac{3}{2} - \tfrac{1}{2} \right) - \tfrac{c_1}{2} = 0 \text{ or } c_2 = \frac{c_1}{7} = \frac{c_0}{70}.$$

One has to proceed in the same way to get the other coefficients of the series. Hence

$$\psi_1(x) = x \left(1 + \frac{x}{5} + \frac{x^2}{70} + \cdots \right), \ x \in \mathbb{R},$$

is a solution to the given equation.

To get the second Frobenius solution put $m = -\frac{1}{2}$ in and repeat the same process as before. On equating the coefficients of x and x^2 to zero, we get $\tilde{c}_1 = -\tilde{c}_0$, and $\tilde{c}_2 = -\frac{1}{2}\tilde{c}_0$, respectively. Therefore

$$\psi_2(x) = |x|^{-\frac{1}{2}} \left(1 - x - \frac{x^2}{2} + \cdots \right), \ x \neq 0,$$

is another Frobenius solution. $\qquad \square$

We now turn our attention toward the case when we have equal roots for the indicial equation $f(m) = 0$. Let $m = m_1$ be the repeated root of $f(m) = 0$. This is the same as $f(m_1) = f'(m_1) = 0$. We now define the operator

$$\mathscr{L}y := y'' + a_1 y' + a_0 y.$$

Let c_n be given by (7.31) for $n \geq 1$. Then we note that c_n's depend on m. We suppose $\sum_{n=0}^{\infty} c_n x^n$, $|x| < \delta$ converges and set $\psi(x; m) = |x|^m \sum_{n=0}^{\infty} c_n x^n$, $0 < |x| < \delta$. As usual, we first consider the case when $x > 0$. In view of (7.32), the following identity holds

$$\mathscr{L}\psi = c_0 f(m) x^{m-2}, \quad (7.40)$$

where we take c_0 to be a constant that does not depend on m.
On differentiating (7.40) with respect to m, we obtain

$$\frac{\partial}{\partial m}\mathscr{L}\psi = \mathscr{L}\left(\frac{\partial \psi}{\partial m}\right) = c_0\big(f'(m) + \log(x)f(m)\big)x^{m-2}. \tag{7.41}$$

We now put $m = m_1$ in (7.41) to arrive at

$$\mathscr{L}\left(x^{m_1}\sum_{n=0}^{\infty}\frac{dc_n}{dm}(m_1)x^n + (\log x)\psi\right) = c_0\big(f'(m_1) + \log(x)f(m_1)\big)x^{m_1-2} = 0.$$

The case when $x < 0$ can be dealt with in the same way to obtain the following result:

The two linearly independent solutions to (7.25) for $0 < |x| < \delta$ are given by

$$\phi(x) = |x|^{m_1}\sum_{n=0}^{\infty}c_n x^n, \quad \tilde{\phi}(x) = |x|^{m_1}\sum_{n=0}^{\infty}\frac{dc_n}{dm}(m_1)x^n + \phi(x)\log|x|,$$

where m_1 is the repeated root of the indicial equation and c_n, are given by (7.31) for $n \geq 1$. Since c_0 is independent of m, the first term in the series in $\tilde{\phi}$ is zero.

On the other hand, suppose that m_1 and m_2 are the zeros of the indicial equation with $m_1 - m_2 \in \mathbb{N}$. In this case also, one can show that a solution to (7.25) which is not given in Theorem 7.4.4 is

$$\tilde{\phi}(x) = |x|^{m_2}\sum_{n=0}^{\infty}\frac{dc_n}{dm}(m_2)x^n + c\phi(x)\log|x|,$$

where c is a constant and ϕ is a Frobenius solution corresponding to the root $m = m_1$. For details see [8].

We conclude the section by summarizing our discussion pertaining to the case $|m_1 - m_2| \in \mathbb{N} \cup \{0\}$ in the following theorem.

Theorem 7.4.9. *Assume that x_0 is a regular singular point of*

$$y''(x) + a_1(x)y'(x) + a_2(x)y(x) = 0,$$

$(x - x_0)a_1(x)$ and $(x - x_0)^2 a_2(x)$ are analytic in $|x - x_0| < \delta$. Let m_1 and m_2 be the roots of the indicial equation. Then the following hold true.

(a) If $m_1 = m_2$, then there exist two solutions in $(x_0 - \delta, x_0 + \delta)\backslash\{x_0\}$ of the form

$$\phi(x) = |x - x_0|^{m_1}\sum_{n=0}^{\infty}c_n(x - x_0)^n,$$

$$\tilde{\phi}(x) = |x - x_0|^{m_1+1}\sum_{n=0}^{\infty}\tilde{c}_n(x - x_0)^n + \phi(x)\log|x - x_0|,$$

where $c_0 \neq 0$, both $\sum_{n=0}^{\infty}c_n(x - x_0)^n$ and $\sum_{n=0}^{\infty}\tilde{c}_n(x - x_0)^n$ converge in J.

(b) If $m_1 - m_2 \in \mathbb{N}$, then there exist two solutions in $(x_0 - \delta, x_0 + \delta) \backslash \{x_0\}$ of the form

$$\phi(x) = |x - x_0|^{m_1} \sum_{n=0}^{\infty} c_n (x - x_0)^n,$$

$$\tilde{\phi}(x) = |x - x_0|^{m_2} \sum_{n=0}^{\infty} \tilde{c}_n (x - x_0)^n + c\phi(x) \log|x - x_0|,$$

where $c_0 \neq 0$, $\tilde{c}_0 \neq 0$, $c \in \mathbb{R}$, both $\sum_{n=0}^{\infty} c_n(x-x_0)^n$ and $\sum_{n=0}^{\infty} \tilde{c}_n(x-x_0)^n$ converge in J. □

7.5 Bessel's equation

The objective of this section is to find the general solution to the Bessel equation and study some properties of its solutions. Bessel's equation appears naturally in many places in mathematical physics. For example, it arises in the study of thin circular vibrating membrane, planetary motion, potential theory etc. Bessel's equation (functions) are treated at length in many books on differential equations, see for instance, [14, 36]. The ODE

$$x^2 y''(x) + xy'(x) + (x^2 - \alpha^2)y(x) = 0 \tag{7.42}$$

is called the Bessel equation of order $\alpha \geq 0$.

Observe that $x = 0$ is a regular singular point of (7.42). From the previous section, it is clear that there exists at least one Frobenius solution. Therefore we find (c_n) such that

$$\phi(x) = x^m \sum_{n=0}^{\infty} c_n x^n.$$

is a solution to (7.42). As usual we compute ϕ', ϕ'', substitute in (7.42) and equate the coefficient of each power of x to zero to obtain

$$m(m - 1) + m - \alpha^2 = 0 \text{ or } m = \pm\alpha, \tag{7.43}$$

$$((m + 1)^2 - \alpha^2)c_1 = 0, \tag{7.44}$$

$$((n + m)^2 - \alpha^2)c_n + c_{n-2} = 0, \ n \geq 2. \tag{7.45}$$

On substituting $m = \alpha$ in (7.44), we find that $c_1 = 0$. Moreover from (7.45) we get

$$c_n = \frac{-c_{n-2}}{n(n + 2\alpha)}, \ n \geq 2.$$

Since $c_1 = 0$, it follows that $c_3 = c_5 = \cdots = 0$. We now consider

$$
\begin{aligned}
c_{2k} &= \frac{-c_{2k-2}}{2^2 k(k+\alpha)} \\
&= \frac{c_{2k-4}}{2^4 k(k-1)(k+\alpha)(k-1+\alpha)} \\
&= \frac{-c_{2k-6}}{2^6 k(k-1)(k-2)(k+\alpha)(k-1+\alpha)(k-2+\alpha)} \\
&\ \vdots \\
&= \frac{(-1)^k c_0}{2^{2k}(k!)(k+\alpha)(k-1+\alpha)\cdots(2+\alpha)(1+\alpha)}.
\end{aligned}
$$

Therefore a solution to (7.42) is given by

$$
\psi(x) = c_0 x^\alpha \sum_{k=0}^\infty \frac{(-1)^k x^{2k}}{2^{2k}(k!)(k+\alpha)(k-1+\alpha)\cdots(2+\alpha)(1+\alpha)}, \quad x > 0. \quad (7.46)
$$

If we take $c_0 = \frac{1}{2^\alpha \Gamma(\alpha+1)}$, where Γ is the Euler gamma function[2] in (7.46), then the solution ψ to (7.42) is called the Bessel function of order α of the first kind and is denoted by J_α. Using the gamma function we write

$$
J_\alpha(x) = \left(\frac{x}{2}\right)^\alpha \sum_{k=0}^\infty \frac{(-1)^k}{k!\Gamma(k+\alpha+1)} \left(\frac{x}{2}\right)^{2k}, \quad x > 0. \quad (7.47)
$$

Suppose 2α is not an integer, then the hypothesis of Theorem 7.4.4 holds true. Therefore another Frobenius solution to (7.42) exists and is given by

$$
J_{-\alpha}(x) = \left(\frac{x}{2}\right)^{-\alpha} \sum_{k=0}^\infty \frac{(-1)^k}{k!\Gamma(k-\alpha+1)} \left(\frac{x}{2}\right)^{2k}, \quad x > 0. \quad (7.48)
$$

If $\alpha \notin \mathbb{N}$ and $2\alpha \in \mathbb{N}$, then we can observe that $J_{-\alpha}$ given in (7.48) is another solution to (7.42). If α is a positive integer, then there is a solution to the Bessel equation apart from the Frobenius solution (7.46), which is called the Bessel function of order α of the second kind. For details see [8, 36].

We now prove some properties of the Bessel function J_α.

Theorem 7.5.1. *The following properties hold for Bessel's functions:*

(i) $J_0'(x) = -J_1(x)$.

(ii)(a) $\dfrac{d}{dx}\left[x^\alpha J_\alpha(x)\right] = x^\alpha J_{\alpha-1}(x)$.

---

[2]The gamma function is defined as $\Gamma(s) := \int_0^\infty e^{-x} x^{s-1}\,dx$, $s > 0$. It is easy to verify that the gamma function satisfies $\Gamma(s+1) = s\Gamma(s)$. This definition can be extended to the set $s \in \mathbb{R}\backslash\{-n : n \in \mathbb{N} \cup \{0\}\}$ using this relation.

$(ii)(b)$ $\dfrac{d}{dx}\left[x^{-\alpha}J_\alpha(x)\right] = -x^{-\alpha}J_{\alpha+1}(x).$

$(iii)(a)$ $J_{\alpha-1}(x) + J_{\alpha+1}(x) = 2\alpha x^{-1}J_\alpha(x).$

$(iii)(b)$ $J_{\alpha-1}(x) - J_{\alpha+1}(x) = 2J'_\alpha(x).$

Proof. (i) From formula (7.47) we get that

$$J_0(x) = \sum_{k=0}^{\infty} \frac{(-1)^k}{(k!)^2}\left(\frac{x}{2}\right)^{2k}, \quad J_1(x) = \sum_{k=0}^{\infty}\frac{(-1)^k}{(k+1)!\,k!}\left(\frac{x}{2}\right)^{2k+1}, \quad x > 0.$$

On differentiating J_0 we obtain

$$J'_0(x) = \sum_{k=1}^{\infty}\frac{(-1)^k}{(k-1)!\,k!}\left(\frac{x}{2}\right)^{2k-1} = \sum_{k=0}^{\infty}\frac{(-1)^{k+1}}{(k+1)!\,k!}\left(\frac{x}{2}\right)^{2k+1} = -J_1(x).$$

This proves (i).

$(ii)(a)$ We notice that

$$x^\alpha J_\alpha(x) = \sum_{k=0}^{\infty}\frac{(-1)^k 2^\alpha}{k!\Gamma(k+\alpha+1)}\left(\frac{x}{2}\right)^{2k+2\alpha}, \quad x > 0.$$

On differentiating both sides with respect to x, we find

$$
\begin{aligned}
\frac{d}{dx}\left[x^\alpha J_\alpha(x)\right] &= \sum_{k=0}^{\infty}\frac{(-1)^k 2^\alpha}{k!}\frac{(k+\alpha)}{\Gamma(k+\alpha+1)}\left(\frac{x}{2}\right)^{2k+2\alpha-1} \\
&= x^\alpha \left(\frac{x}{2}\right)^{\alpha-1}\sum_{k=0}^{\infty}\frac{(-1)^k}{k!\Gamma(k+\alpha)}\left(\frac{x}{2}\right)^{2k} \\
&= x^\alpha J_{\alpha-1}(x).
\end{aligned}
$$

This completes the proof of $(ii)(a)$. Using the similar arguments. one can easily prove $(ii)(b)$.

$(iii)(a)$ After canceling the common factor $x^{\alpha-1}$ in $(ii)(a)$ we obtain

$$xJ'_\alpha(x) + \alpha J_\alpha(x) = xJ_{\alpha-1}(x). \tag{7.49}$$

Similarly from $(ii)(b)$ we get

$$xJ'_\alpha(x) - \alpha J_\alpha(x) = -xJ_{\alpha+1}(x). \tag{7.50}$$

On subtracting (7.50) from (7.49), we obtain that $(iii)(a)$ holds.

On adding (7.49) and (7.50) we get $(iii)(b)$. $\qquad\square$

Theorem 7.5.2. *For $\alpha > 0$, the following statements are true.*

(i) *Between any two positive zeros of $J_{\alpha+1}$ (resp. J_α) there is a zero of J_α (resp. $J_{\alpha+1}$).*

(ii) *The function J_α has infinitely many positive zeros.*

Proof. (*i*) Let x_1 and x_2 be two zeros of $J_{\alpha+1}$ in $(0, \infty)$. We set $g(x) := x^{\alpha+1} J_{\alpha+1}(x)$, $x > 0$, then $g(x_1) = g(x_2) = 0$. By Rolle's theorem there exists $\eta \in (x_1, x_2)$ such that $g'(\eta) = 0$. In view of Theorem 7.5.1 (*ii*)(*a*), we have $J_\alpha(\eta) = 0$. This shows that between any two positive zeros of $J_{\alpha+1}$ there is a zero of J_α.

In order to prove the remaining part, we use the same argument with $g(x) = x^{-\alpha} J_\alpha(x)$, $x > 0$ and Theorem 7.5.1 (*ii*)(*b*).

(*ii*) We first prove that J_α has infinitely many positive zeros for $-\frac{1}{2} < \alpha \le \frac{1}{2}$. For, we set $u(x) := \sqrt{x} J_\alpha(x)$. (See Chapter 3, Section 3.2.4 for motivation to consider this u.) Then u satisfies

$$u''(x) + \left(1 + \frac{1 - 4\alpha^2}{4x^2}\right) u = 0, \ x > 0. \tag{7.51}$$

Now comparing (7.51) with $y'' + y = 0$ and using the Sturm comparison theorem, we conclude that J_α has infinitely many positive zeros if α lies in $\left(-\frac{1}{2}, \frac{1}{2}\right]$.

For any $\alpha > \frac{1}{2}$, choose $m \in \mathbb{N} \cup \{0\}$ such that $-\frac{1}{2} < \alpha - m \le \frac{1}{2}$. Since $J_{\alpha-m}$ has infinitely many positive zeros, in view of (*i*) of this Theorem, $J_{\alpha+1-m}$ has infinitely many zeros. Proceeding in the same way for m times, we can conclude that J_α has infinitely many positive zeros. $\qquad \square$

We conclude this section by providing the orthogonality result for Bessel functions. To this end, let $\lambda \in \mathbb{R}$, $\psi(x) := J_\alpha(\lambda x)$. Then we have

$$\psi'(x) = \lambda J'_\alpha(\lambda x), \quad \psi''(x) = \lambda^2 J''_\alpha(\lambda x).$$

We now consider the identity

$$\lambda^2 x^2 J''_\alpha(\lambda x) + \lambda x J'_\alpha(\lambda x) + (\lambda^2 x^2 - \alpha^2) J_\alpha(\lambda x) = 0,$$

which is the same as

$$x^2 \psi''(x) + x\psi'(x) + (\lambda^2 x^2 - \alpha^2)\psi(x) = 0.$$

Therefore ψ satisfies

$$x\big(x\psi'(x)\big)' + (\lambda^2 x^2 - \alpha^2)\psi(x) = 0. \tag{7.52}$$

Similarly, if $\phi(x) := J_\alpha(\mu x)$ then we obtain

$$x\big(x\phi'(x)\big)' + (\mu^2 x^2 - \alpha^2)\phi(x) = 0. \tag{7.53}$$

On multiplying (7.52) and (7.53) with ϕ and ψ, respectively, and subtracting one from the other, we get

$$x\big(x\phi(x)\psi'(x) - x\psi(x)\phi'(x)\big)' + x^2(\lambda^2 - \mu^2)\psi(x)\phi(x) = 0.$$

Now integrating from 0 to a, we find

$$(\lambda^2 - \mu^2) \int_0^a x\psi(x)\phi(x)dx = a\big(\psi(a)\phi'(a) - \phi(a)\psi'(a)\big). \tag{7.54}$$

We now state the orthogonality property of the Bessel function in the following theorem.

Theorem 7.5.3. *Let $a > 0, \alpha \geq 0$. Suppose λ, μ are such that $\lambda \neq \mu$ and $J_\alpha(\lambda a) = J_\alpha(\mu a) = 0$, then the following orthogonality property holds:*

$$\int_0^a xJ_\alpha(\lambda x)J_\alpha(\mu x)dx = 0. \tag{7.55}$$

Proof. Proof follows immediately from (7.54). □

7.6 Regular singular points at infinity

In this section, we study the behavior of the solution to

$$y''(x) + a_1(x)y'(x) + a_2(x)y(x) = 0, \tag{7.56}$$

for $|x|$ large enough. Let ϕ be a solution to (7.56) in $|x| > r$ for some $r > 0$. By putting $t = \frac{1}{x}$ and changing the independent variable in (7.56) to t, we get another equation. We want to say that $x = \infty$ is a regular singular point of (7.56) whenever $t = 0$ is a regular singular point of the new equation. To be more precise, we set $\psi(t) := \phi(\frac{1}{t})$. Then it follows that

$$\frac{d\psi}{dt}(t) = -\frac{1}{t^2}\frac{d\phi}{dx}\left(\frac{1}{t}\right), \quad \frac{d^2\psi}{dt^2}(t) = \frac{1}{t^4}\frac{d^2\phi}{dx^2}\left(\frac{1}{t}\right) + \frac{2}{t^3}\frac{d\phi}{dx}\left(\frac{1}{t}\right), \quad 0 < |t| < \frac{1}{r}.$$

Hence we get

$$\frac{d\phi}{dx}\left(\frac{1}{t}\right) = -t^2\frac{d\psi}{dt}(t), \quad \frac{d^2\phi}{dx^2}\left(\frac{1}{t}\right) = t^4\frac{d^2\psi}{dt^2}(t) + 2t^3\frac{d\psi}{dt}(t). \tag{7.57}$$

In view of (7.56), we have

$$\frac{d^2\phi}{dx^2}\left(\frac{1}{t}\right) + a_1\left(\frac{1}{t}\right)\frac{d\phi}{dx}\left(\frac{1}{t}\right) + a_0\left(\frac{1}{t}\right)\phi\left(\frac{1}{t}\right) = 0, \quad 0 < |t| < \frac{1}{r}. \tag{7.58}$$

From (7.57)–(7.58), we obtain

$$t^4\frac{d^2\psi}{dt^2}(t) + \left(2t^3 - t^2a_1\left(\frac{1}{t}\right)\right)\frac{d\psi}{dt}(t) + a_0\left(\frac{1}{t}\right)\psi(t) = 0, \quad 0 < |t| < \frac{1}{r}.$$

Therefore the transformed equation near $t = 0$ is given by

$$\frac{d^2\psi}{dt^2}(t) + \left(\frac{2}{t} - \frac{1}{t^2}a_1\left(\frac{1}{t}\right)\right)\frac{d\psi}{dt}(t) + \frac{1}{t^4}a_0\left(\frac{1}{t}\right)\psi(t) = 0, \quad 0 < |t| < \frac{1}{r}. \tag{7.59}$$

Definition 7.6.1. We say that $x = \infty$ (or infinity) is a regular singular point of (7.56) if $t = 0$ is a regular singular point of (7.59).

Example 7.6.2. Find a condition that integers β and γ satisfy such that $x = \infty$ is a regular singular point of

$$y''(x) + ax^\beta y'(x) + bx^\gamma y(x) = 0.$$

Solution: Here we have $a_1(x) = ax^\beta$, and $a_2(x) = bx^\gamma$. In this case the transformed equation (7.59) becomes

$$\frac{d^2\psi}{dt^2}(t) + \left(\frac{2}{t} - \frac{a}{t^{2+\beta}}\right)\frac{d\psi}{dt}(t) + \frac{b}{t^{4+\gamma}}\psi(t) = 0. \tag{7.60}$$

Thus $t = 0$ is a regular singular point of (7.60) if and only if both

$$2 - \frac{a}{t^{1+\beta}} \text{ and } \frac{b}{t^{2+\gamma}}$$

are analytic in a neighborhood of $t = 0$. Therefore $t = 0$ is a regular singular point of (7.60) or $x = \infty$ is a regular singular point of the given equation if and only if $\beta \le -1$ and $\gamma \le -2$. □

On the other hand, $x = 0$ is a regular singular point for the equation given in Example 7.6.2 if and only if $\beta \ge -1$ and $\gamma \ge -2$. Hence both $x = 0$ and $x = \infty$ are regular singular points for the problem given in Example 7.6.2 if and only if $\beta = -1$ and $\gamma = -2$. Therefore we have proved that the Euler equation has regular singular points at $x = 0$ and $x = \infty$.

Example 7.6.3. Show that $x = \infty$ is not a regular singular point of the Bessel equation

$$x^2 y''(x) + xy'(x) + (x^2 - \alpha^2)y(x) = 0.$$

Solution. We write the Bessel equation in the standard form (7.56) to find that

$$a_1(x) = \frac{1}{x}, \ a_2(x) = 1 - \frac{\alpha^2}{x^2}.$$

Then the transformed equation near $t = 0$ becomes (see (7.59))

$$\frac{d^2\psi}{dt^2}(t) + \frac{1}{t}\frac{d\psi}{dt}(t) + \frac{1}{t^4}(1 - \alpha^2 t^2)\psi(t) = 0. \tag{7.61}$$

Clearly $t = 0$ is not a regular singular point of (7.61) and hence $x = \infty$ is not a regular singular point of the Bessel equation. □

Exercise 7.1. Solve the following equations using the power series method:

(i) $y'' - 2y' + y = 0$.

(ii) $y'' - 3y' + 2y = 0$.

(*iii*) $y'' + x^2 y' + x^2 y = 0.$

(*iv*) $(1 + x^2)y'' + y = 0.$

(*v*) $y'' + (x-1)^2 y' - (x-1)y = 0.$

Exercise 7.2. For $n \in \mathbb{N}$, show that P_n has exactly n real roots in $[-1, 1]$.

Hint: Define $h(x) := (x^2 - 1)^n$. From the proof of Lemma 7.3.5 we have $h^{(k)}(\pm 1) = 0$, $0 \le k \le n-1$. Use Rolle's theorem repeatedly n times to get that $h^{(n)}$ has n roots in $[-1, 1]$.

Exercise 7.3. For $n \in \mathbb{N}$, show that:

(*i*) $\displaystyle \int_{-1}^{1} x P_{n-1}(x) P_n(x) dx = \frac{2n}{4n^2 - 1}.$

(*ii*) $\displaystyle \int_{-1}^{1} x P_n'(x) P_n(x) dx = \frac{2n}{2n + 1}.$

(*iii*) $\displaystyle \int_{-1}^{1} P_{n+1}'(x) P_n(x) dx = 2.$

Hint: Use Lemma 7.3.5 with $f = x P_{n-1}$, $x P_n'$, and P_{n+1}' to prove (*i*), (*ii*), and (*iii*), respectively. Use (7.19) and Claim in Theorem 7.3.6 if necessary.

Exercise 7.4. For $n \in \mathbb{N}$, show that

$$(n+1)P_{n+1}(x) - (2n+1)x P_n(x) + n P_{n-1}(x) = 0, \ x \in [-1, 1].$$

Hint: Since $x P_n$ is a polynomial of degree $(n+1)$, write $x P_n = \sum_{k=0}^{n+1} \alpha_k P_k$. To find α_k, we multiply on both sides with P_k, use Theorem 7.3.6(*ii*) and Exercise 7.3 (*i*).

Exercise 7.5. (*i*) For $n \in \mathbb{N}$, show that

$$P_{n+1}'(x) - P_{n-1}'(x) = (2n+1)P_n(x), \ x \in [-1, 1].$$

(*ii*) For $n \in \mathbb{N} \cup \{0\}$, show that

$$P_{n+1}'(x) = x P_n'(x) + (n+1)P_n(x), \ x \in [-1, 1].$$

(*iii*) For $n \in \mathbb{N}$, show that

$$n P_n(x) = x P_n'(x) - P_{n-1}'(x), \ x \in [-1, 1].$$

(*iv*) For $n \in \mathbb{N}$, show that

$$(1 - x^2)P_n'(x) = n P_{n-1}(x) - n x P_n(x), \ x \in [-1, 1].$$

Hint: To prove (*i*), let $f(x) := (x^2 - 1)^{n+1}$. Then we find that

$$f''(x) = 2(n+1)(2n+1)(x^2 - 1)^n + 4n(n+1)(x^2 - 1)^{n-1}.$$

On dividing both sides with $(n+1)!2^{n+1}$ and differentiating n times we get (*i*). We use (*i*) and (*ii*) in order to prove (*iii*). For (*iv*), we multiply (*iii*) with x to get a new identity, replace n in (*ii*) by $(n-1)$ and subtract from the new identity.

Exercise 7.6. Check whether $x = 0$ is a regular singular point of the following ODEs. If so, check whether any Frobenius solutions exist. Moreover, find the Frobenius solutions whenever they exist.

(i) $xy'' + 9y = 0$.

(ii) $16x^2y'' + 16xy' + (16x^2 - 1)y = 0$.

(iii) $x^2y'' + (x^2 - 5x)y' + y = 0$.

(iv) $x^2y'' - xy' + y = 0$.

(v) $x^2y'' + (\cos x)y' + (\sin x)y = 0$.

(vi) $x^2y'' + (\sin 2x)y' + (\cos 2x)y = 0$.

Exercise 7.7. Suppose $\alpha \geq 0$, $a > 0, \mu > 0$. If $J(\mu a) = 0$ then show that

$$\int_0^a xJ_\alpha^2(\mu x)dx = \frac{a^2}{2}J_{\alpha+1}^2(\mu a).$$

Hint: Multiply (7.53) with ϕ' to get

$$\frac{d}{dx}\left(x\phi'(x)\right)^2 + (\mu^2x^2 - \alpha^2)\frac{d}{dx}\left(\phi^2(x)\right) = 0.$$

On integrating this equation over $[0, a]$, using the integration by parts and (7.49) one gets the required result.

Exercise 7.8. Check whether $x = \infty$ is a regular singular point for the following ODEs:

(i) $(1 - x^2)y'' - 2xy' + 6y = 0$.

(ii) $y'' - 20y' + 7y = 0$.

(iii) $(1 - x^3)y'' - x^2y' + y = 0$.

Chapter 8

The Laplace transforms

8.1 Introduction

In this chapter, we introduce a linear transformation called the Laplace transform on a 'large' class of functions. In view of its very interesting properties, the Laplace transform is very useful in solving linear ODEs. In particular, Laplace transform methods are widely used to solve linear the initial value problems and boundary value problems. In fact, methods of dealing with different types of differential/difference equations using the Laplace transforms, Fourier transforms, Mellin transforms, Z-transforms etc. have many applications in various branches of engineering sciences. Here we focus only on Laplace transforms.

8.2 Definition and properties

In this section, we define the Laplace transform and prove some of its properties which are useful in solving linear ODEs. The Laplace transform of the function $f(t)$, which is denoted by $F(s)$ or $\mathcal{L}[f(t)]$ is defined as

$$\mathcal{L}[f(t)] = F(s) = \int_0^\infty e^{-st} f(t)dt, \quad s > 0, \tag{8.1}$$

whenever the improper integral[1] exists.

Throughout the chapter:

(i) we take t as the variable in the functions before taking the Laplace transform and their Laplace transforms are functions of s.

(ii) unless specified otherwise we assume that $f(t)$ is a piecewise continuous function[2] on $[0, \infty)$.

---

[1] The improper integral $\displaystyle\int_0^\infty g(t)dt$ is defined as $\displaystyle\lim_{T\to\infty} \int_0^T f(t)dt$ whenever the limit exists.

[2] A function $g : [a, b] \to \mathbb{R}$ is said to be piecewise continuous, if (i) g is continuous at every point except at possibly finitely many points, (ii) at every point in (a, b) both the left

A very important fact which will be used often to compute the Laplace transform is the following: If $a > 0$ and $f(t)$ is a polynomial function then

$$\lim_{t\to\infty} e^{-at} f(t) = 0.$$

We now compute the Laplace transform of some elementary functions in the following example.

Example 8.2.1. Find the Laplace transform of each of the following functions defined on $(0,\infty)$:

(i) $f(t) = 1$. (ii) $f(t) = t$. (iii) $f(t) = t^n$, $n \in \mathbb{N}$. (iv) $f(t) = e^t$. (v) $f(t) = \sin t$. (vi) $f(t) = \cos t$.

Solution. (i) If $f(t) = 1$, $t > 0$, then

$$\int_0^T f(t)e^{-st}dt = \int_0^T e^{-st}dt = \frac{1}{s} - \frac{e^{-sT}}{s}.$$

Thus we have

$$F(s) = \int_0^\infty e^{-st}dt = \lim_{T\to\infty}\int_0^T e^{-st}dt = \frac{1}{s} - \lim_{T\to\infty}\frac{e^{-sT}}{s} = \frac{1}{s}.$$

(ii) If $f(t) = t$, $t > 0$, then consider

$$\int_0^T te^{-st}dt = -\frac{Te^{-sT}}{s} + \frac{1}{s^2} - \frac{e^{-sT}}{s^2}.$$

Hence we get

$$\mathcal{L}[t] = \int_0^\infty te^{-st}dt = \lim_{T\to\infty}\int_0^T te^{-st}dt = \frac{1}{s^2}.$$

(iii) For $n > 1$, we observe that

$$\int_0^T t^n e^{-st}dt = -\frac{T^n e^{-sT}}{s} + \frac{n}{s}\int_0^T t^{n-1}e^{-st}dt.$$

Let $T \to \infty$ to obtain

$$\mathcal{L}[t^n] = \frac{n}{s}\mathcal{L}[t^{n-1}] = \frac{n(n-1)}{s^2}\mathcal{L}[t^{n-2}] = \cdots = \frac{n!}{s^{n-1}}\mathcal{L}[t] = \frac{n!}{s^{n+1}}.$$

(iv) We now notice that

$$\int_0^T e^t e^{-st}dt = \frac{1 - e^{(1-s)T}}{s - 1}.$$

Suppose $s > 1$ then, after letting $T \to \infty$, we get

and the right limits exist, (iii) the left and the right limits exist at b and a, respectively. A function defined on $[0,\infty)$ is said to be piecewise continuous if it is piecewise continuous in $[0,T]$ for every $T > 0$.

$$\mathcal{L}[e^t] = \frac{1}{s-1}.$$

(*v*) A straightforward computation yields

$$\int_0^T e^{-st} \sin t dt = 1 - e^{-sT} \cos T - s \int_0^T e^{-st} \cos t dt. \tag{8.2}$$

Applying the integration by parts formula again to the integral on the right hand side and rearranging the terms we obtain

$$\int_0^T e^{-st} \sin t dt = \frac{1}{s^2+1} \left(1 - e^{-sT} \cos T - s e^{-sT} \sin T \right).$$

We now let $T \to \infty$ to find

$$\mathcal{L}[\sin t] = \frac{1}{s^2+1}.$$

(*vi*) Let $T \to \infty$ in (8.2) to observe that

$$\mathcal{L}[\sin t] = 1 - s\mathcal{L}[\cos t].$$

Hence we have

$$\mathcal{L}[\cos t] = \frac{s}{s^2+1}.$$

This completes the solution. □

The reader is advised to remember the formulae of the Laplace transforms of all the functions that are considered in Example 8.2.1.

Definition 8.2.2. A piecewise continuous function $f : [0, \infty) \to \mathbb{R}$ is said to be of exponential order if there exists $\alpha > 0$ and $M > 0$ such that $|f(t)| \le Me^{\alpha t}$, for every $t > 0$.

Example 8.2.3. Every bounded and piecewise continuous function is a function of exponential order, e.g., $\sin x$, $\cos x$, $\frac{1}{1+x^2}$. Furthermore, all polynomial functions are functions of exponential order. □

Remark 8.2.4. If f and g are functions of exponential order, then the functions (*i*) $f + g$ (*ii*) fg (*iii*) af, $a \in \mathbb{R}$ (*iv*) $\int_0^t f(\tau)d\tau$ are functions of exponential order.

On the other hand, there are many functions which grow faster than exponential functions at infinity. For instance, e^{at^2}, where $a > 0$, is not a function of exponential order (why?).

We now provide a sufficient condition for existence of the improper integral in (8.1).

Proposition 8.2.5. *Assume that $f : [0, \infty) \to \mathbb{R}$ is a piecewise continuous function with $|f(t)| \le Me^{\alpha t}$, $t > 0$. Then the Laplace transform of f exists, i.e., the improper integral in (8.1) exists for $s > \alpha$.*

Proof. The main result we use in this proof is the following.

If $h : [0, \infty) \to \mathbb{R}$ is integrable in every closed and bounded subinterval of $[0, \infty)$, and $\int_0^\infty |h(x)| dx$ exists, then $\int_0^\infty h(x) dx$ also exists.

Since f is a piecewise continuous function, $e^{-st} f(t)$ is integrable in $[T_1, T_2]$ for every $T_2 > T_1 > 0$. To complete the proof of the proposition it is enough to prove that

$$\int_0^\infty |f(t)| e^{-st} dt < \infty, \ s > \alpha. \tag{8.3}$$

To that end, we fix $s > \alpha$ and set

$$g(T) := \int_0^T e^{-st} |f(t)| dt.$$

Since the integrand in $g(T)$ is nonnegative, it follows that g is nondecreasing. In order to prove (8.3), we show that g is bounded. For, we consider

$$g(T) \leq M \int_0^T e^{-st} e^{\alpha t} dt = \frac{M}{s - \alpha} (1 - e^{-(s-\alpha)T}) \leq \frac{M}{s - \alpha}. \tag{8.4}$$

Since g is bounded, $\lim_{T \to \infty} g(T)$ exists. Hence the Laplace transform of f exists. $\square$

Lemma 8.2.6. *Suppose f is as in Proposition 8.2.5, then we get that*

$$\lim_{s \to \infty} F(s) = 0. \tag{8.5}$$

Proof. We consider

$$|F(s)| \leq \int_0^\infty |f(t)| e^{-st} dt \leq M \int_0^\infty e^{\alpha t} e^{-st} dt = \frac{M}{s - \alpha}.$$

We now let $s \to \infty$ to get the required result. $\square$

Now onward, unless stated otherwise, we assume that all the functions that are considered in this chapter are the functions which satisfy the hypotheses of Proposition 8.2.5.

There are some functions which do not satisfy the hypotheses of Proposition 8.2.5, but their Laplace transforms exist. In the next example, we present a function f such that (i) it is piecewise continuous (ii) it has Laplace transform (iii) it is not of exponential order.

Example 8.2.7. Let $f : [0, \infty) \to \mathbb{R}$ be defined by

$$f(t) = \begin{cases} e^{n^2}, \ n \leq t \leq n + e^{-n^2}, \ n \in \mathbb{N}, \\ 0, \ \text{otherwise.} \end{cases}$$

Then clearly f is a piecewise continuous map which is not of exponential order. We now show that the Laplace transform of f exists. For, since f is nonnegative, it is enough to show that $\int_0^\infty f(t) e^{-st} dt < \infty$. An important inequality which is used to show this is $1 - e^{-x} \leq x$, $x > 0$. Now for $s > 0$, consider

$$\int_0^\infty e^{-st} f(t)dt = \sum_{n=1}^\infty \int_n^{n+e^{-n^2}} e^{n^2} e^{-st} dt$$

$$= \frac{1}{s} \sum_{n=1}^\infty e^{n^2} e^{-sn} \left(1 - e^{-se^{-n^2}}\right)$$

$$\leq \sum_{n=1}^\infty e^{-sn}$$

$$= \frac{e^{-s}}{1 - e^{-s}} < \infty.$$

Thus the Laplace transform of f exists. □

We compute the Laplace transforms of some functions which are not differentiable/ continuous in the next few examples.

Example 8.2.8. Find the Laplace transform of

$$f(t) = \begin{cases} 0, & 0 \leq t \leq t_0, \\ t - t_0, & t_0 < t \leq t_1, \\ 2t_1 - t_0 - t, & t_1 < t \leq 2t_1 - t_0, \\ 0, & t > 2t_1 - t_0. \end{cases}$$

Solution. We first note that the given function is bounded, continuous but not differentiable. Since f vanishes in $[0, t_0) \cup [2t_1 - t_0, \infty)$, we obtain

$$F(s) = \int_{t_0}^{t_1} (t - t_0)e^{-st}dt + \int_{t_1}^{2t_1-t_0} (2t_1 - t_0 - t)e^{-st}dt$$

$$= -(t - t_0)\frac{e^{-st}}{s}\Big|_{t=t_0}^{t_1} + \int_{t_0}^{t_1} \frac{e^{-st}}{s}dt$$

$$-(2t_1 - t_0 - t)\frac{e^{-st}}{s}\Big|_{t=t_1}^{2t_1-t_0} - \int_{t_1}^{2t_1-t_0} \frac{e^{-st}}{s}dt$$

$$= -\frac{e^{-st}}{s^2}\Big|_{t=t_0}^{t_1} + \frac{e^{-st}}{s^2}\Big|_{t=t_1}^{2t_1-t_0}$$

$$= \frac{1}{s^2}\left(e^{-2st_1+st_0} - 2e^{-st_1} + e^{-st_0}\right)$$

$$= \frac{e^{st_0}}{s^2}\left(e^{-st_1} - e^{-st_0}\right)^2.$$

This completes the solution. □

Example 8.2.9. Find the Laplace transform of

$$f(t) = \begin{cases} a, & 0 \leq t \leq t_0, \\ b, & t > t_0. \end{cases}$$

Solution. We notice that the given function is bounded but discontinuous. We compute

$$
\begin{aligned}
F(s) &= \int_0^{t_0} ae^{-st}dt + \lim_{T\to\infty}\int_{t_0}^{T} be^{-st}dt \\
&= \frac{a}{s}\left(1 - e^{-st_0}\right) + \lim_{T\to\infty}\frac{b}{s}\left(e^{-st_0} - e^{-sT}\right) \\
&= \frac{a}{s} + e^{-st_0}\left(\frac{b}{s} - \frac{a}{s}\right).
\end{aligned}
$$

This completes the solution. $\qquad\square$

Example 8.2.10. Find the Laplace transform of

$$
f(t) = \begin{cases} a, & 0 \le t \le t_0, \\ b, & t_0 < t < t_1, \\ c, & t > t_1. \end{cases}
$$

Solution. As in the previous example, we consider

$$
\begin{aligned}
F(s) &= \int_0^{t_0} ae^{-st}dt + \int_{t_0}^{t_1} be^{-st}dt + \lim_{T\to\infty}\int_{t_1}^{T} ce^{-st}dt \\
&= \frac{a}{s}\left(1 - e^{-st_0}\right) + \frac{b}{s}\left(e^{-st_0} - e^{-st_1}\right) + \lim_{T\to\infty}\frac{c}{s}\left(e^{-st_1} - e^{-sT}\right) \\
&= \frac{a}{s} + e^{-st_0}\left(\frac{b}{s} - \frac{a}{s}\right) + e^{-st_1}\left(\frac{c}{s} - \frac{b}{s}\right).
\end{aligned}
$$

This completes the solution. $\qquad\square$

Remark 8.2.11. If we assume that $\epsilon, \alpha > 0$, $a = 0 = c$, $b = \frac{1}{2\epsilon}$, $0 < t_0 = \alpha - \epsilon$ and $t_1 = \alpha + \epsilon$ in the function given in Example 8.2.10 (such a function is called an impulse function) then the Laplace transform of that function becomes

$$
F(s) = e^{-s\alpha}\left(\frac{e^{s\epsilon} - e^{-s\epsilon}}{2\epsilon s}\right).
$$

In the next lemma, we derive a formula for the Laplace transform of periodic functions.

Lemma 8.2.12. *Assume that $f : [0, \infty) \to \mathbb{R}$ be a piecewise continuous function. If f is a periodic function with period T_0, i.e., $f(t + T_0) = f(t)$, $t \ge 0$, then the Laplace transform of f is given by*

$$
F(s) = \frac{\displaystyle\int_0^{T_0} e^{-st} f(t)dt}{1 - e^{-sT_0}}.
$$

Proof. From the periodicity of f, it follows that $f(t + nT_0) = f(t)$, $t \geq 0$, $n \in \mathbb{N}$. Moreover, f is bounded and hence of exponential order. We consider

$$
\begin{aligned}
F(s) &= \sum_{k=0}^{\infty} \int_{kT_0}^{(k+1)T_0} f(t)e^{-st}dt \\
&= \sum_{k=0}^{\infty} \int_0^{T_0} f(\tau + kT_0)e^{-s(\tau+kT_0)}d\tau \\
&= \sum_{k=0}^{\infty} e^{-ksT_0} \int_0^{T_0} f(\tau)e^{-s\tau}d\tau \\
&= \frac{\int_0^{T_0} e^{-st}f(t)dt}{1 - e^{-sT_0}}.
\end{aligned}
$$

This completes the proof of the lemma. $\qquad\qquad\qquad\qquad\qquad\qquad\square$

We now use Lemma 8.2.12 to compute the Laplace transform of some periodic functions.

Example 8.2.13. Use Lemma 8.2.12 to compute the Laplace transform of $\sin t$.

Solution. Since $\sin t$ is a periodic function of period 2π, hence we need to take $T_0 = 2\pi$ in Lemma 8.2.12. We consider

$$
\begin{aligned}
\int_0^{2\pi} e^{-st}\sin(t)dt &= -\cos(t)e^{-st}\Big|_{t=0}^{2\pi} - s\int_0^{2\pi}\cos(t)e^{-st}dt \\
&= 1 - e^{-2\pi s} - s(e^{-st}\sin(t))\Big|_{t=0}^{2\pi} - s^2\int_0^{2\pi}e^{-st}\sin(t)dt \\
&= 1 - e^{-2\pi s} - s^2\int_0^{2\pi}e^{-st}\sin(t)dt.
\end{aligned}
$$

Thus we get

$$
\int_0^{2\pi} e^{-st}\sin(t)dt = \frac{1 - e^{-2\pi s}}{1 + s^2}.
$$

Therefore in view of Lemma 8.2.12 we have

$$
\mathcal{L}[\sin t] = \frac{\displaystyle\int_0^{2\pi} e^{-st}\sin(t)dt}{1 - e^{-2\pi s}} = \frac{1}{1 + s^2},
$$

which is the same as what we obtained in Example 8.2.1(*v*). $\qquad\qquad\square$

Example 8.2.14. Find the Laplace transform of the function $f : [0, \infty) \to \mathbb{R}$ given by

$$f(t) = \begin{cases} a, & 0 \le t < T_1, \\ b, & T_1 \le t < 2T_1, \end{cases}$$

and $f(t + 2T_1) = f(t)$, $t > 0$.

Solution. We consider

$$
\begin{aligned}
\int_0^{2T_1} f(t) e^{-st} dt &= a \int_0^{T_1} e^{-st} dt + b \int_{T_1}^{2T_1} e^{-st} dt \\
&= \frac{a}{s}(1 - e^{-sT_1}) + \frac{be^{-sT_1}}{s}(1 - e^{-sT_1}).
\end{aligned}
$$

Then the function

$$
F(s) = \frac{\int_0^{2T_1} e^{-st} f(t) dt}{1 - e^{-2T_1 s}} = \frac{a + be^{-sT_1}}{s(1 + e^{-sT_1})}, \quad s > 0,
$$

is the required Laplace transform.   □

So far, we have computed the Laplace transform of functions from the first principle. We next prove some results which enable us to find the Laplace transform of more functions very efficiently. We now begin with the following elementary properties.

Theorem 8.2.15. *Let f, g satisfy hypotheses of Proposition 8.2.5. Then the following properties hold true.*

(i) (Linearity) $\mathcal{L}[af(t) + bg(t)] = a\mathcal{L}[f(t)] + b\mathcal{L}[g(t)]$, $a, b \in \mathbb{R}$.

(ii) (Shifting) $\mathcal{L}[e^{at} f(t)] = F(s - a)$, $s > a$.

(iii) (Scaling) If $c > 0$, then we have $\mathcal{L}[f(ct)] = \dfrac{1}{c}F(\dfrac{s}{c})$.

Proof. (i) Since f, g satisfy hypotheses of Proposition 8.2.5, the Laplace transforms of f, g and $af + bg$ exist. In view of the linearity of improper integrals, (i) follows immediately.

(ii) Since $f(t)$ is a function of exponential order, so is the function $e^{at} f(t)$. We consider

$$
\int_0^T e^{-st} e^{at} f(t) dt = \int_0^T e^{-(s-a)t} f(t) dt.
$$

Fix $s > a$, and let $T \to \infty$ to get

$$
\int_0^\infty e^{-st} e^{at} f(t) dt = \int_0^\infty e^{-(s-a)t} f(t) dt = F(s - a).
$$

(iii) We next consider

$$\mathcal{L}[f(ct)] = \lim_{T \to \infty} \int_0^T e^{-st} f(ct) dt = \lim_{T \to \infty} \int_0^{cT} \frac{1}{c} e^{-\frac{s}{c}\tau} f(\tau) d\tau = \frac{1}{c} F\left(\frac{s}{c}\right).$$

This proves (*iii*) and hence the theorem. □

In the following example, we use Theorem 8.2.15 and the formulae of Laplace transforms found in Example 8.2.1.

Example 8.2.16. Find the Laplace transform of each of the following functions:

(*i*) $\sin(\omega t)$. (*ii*) $\cos(\omega t)$. (*iii*) $e^{-at} \sin(\omega t)$, $a > 0$. (*iv*) $e^{5t} \cos(\omega t)$.

Solution. (*i*) We first recall that $\mathcal{L}[\sin t] = \dfrac{1}{1 + s^2}$. We first assume that $\omega > 0$. In view of the scaling property (see Theorem 8.2.15 (*iii*)), we get

$$\mathcal{L}[\sin(\omega t)] = \frac{1}{\omega} \frac{1}{\left(1 + \frac{s^2}{\omega^2}\right)} = \frac{\omega}{\omega^2 + s^2}, \quad s > 0. \tag{8.6}$$

For $\omega < 0$ also (8.6) holds due to $\sin(-\omega t) = -\sin(\omega t)$ and Theorem 8.2.15 (*i*).

(*ii*) As before we recall $\mathcal{L}[\cos t] = \dfrac{s}{1 + s^2}$. From the scaling property, we have

$$\mathcal{L}[\cos(\omega t)] = \frac{1}{\omega} \frac{\frac{s}{\omega}}{\left(1 + \frac{s^2}{\omega^2}\right)} = \frac{s}{\omega^2 + s^2}, \quad s > 0. \tag{8.7}$$

(*iii*) From the shifting property proved in Theorem 8.2.15 (*ii*) and (8.6) it follows that

$$\mathcal{L}[e^{-at} \sin(\omega t)] = \frac{\omega}{\omega^2 + (s + a)^2}, \quad s > 0.$$

(*iv*) Again, using the shifting property and (8.7) we obtain

$$\mathcal{L}[e^{5t} \cos(\omega t)] = \frac{s - 5}{\omega^2 + (s - 5)^2}, \quad s > 5.$$

This completes the solution. □

We now state and prove some useful properties of the Laplace transform which connect differentiation (integration) and multiplication (division) with polynomials. As a first result, we show that the Laplace transform of the derivative of a function is equal to the product of s and the Laplace transform of the function (up to a constant).

Theorem 8.2.17. *Assume that $f \in C^1([0, \infty))$ satisfies $|f(t)| \le M e^{\alpha t}$, $t > 0$. Then the Laplace transform of f' exists and is given by*

$$\mathcal{L}[f'(t)] = sF(s) - f(0), \quad s > \alpha. \tag{8.8}$$

Besides, if $f \in C^n([0,\infty))$ and $|f^{(k)}(t)| \leq Me^{\alpha t}$, $t > 0$, $0 \leq k \leq n-1$, then the Laplace transform of $f^{(n)}$ exists and

$$\mathcal{L}[f^{(n)}(t)] = s^n F(s) - s^{n-1} f(0) - s^{n-2} f'(0) - \cdots - s f^{(n-2)}(0) - f^{(n-1)}(0), \quad (8.9)$$

where $s > \alpha$.

Proof. From the integration by parts formula, we get

$$\int_0^T e^{-st} f'(t) dt = e^{-sT} f(T) - f(0) + s \int_0^T e^{-st} f(t) dt. \qquad (8.10)$$

For $s > \alpha$, we have $|e^{-sT} f(T)| \leq M e^{-(s-\alpha)T}$, $T > 0$ and thus $e^{-sT} f(T) \to 0$ as $T \to \infty$. Letting $T \to \infty$ on both sides of (8.10) we obtain (8.8).

We prove (8.9) by induction. Let $f \in C^{n+1}([0,\infty))$ be such that $|f^{(k)}(t)| \leq Me^{\alpha t}$, $t > 0$, $0 \leq k \leq n$. Furthermore, we assume that

$$\mathcal{L}[f^{(n)}](s) = s^n F(s) - \sum_{k=0}^{n-1} s^k f^{(n-1-k)}(0), \quad s > \alpha.$$

In view of (8.8) and the induction hypothesis, we get

$$
\begin{aligned}
\mathcal{L}[f^{(n+1)}(t)] &= s\mathcal{L}[f^{(n)}(t)] - f^{(n)}(0) \\
&= s\left(s^n F(s) - \sum_{k=0}^{n-1} s^k f^{(n-1-k)}(0) \right) - f^{(n)}(0) \\
&= s^{n+1} F(s) - \sum_{k=0}^{n} s^k f^{(n-k)}(0).
\end{aligned}
$$

This completes the proof of the theorem. □

In the next result, we establish a relation between the Laplace transform of the product of t and f in terms of the derivative of the Laplace transform of f.

Theorem 8.2.18. *Let $f : [0,\infty) \to \mathbb{R}$ be a continuous function. Assume that $M, \alpha > 0$ are such that $|f(t)|, |tf(t)| \leq Me^{\alpha t}$, $t > 0$. Then*

$$\mathcal{L}[tf(t)] = -F'(s), \quad s > \alpha. \qquad (8.11)$$

In general, for $n \in \mathbb{N}$, if $|t^k f(t)| \leq Me^{\alpha t}$, $t > 0$, $0 \leq k \leq n$, then

$$\mathcal{L}[t^n f(t)] = (-1)^n F^{(n)}(s), \quad s > \alpha. \qquad (8.12)$$

Proof. For $s > \alpha$ we consider

$$\mathcal{L}[tf(t)] = -\int_0^\infty \frac{\partial}{\partial s}(e^{-st}) f(t) dt = -\frac{d}{ds} \int_0^\infty e^{-st} f(t) dt.$$

This proves (8.11). For justification for interchanging the improper integral and the derivative (see [18]).

One can prove (8.12) using induction. Let $M, \alpha > 0$ be such that $|t^k f(t)| < Me^{\alpha t}$, $t > 0$, $1 \leq k \leq n+1$. Moreover we assume

$$\mathcal{L}[t^n f(t)] = (-1)^n F^{(n)}(s).$$

In view of the induction hypothesis and (8.11) we obtain

$$\mathcal{L}[t^{n+1} f(t)] = -\frac{d}{ds}\mathcal{L}[t^n f(t)] = (-1)^{n+1}\frac{d^{n+1}}{ds^{n+1}}F(s).$$

This completes the proof of the theorem. $\qquad\qquad\square$

We now state and prove a couple of results analogous to Theorems 8.2.17 –8.2.18 where we replace the derivative with the integral and multiplication with the division in the following sense.

Theorem 8.2.19. *Let f be a continuous function satisfying the hypotheses of Proposition 8.2.5. Then we have*

$$\mathcal{L}[\int_0^t f(\tau)d\tau] = \frac{F(s)}{s}, \quad s > \alpha. \tag{8.13}$$

Proof. We use Theorem 8.2.17 to prove this result. We set $g(t) = \int_0^t f(\tau)d\tau$. Then g is continuously differentiable. Moreover, we have

$$g'(t) = f(t), \, t > 0,$$

and

$$|g(t)| \leq \int_0^t |f(\tau)|d\tau \leq \int_0^t Me^{\alpha\tau}d\tau = \frac{M}{\alpha}e^{\alpha t} - \frac{M}{\alpha} \leq \frac{M}{\alpha}e^{\alpha t}.$$

Hence g is of exponential order. Therefore g satisfies the hypotheses of Theorem 8.2.17. In view of (8.8) we obtain

$$F(s) = \mathcal{L}[g'(t)] = s\mathcal{L}[g(t)] - g(0) = s\mathcal{L}[g(t)], \, s > \alpha.$$

Therefore we get

$$\mathcal{L}[g(t)] = \frac{F(s)}{s}, \quad s > \alpha.$$

This completes the proof of the theorem. $\qquad\qquad\square$

Theorem 8.2.20. *Let f be a continuous function satisfying the hypotheses of Proposition 8.2.5. Further we assume that $\frac{f(t)}{t}$ is bounded in $(0,1)$. Then*

$$\mathcal{L}\left[\frac{f(t)}{t}\right] = \int_s^\infty F(\sigma)d\sigma, \, s > \alpha. \tag{8.14}$$

Proof. We use Theorem 8.2.18 to prove this result. Let $v(t) = \frac{f(t)}{t}$ and V denote the Laplace transform of v. Observe that v satisfies the hypotheses of Theorem 8.2.18. Now it follows that

$$F(s) = \mathcal{L}[tv(t)] = -\frac{d}{ds}V(s), \quad s > \alpha. \tag{8.15}$$

Fixing $s > \alpha$ and integrating (8.15) over $[s, \sigma]$, we get

$$\int_s^\sigma F(\tilde{s})d\tilde{s} = -\int_s^\sigma \frac{d}{d\tilde{s}}V(\tilde{s})d\tilde{s} = V(s) - V(\sigma).$$

Letting $\sigma \to \infty$ and owing to Lemma 8.2.6, we obtain (8.14).   □

We now find the Laplace transforms of some functions using the results proved in Theorems 8.2.17–8.2.20.

Example 8.2.21. Find the Laplace transform of $f(t) = t\sin(3t)$.

Solution. In view of Theorem 8.2.18, we have

$$\mathcal{L}[t\sin(3t)] = -\frac{d}{ds}\mathcal{L}[\sin(3t)] = -\frac{d}{ds}\left(\frac{3}{s^2+9}\right) = \frac{6s}{(s^2+9)^2}.$$

This completes the solution.   □

Example 8.2.22. Find the Laplace transform of each of the following functions:

(i) $\dfrac{\sin(10t)}{t}$. (ii) $e^{-t}\dfrac{\sin(4t)}{t}$. (iii) $te^{-2t}\cos^2(5t)$. (iv) $\displaystyle\int_0^t \frac{1-e^\tau}{\tau}d\tau$.

Solution. (i) From the scaling property of the Laplace transform and Theorem 8.2.20, we have

$$\mathcal{L}\left[\frac{\sin(10t)}{t}\right] = \int_s^\infty \mathcal{L}[\sin(10t)]d\sigma$$

$$= \int_s^\infty \frac{10d\sigma}{\sigma^2 + 100}$$

$$= \frac{\pi}{2} - \tan^{-1}(\frac{s}{10}), \quad s > 0.$$

(ii) From the shifting, scaling results and Theorem 8.2.20, we obtain

$$\mathcal{L}\left[e^{-t}\frac{\sin(4t)}{t}\right] = \int_s^\infty \mathcal{L}[e^{-t}\sin(4t)]d\sigma$$

$$= \int_s^\infty \frac{4d\sigma}{(\sigma+1)^2 + 16}$$

$$= \frac{\pi}{2} - \tan^{-1}(\frac{s+1}{4}), \quad s > 0.$$

(*iii*) By the shifting, the scaling properties and Theorem 8.2.18, one has

$$
\mathcal{L}\left[te^{-2t}\cos^2(5t)\right] = -\frac{d}{ds}\mathcal{L}\left[\frac{e^{-2t}}{2}\big(1+\cos(10t)\big)\right]
$$

$$
= -\frac{1}{2}\frac{d}{ds}\left(\frac{1}{s+2}\right) - \frac{1}{2}\frac{d}{ds}\left(\frac{s+2}{(s+2)^2+100}\right)
$$

$$
= \frac{1}{2(s+2)^2} + \frac{(s+2)^2-100}{2((s+2)^2+100)^2}.
$$

(*iv*) Using Theorem 8.2.19 we get

$$
\mathcal{L}\left[\int_0^t \frac{1-e^\tau}{\tau}\,d\tau\right] = \frac{1}{s}\mathcal{L}\left[\frac{1-e^t}{t}\right]
$$

$$
= \frac{1}{s}\int_s^\infty \mathcal{L}[1-e^t]\,d\sigma
$$

$$
= \frac{1}{s}\int_s^\infty \left(\frac{1}{\sigma}-\frac{1}{\sigma-1}\right)d\sigma
$$

$$
= -\frac{1}{s}\log\left(\frac{s}{s-1}\right).
$$

This completes the solution. □

Example 8.2.23. Find the Laplace transform of $\mathrm{Si}(t) = \displaystyle\int_0^t \frac{\sin\tau}{\tau}\,d\tau$.

Solution. In view of Theorems 8.2.19–8.2.20, we have

$$
\mathcal{L}\left[\int_0^t \frac{\sin\tau}{\tau}\,d\tau\right] = \frac{1}{s}\mathcal{L}\left[\frac{\sin t}{t}\right]
$$

$$
= \frac{1}{s}\int_s^\infty \mathcal{L}[\sin t](\sigma)\,d\sigma
$$

$$
= \frac{1}{s}\int_s^\infty \frac{1}{\sigma^2+1}\,d\sigma
$$

$$
= \frac{1}{s}\left(\frac{\pi}{2}-\tan^{-1}s\right)
$$

$$
= \frac{1}{s}\cot^{-1}\left(\frac{1}{s}\right).
$$

This completes the solution. □

Another useful result in the computation of the Laplace transform of many functions is

$$
\int_0^\infty e^{-\sigma t^2}\,dt = \frac{\sqrt{\pi}}{2\sqrt{\sigma}}, \quad \sigma > 0. \tag{8.16}
$$

So far, all the functions for which the Laplace transforms are found are bounded in $(0,1)$. In the next example, we consider a function which is unbounded in $(0,1)$ and find its Laplace transform using (8.16).

Example 8.2.24. Find the Laplace transform of $\dfrac{1}{\sqrt{t}}$. Hence find the Laplace transform of $t^{n+\frac{1}{2}}$, where n is a nonnegative integer.

Solution. We begin with the definition

$$\mathcal{L}[\frac{1}{\sqrt{t}}] = \int_0^\infty \frac{e^{-st}}{\sqrt{t}}\,dt.$$

On putting $u = \sqrt{t}$, we get $du = \dfrac{dt}{2\sqrt{t}}$

$$\mathcal{L}[\frac{1}{\sqrt{t}}] = 2\int_0^\infty e^{-su^2}\,du = \frac{\sqrt{\pi}}{\sqrt{s}},\quad s > 0.$$

In the general case, for $n \in \mathbb{N} \cup \{0\}$, let

$$I_n(s) := \mathcal{L}[t^{n+\frac{1}{2}}] = \int_0^\infty e^{-st}t^{n+\frac{1}{2}}\,dt.$$

As before, we set $t = u^2$, so that $dt = 2u\,du$ and

$$
\begin{aligned}
I_n(s) &= 2\int_0^\infty e^{-su^2}u^{2n+2}\,du & (8.17)\\
&= \lim_{U\to\infty} \frac{-1}{s}\int_0^U \frac{d}{du}\left(e^{-su^2}\right)u^{2n+1}\,du \\
&= \lim_{U\to\infty}\left(\frac{-1}{s}e^{-su^2}u^{2n+1}\Big|_0^U + \frac{2n+1}{s}\int_0^U e^{-su^2}u^{2n}\,du\right) \\
&= \frac{2n+1}{s}\int_0^\infty e^{-su^2}u^{2n}\,du \\
&= \frac{2n+1}{2s}I_{n-1}(s). & (8.18)
\end{aligned}
$$

The last equation follows from (8.17). Hence we find that

$$
\begin{aligned}
\mathcal{L}[t^{n+\frac{1}{2}}] &= \frac{2n+1}{2s}\mathcal{L}[t^{n-\frac{1}{2}}] \\
&= \frac{(2n+1)(2n-1)}{2^2 s^2}\mathcal{L}[t^{n-\frac{3}{2}}] \\
&\;\;\vdots \\
&= \frac{(2n+1)(2n-1)\cdots 1}{2^{n+1}s^{n+1}}\mathcal{L}[\frac{1}{\sqrt{t}}] \\
&= \frac{\sqrt{\pi}(2n+1)!}{n!2^{2n+1}s^{n+\frac{3}{2}}},\quad s > 0.
\end{aligned}
$$

This completes the solution. $\qquad\square$

8.2.1 The Heaviside function

The function $H : \mathbb{R} \to \mathbb{R}$ defined by

$$H(t) = \begin{cases} 0, \ t < 0, \\ 1, \ t \geq 0, \end{cases} \tag{8.19}$$

is called as the Heaviside unit step function or the Heaviside function.

Notation. Let $f : [0, \infty) \to \mathbb{R}$ be a function and $a > 0$. Then $H(t - a)f(t)$ denotes the function defined on $[0, \infty]$ and is given by

$$H(t - a)f(t) = \begin{cases} 0, \ 0 \leq t < a, \\ f(t), \ t \geq a. \end{cases}$$

We now prove another shifting theorem involving the Heaviside function.

Theorem 8.2.25. *Assume that f is a piecewise continuous function which is of exponential order. If $a > 0$, then*

$$\mathcal{L}[f(t - a)H(t - a)] = e^{-as}F(s), \ s > 0. \tag{8.20}$$

Proof. We consider

$$\mathcal{L}[f(t-a)H(t-a)] = \int_0^\infty e^{-st} f(t-a)H(t-a)dt$$

$$= \int_a^\infty e^{-st} f(t-a)dt.$$

By setting $\xi = t - a$, we obtain

$$\mathcal{L}[f(t-a)H(t-a)] = \int_0^\infty e^{-s(\xi+a)} f(\xi)d\xi = e^{-as}F(s), \ s > 0.$$

This completes the proof of the theorem. $\qquad\square$

Example 8.2.26. Let $a, b > 0$. Find the Laplace transform of

$$f(t) = \begin{cases} 0, & 0 \leq t \leq b, \\ \sin(t - a), & t > b. \end{cases}$$

Solution. We first note that

$$\begin{aligned} f(t) &= \sin(t-a)H(t-b) \\ &= \sin\big((t-b) + (b-a)\big)H(t-b) \\ &= \big(\cos(b-a)\sin(t-b) + \sin(b-a)\cos(t-b)\big)H(t-b). \end{aligned}$$

Thus in view of Theorem 8.2.25, we obtain

$$\begin{aligned} F(s) &= e^{-bs}\big(\cos(b-a)\mathcal{L}[\sin t] + \sin(b-a)\mathcal{L}[\cos t]\big) \\ &= e^{-bs}\Big(\cos(b-a)\frac{1}{s^2+1} + \sin(b-a)\frac{s}{s^2+1}\Big), \ s > 0. \end{aligned}$$

This completes the solution. $\qquad\square$

Example 8.2.27. Let $a, b > 0$. Find the Laplace transform of

$$f(t) = \begin{cases} 0, & 0 \le t \le b, \\ (t-a)^2, & t > b. \end{cases}$$

Solution. As before, we observe that

$$\begin{aligned} f(t) &= (t-a)^2 H(t-b) \\ &= \big((t-b) + (b-a)\big)^2 H(t-b) \\ &= \big((t-b)^2 + 2(t-b)(b-a) + (b-a)^2\big) H(t-b). \end{aligned}$$

Using Theorem 8.2.25, we get

$$\begin{aligned} F(s) &= e^{-bs}\big(\mathcal{L}[t^2] + (b-a)\mathcal{L}[2t] + (b-a)^2\mathcal{L}[1]\big) \\ &= e^{-bs}\left(\frac{2}{s^3} + \frac{2(b-a)}{s^2} + \frac{(b-a)^2}{s}\right), \quad s > 0. \end{aligned}$$

This completes the solution. $\square$

8.2.2　The convolution

The convolution of two functions $u, v : \mathbb{R} \to \mathbb{R}$ is (formally) defined as

$$(u * v)(x) = \int_{-\infty}^{\infty} u(x - \tau)v(\tau)d\tau, \quad x \in \mathbb{R}.$$

Since all functions that are considered in this chapter are defined on $[0, \infty)$, we define the convolution of such functions using the following strategy. Let f and g be real valued functions defined on $[0, \infty)$. We extend f and g for negative real numbers by defining new functions

$$\bar{f}(x) = \begin{cases} f(x), & x \ge 0, \\ 0, & x < 0, \end{cases} \qquad \bar{g}(x) = \begin{cases} g(x), & x \ge 0, \\ 0, & x < 0, \end{cases}$$

respectively. A natural choice of defining the convolution of f and g is $\bar{f} * \bar{g}$. In that case, using the definition of $\bar{f}$ and $\bar{g}$, we have

$$(\bar{f} * \bar{g})(t) = \int_{-\infty}^{\infty} \bar{f}(t-\tau)\bar{g}(\tau)d\tau = \int_{0}^{\infty} \bar{f}(t-\tau)g(\tau)d\tau = \int_{0}^{t} f(t-\tau)g(\tau)d\tau.$$

This motivates us to define the convolution of functions on $[0, \infty)$ as follows.

Definition 8.2.28. Assume that $f, g : [0, \infty) \to \mathbb{R}$ are piecewise continuous functions. The convolution of f and g is a function on $[0, \infty)$ defined as

$$(f * g)(t) = \int_{0}^{t} f(t-\tau)g(\tau)d\tau, \quad t \ge 0. \tag{8.21}$$

The following properties of convolution immediately follow from the definition.

Proposition 8.2.29. *Assume that $f, g, h : [0, \infty) \to \mathbb{R}$ are piecewise continuous functions. Then the following hold:*

(i) (Commutativity) $f * g = g * f$,

(ii) (Distributivity) $f * (g + h) = f * g + f * h$,

(iii) (Associativity) $f * (g * h) = (f * g) * h$.

Proof. (i) From the definition of the convolution, we have

$$(f * g)(t) = \int_0^t f(t - \tau)g(\tau)d\tau, \ t \geq 0.$$

By setting $\xi = t - \tau$, we get $d\xi = -d\tau$. Moreover, the upper and the lower limits of ξ become t and zero, respectively. Hence, we obtain

$$(f * g)(t) = -\int_t^0 f(\xi)g(t - \xi)d\xi = \int_0^t f(\xi)g(t - \xi)d\xi = (g * f)(t), \ t \geq 0.$$

(ii) The proof of (ii) is an immediate consequence of the definition of the convolution and linearity of the integral.

(iii) It is also straightforward and is left to the reader as a simple exercise. $\square$

Example 8.2.30. Compute the following convolutions:

(i) $t * e^t$. (ii) $\sin(mt) * \cos(nt)$.

Solution. (i) For $t \geq 0$, we have

$$t * e^t = \int_0^t \tau e^{t-\tau}d\tau = e^t \left(-\tau e^{-\tau} \Big|_{\tau=0}^{t} + \int_0^t e^{-\tau}d\tau \right) = -t - 1 + e^t.$$

(ii) For $m \neq n$, we readily get

$$
\begin{aligned}
\sin(mt) * \cos(nt) &= \int_0^t \sin(m\tau)\cos(nt - n\tau)d\tau \\
&= \frac{1}{2}\int_0^t \Big(\sin((m - n)\tau + nt) + \sin((m + n)\tau - nt) \Big)d\tau \\
&= \frac{m}{m^2 - n^2}\Big(\cos(nt) - \cos(mt) \Big), \ t \geq 0.
\end{aligned}
$$

Using the same method, we compute

$$
\begin{aligned}
\sin(nt) * \cos(nt) &= \int_0^t \sin(n\tau)\cos(nt - n\tau)d\tau \\
&= \frac{1}{2}\int_0^t \Big(\sin(nt) + \sin(2n\tau - nt) \Big)d\tau \\
&= \frac{t}{2}\sin(nt), \ t \geq 0.
\end{aligned}
$$

This completes the solution. $\square$

We now state a result pertaining to the Laplace transform of the convolution of two functions.

Theorem 8.2.31. *Suppose f and g are piecewise continuous functions which are of exponential order. Then the Laplace transform of $f * g$ exists and*

$$\mathcal{L}[(f * g)(t)] = \mathcal{L}[f(t)]\mathcal{L}[g(t)]. \tag{8.22}$$

Proof. Let $M, \alpha > 0$ be such that $|f(t)|, |g(t)| \leq Me^{\alpha t}$, $t > 0$. Then we have

$$|(f * g)(t)| \leq \int_0^t |f(\tau)||g(t - \tau)|d\tau$$

$$\leq M^2 \int_0^t e^{\alpha \tau} e^{\alpha(t-\tau)} d\tau$$

$$= M^2 t e^{\alpha t}$$

$$\leq M^2 e^{(\alpha+1)t}, \ t \geq 0.$$

Thus $f * g$ is a function of exponential order. The main idea in the proof is the conversion of a double integral to successive integrals[3] and vice versa. We consider

$$\mathcal{L}[(f * g)(t)] = \int_{t=0}^{\infty} e^{-st} \left(\int_{\tau=0}^{t} f(\tau)g(t - \tau)d\tau \right) dt$$

$$= \iint_S e^{-st} f(\tau)g(t - \tau)d\tau dt, \tag{8.23}$$

where $S = \{(\tau, t) : t \geq 0, 0 \leq \tau \leq t\}$. We now use the coordinate transformation $\eta = \tau$ and $\xi = t - \tau$ to evaluate the double integral in equation (8.23). Under this transformation, S gets mapped to $R := \{(\xi, \eta) : \xi \geq 0, \eta \geq 0\}$ (see Figures 8.1(a)–8.1(b)).
The Jacobian of this transformation is

$$\frac{\partial(\tau, t)}{\partial(\xi, \eta)} = \begin{vmatrix} \frac{\partial \tau}{\partial \xi} & \frac{\partial \tau}{\partial \eta} \\ \frac{\partial t}{\partial \xi} & \frac{\partial t}{\partial \eta} \end{vmatrix} = \begin{vmatrix} \frac{\partial \eta}{\partial \xi} & \frac{\partial \eta}{\partial \eta} \\ \frac{\partial}{\partial \xi}(\xi + \eta) & \frac{\partial}{\partial \eta}(\xi + \eta) \end{vmatrix} = \begin{vmatrix} 0 & 1 \\ 1 & 1 \end{vmatrix} = -1.$$

Thus we obtain

$$\mathcal{L}[(f * g)(t)] = \iint_R e^{-s(\xi+\eta)} f(\xi)g(\eta) \left| \frac{\partial(\tau, t)}{\partial(\xi, \eta)} \right| d\xi d\eta$$

$$= \int_{\xi=0}^{\infty} e^{-s\xi} f(\xi)d\xi \int_{\eta=0}^{\infty} e^{-s\eta} g(\eta)d\eta$$

$$= \mathcal{L}[f(t)]\mathcal{L}[g(t)].$$

This completes the proof. □

---

[3]The reader is advised to learn Fubini's theorem which allows us change the order of integration. See for instance [31].

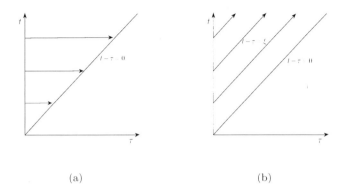

(a)   (b)

FIGURE 8.1: (a) The region S in the τt-plane over which the double integral is evaluated; (b) After introducing ξ and η coordinates, S is covered in a different manner.

It is easy to verify that $\int_0^t f(\tau)d\tau = (1 * f)(t)$. By taking the Laplace transform on both sides, we immediately get (8.13).

8.3   Inverse Laplace transforms

If the Laplace transform of f is F, then we say that an inverse Laplace transform of F is f. Suppose f_1 and f_2 are such that $\mathcal{L}[f_1(t)] = \mathcal{L}[f_2(t)]$, then due to the Lerch theorem, we have $f_1 = f_2$ (see [15, 19]). This establishes the uniqueness of an inverse Laplace transform. We do not prove that result here. From the linearity of the Laplace transform we immediately get that the uniqueness result is equivalent to the following. If $\mathcal{L}[f(t)] = 0$, then $f(t) = 0$, $t > 0$.

In this section, we present some methods to find the inverse Laplace transform of functions. In other words, for a given $F(s)$, our objective is to find $f(t)$ such that $\mathcal{L}[f(t)] = F(s)$. We denote by $\mathcal{L}^{-1}[F(s)]$ the inverse Laplace transform of f, i.e., $\mathcal{L}[\mathcal{L}^{-1}[F]] = F$.

From the examples presented in the previous section, we have the following formulae:

1. $\mathcal{L}^{-1}\left[\dfrac{1}{s^{n+1}}\right] = \dfrac{t^n}{n!}$, $n \in \mathbb{N} \cup \{0\}$,

2. $\mathcal{L}^{-1}\left[\dfrac{a}{s^2 + a^2}\right] = \sin(at)$, $a \in \mathbb{R}$,

3. $\mathcal{L}^{-1}\left[\dfrac{s}{s^2 + a^2}\right] = \cos(at), a \in \mathbb{R}$.

Let $\mathcal{L}[f(t)] = F(s)$ and $\mathcal{L}[f_i(t)] = F_i(s)$, $i = 1, 2$. We assume that f and F satisfy the hypotheses of the theorems proved in the previous section. Then we have the following results:

$$\mathcal{L}^{-1}[aF_1(s) + bF_2(s)] = a\mathcal{L}^{-1}[F_1(s)] + b\mathcal{L}^{-1}[F_2(s)], \quad a, b \in \mathbb{R}, \qquad (8.24)$$

$$\mathcal{L}^{-1}[F(s + a)] = e^{-at} f(t), \quad a \in \mathbb{R}, \qquad (8.25)$$

$$\mathcal{L}^{-1}\left[e^{-as} F(s)\right] = H(t - a)f(t - a), \quad a > 0, \qquad (8.26)$$

$$\mathcal{L}^{-1}\left[F\left(\frac{s}{a}\right)\right] = af(at), \quad a > 0, \qquad (8.27)$$

$$\mathcal{L}^{-1}\left[\frac{d^n}{ds^n}F(s)\right] = (-1)^n t^n f(t), \qquad (8.28)$$

$$\mathcal{L}^{-1}\left[\int_s^\infty F(\sigma)d\sigma\right] = \frac{1}{t}f(t), \qquad (8.29)$$

$$\mathcal{L}^{-1}\left[\frac{1}{s}F(s)\right] = \int_0^t f(\tau)d\tau, \qquad (8.30)$$

$$\mathcal{L}^{-1}[F_1(s)F_2(s)] = (f_1 * f_2)(t). \qquad (8.31)$$

The proofs of the above results are left to the reader as an easy exercise. We present some examples in which we compute the inverse Laplace transform of the functions using the above mentioned formulae and results.

Example 8.3.1. Find the inverse Laplace transform of each of the following functions:

(i) $\dfrac{as + b}{l^2 s^2 + m^2}$, $l, m \neq 0$. (ii) $P\left(\dfrac{1}{s}\right)$, where P is a polynomial with $P(0) = 0$.

Solution. (i) From the linearity of the inverse Laplace transform, we have

$$\mathcal{L}^{-1}\left[\frac{as + b}{l^2 s^2 + m^2}\right] = \frac{a}{l^2}\left[\mathcal{L}^{-1}\frac{s}{s^2 + (m/l)^2}\right] + \frac{b}{l^2}\mathcal{L}^{-1}\left[\frac{1}{s^2 + (m/l)^2}\right]$$

$$= \frac{a}{l^2}\cos\left(\frac{mt}{l}\right) + \frac{b}{lm}\sin\left(\frac{mt}{l}\right).$$

(ii) Let $P(X) = a_n X^n + a_{n-1} X^{n-1} + \cdots + a_1 X$ be the given polynomial. Then we have

$$\mathcal{L}^{-1}[P(\tfrac{1}{s})] = a_n \mathcal{L}^{-1}[\tfrac{1}{s^n}] + \cdots + a_1 \mathcal{L}^{-1}[\tfrac{1}{s}] = a_n \frac{t^{n-1}}{(n-1)!} + a_{n-1}\frac{t^{n-2}}{(n-2)!} + \cdots + a_1.$$

This completes the solution. □

We now present the computation of the inverse Laplace transforms using the shifting results.

Example 8.3.2. Find the inverse Laplace transform of each of the following functions:

(i) $\dfrac{1}{s+a} - \dfrac{1}{(s+b)^2}$, $a, b \in \mathbb{R}$. (ii) $\dfrac{s+6}{s^2+2s+50}$. (iii) $\dfrac{(s+6)e^{-3s}}{s^2-8s+20}$.

Solution. (i) Using the shifting result (8.25) we readily obtain that

$$\mathcal{L}^{-1}\left[\frac{1}{s+a} - \frac{1}{(s+b)^2}\right] = e^{-at}\mathcal{L}^{-1}\left[\frac{1}{s}\right] - e^{-bt}\mathcal{L}^{-1}\left[\frac{1}{s^2}\right] = e^{-at} - te^{-bt}.$$

(ii) We first write

$$\frac{s+6}{s^2+2s+50} = \frac{s+1}{(s+1)^2+7^2} + \frac{5}{(s+1)^2+7^2}.$$

Using the shifting result for the inverse Laplace transform (8.25), we get

$$\begin{aligned}
\mathcal{L}^{-1}\left[\frac{s+6}{s^2+2s+50}\right] &= \mathcal{L}^{-1}\left[\frac{s+1}{(s+1)^2+7^2}\right] + \mathcal{L}^{-1}\left[\frac{5}{(s+1)^2+7^2}\right] \\
&= e^{-t}\mathcal{L}^{-1}\left[\frac{s}{s^2+7^2}\right] + e^{-t}\mathcal{L}^{-1}\left[\frac{5}{s^2+7^2}\right] \\
&= e^{-t}\cos(7t) + \frac{5}{7}e^{-t}\sin(7t).
\end{aligned}$$

(iii) We first consider

$$\mathcal{L}^{-1}\left[\frac{s+6}{s^2-8s+20}\right] = \mathcal{L}^{-1}\left[\frac{(s-4)+10}{(s-4)^2+4}\right] = e^{4t}\big(\cos(2t) + 5\sin(2t)\big).$$

We now use (8.26) to arrive at

$$\mathcal{L}^{-1}\left[\frac{(s+6)e^{-3s}}{s^2-8s+20}\right] = H(t-3)e^{4(t-3)}\big(\cos(2t-6) + 5\sin(2t-6)\big).$$

This completes the solution. $\quad\square$

Another important technique to find the inverse Laplace transform of a rational function is to decompose it using partial fractions. This method is demonstrated in the next couple of examples.

Example 8.3.3. Find the inverse Laplace transform of

$$F(s) = \frac{1}{s^2-9s+18}.$$

Solution. We begin with factorizing the denominator of the given fraction to write

$$F(s) = \frac{1}{(s-6)(s-3)}.$$

By the partial fraction decomposition, we get

$$F(s) = \frac{1}{3(s-6)} - \frac{1}{3(s-3)}.$$

We now apply the inverse Laplace transform on both sides to obtain

$$f(t) = \mathcal{L}^{-1}\left[\frac{1}{3(s-6)}\right] - \mathcal{L}^{-1}\left[\frac{1}{3(s-3)}\right] = \frac{1}{3}(e^{6t} - e^{3t}).$$

This completes the solution of the problem. □

Example 8.3.4. Find the inverse Laplace transform of

$$F(s) = \frac{s+1}{s^3 - 9s^2 + 9s - 81}.$$

Solution. We first factorize the denominator of the given fraction to write

$$F(s) = \frac{s+1}{(s-9)(s^2+9)}.$$

Using the partial fraction decomposition, we get

$$F(s) = \frac{1}{9(s-9)} - \frac{s}{9(s^2+9)}.$$

Hence it follows that

$$f(t) = \mathcal{L}^{-1}\left[\frac{1}{9(s-9)}\right] - \mathcal{L}^{-1}\left[\frac{s}{9(s^2+9)}\right] = \frac{1}{9}(e^{9t} - \cos(3t)).$$

This completes the solution of the problem. □

In some cases, it is easier to find the inverse Laplace transform of the derivative/integral of the given function than that of the original function. Then one can use (8.28)–(8.29) to obtain the required inverse Laplace transform. The following example demonstrates this technique.

Example 8.3.5. Find the inverse Laplace transform of each of the following functions.

(i) $\log\left(\dfrac{s^2+a^2}{s^2+b^2}\right)$, $a, b \in \mathbb{R}$. (ii) $\cot^{-1}(as)$, $a \neq 0$. (iii) $\dfrac{e^{-s}}{s}\cot^{-1}(s)$.

Solution. (i) Let $F(s) = \log\left(\dfrac{s^2+a^2}{s^2+b^2}\right)$, then

$$F'(s) = \frac{2s}{s^2+a^2} - \frac{2s}{s^2+b^2}.$$

Therefore, we have

$$\mathcal{L}^{-1}[F'(s)] = 2(\cos(at) - \cos(bt)).$$

In view of (8.28), it follows that

$$\mathcal{L}^{-1}[F(s)] = -\frac{1}{t}\mathcal{L}^{-1}[F'(s)] = -\frac{2}{t}(\cos(at) - \cos(bt)).$$

(*ii*) By setting $F(s) = \cot^{-1}(as)$, we get $F'(s) = \dfrac{-a}{1 + a^2 s^2}$.

Hence we readily obtain

$$\mathcal{L}^{-1}[F'(s)] = -\sin(\frac{t}{a}),$$

Thus from (8.28) we find that

$$\mathcal{L}^{-1}[F(s)] = \frac{1}{t}\sin(\frac{t}{a}).$$

(*iii*) From the result proved in (*ii*), we get $\mathcal{L}^{-1}[\cot^{-1}(s)] = \dfrac{\sin t}{t}$.
Due to (8.30) it follows that

$$\mathcal{L}^{-1}[\frac{1}{s}\cot^{-1}(s)] = \int_0^t \frac{\sin \tau}{\tau}d\tau.$$

Therefore from (8.26) we have

$$\mathcal{L}^{-1}\left[\frac{e^{-s}}{s}\cot^{-1}(s)\right] = H(t-1)\int_0^{t-1}\frac{\sin \tau}{\tau}d\tau.$$

This completes the solution. □

The convolution result for the inverse Laplace transform (see (8.31)) is used in the next example.

Example 8.3.6. Find the inverse Laplace transform of each of the following functions:

(*i*) $\dfrac{1}{(s^2 + 1)^2}$. (*ii*) $\dfrac{s}{(s^2 + 1)^2}$.

Solution. (*i*) We first consider

$$\begin{aligned}
\mathcal{L}^{-1}\left[\frac{1}{(s^2+1)^2}\right] &= \mathcal{L}^{-1}\left[\frac{1}{(s^2+1)}\right] * \mathcal{L}^{-1}\left[\frac{1}{(s^2+1)}\right]\\
&= (\sin t) * (\sin t)\\
&= \int_0^t \sin(t-\tau)\sin(\tau)d\tau\\
&= \frac{1}{2}\big(\sin(t) - t\cos(t)\big).
\end{aligned}$$

(*ii*) We next consider

$$\begin{aligned}
\mathcal{L}^{-1}\left[\frac{s}{(s^2+1)^2}\right] &= \mathcal{L}^{-1}\left[\frac{s}{(s^2+1)}\right] * \mathcal{L}^{-1}\left[\frac{1}{(s^2+1)}\right]\\
&= \cos t * \sin t\\
&= \frac{t}{2}\sin t,
\end{aligned}$$

where the last step follows from Example 8.2.30.

Example 8.3.7. Find the inverse Laplace transform of

$$F(s) = \frac{s^5}{(s^2+1)^2(s^2+2)}.$$

Solution. We first decompose the given fraction using the method of partial fractions as follows

$$\frac{s^5}{(s^2+1)^2(s^2+2)} = \frac{s}{(s^2+1)^2} - \frac{3s}{(s^2+1)} + \frac{4s}{(s^2+2)}.$$

By taking the inverse Laplace transform, we get

$$
\begin{aligned}
\mathcal{L}^{-1}[F(s)] &= \mathcal{L}^{-1}\left[\frac{s}{(s^2+1)^2}\right] - \mathcal{L}^{-1}\left[\frac{3s}{(s^2+1)}\right] + \mathcal{L}^{-1}\left[\frac{4s}{(s^2+2)}\right] \\
&= ((\cos t) * (\sin t)) - 3\cos(t) + 4\cos(\sqrt{2}t) \\
&= \frac{t}{2}\sin(t) - 3\cos(t) + 4\cos(\sqrt{2}t),
\end{aligned}
$$

owing to Example 8.2.30. □

8.4 Applications to ODEs

In this section, we show how to use the Laplace transforms to solve (system of) ODEs with constant coefficients. Upon taking the Laplace transforms of ODEs, they are reduced to algebraic equations. After solving the algebraic equations, we take the inverse Laplace transform to obtain the solution of ODEs. We explain this method by presenting some examples where we consider initial value problems and boundary value problems.

Example 8.4.1. Solve the initial value problem:

$$
\begin{cases}
y'' + 4y' + 4y = e^{-2t}, \ t > 0, \\
y(0) = 1, \ y'(0) = 2.
\end{cases}
$$

Solution. In view of the results proved in Section 3.1, it follows that the general solution to the given ODE, and its derivatives are of exponential order. This allows us to take the Laplace transform of y, y', and y''. Let $Y(s)$ denote the Laplace transform of $y(t)$. Taking the Laplace transform of the given equation and using Theorem 8.2.17, we get

$$s^2 Y(s) - s - 2 + 4(sY(s) - 1) + 4Y(s) = \frac{1}{s+2}.$$

After rearranging the terms we obtain

$$Y(s) = \frac{s^2 + 8s + 13}{(s+2)^3}.$$

Using the partial fractions, we rewrite the previous equation as

$$Y(s) = \frac{1}{s+2} + \frac{4}{(s+2)^2} + \frac{1}{(s+2)^3}$$

By taking the inverse Laplace transform on both sides to have

$$y(t) = e^{-2t} + \left((4e^{-2t}) * (e^{-2t}) \right) + e^{-2t} * e^{-2t} * e^{-2t}.$$

We now compute

$$e^{-2t} * e^{-2t} = \int_0^t e^{-2(t-\tau)} e^{-2\tau} d\tau = te^{-2t},$$

and

$$e^{-2t} * e^{-2t} * e^{-2t} = e^{-2t} * (te^{-2t}) = \int_0^t \tau e^{-2(t-\tau)} e^{-2\tau} d\tau = \frac{t^2 e^{-2t}}{2}.$$

Therefore

$$y(t) = \left(1 + 4t + \frac{t^2}{2} \right) e^{-2t}, \ t \geq 0$$

is the solution to the given problem. $\square$

Example 8.4.2. Solve the initial value problem:

$$\begin{cases} y'' + y = 1, \ t > 0, \\ y(0) = 1, \ y'(0) = -2. \end{cases}$$

Solution. Let $Y(s)$ denote the Laplace transform of $y(t)$. As before, we take the Laplace transform of the given equation to get

$$s^2 Y(s) - s + 2 + Y(s) = \frac{1}{s}.$$

Thus

$$Y(s) = \frac{s^2 - 2s + 1}{s(s^2 + 1)},$$

and using partial fractions we obtain

$$Y(s) = \frac{1}{s} - \frac{2}{s^2 + 1}.$$

By taking the inverse Laplace transform on both sides, we find that

$$y(t) = 1 - 2\sin t, \ t \geq 0$$

is the solution to the given problem. $\square$

Example 8.4.3. Solve the initial value problem:

$$\begin{cases} y'' + y = e^{-2t} \sin t, \ t > 0, \\ y(0) = 0, \ y'(0) = 0. \end{cases}$$

Solution. Let $Y(s)$ denote the Laplace transform of $y(t)$. After taking the

Laplace transform of the given equation, we get

$$(s^2 + 1)Y(s) = \frac{1}{(s+2)^2 + 1},$$

or

$$Y(s) = \frac{1}{(s^2 + 1)(s^2 + 4s + 5)}.$$

Using the partial fractions, we write

$$
\begin{aligned}
Y(s) &= \frac{-s+1}{8(s^2 + 1)} + \frac{s+3}{8(s^2 + 4s + 5)} \\
&= \frac{1}{8(s^2 + 1)} - \frac{s}{8(s^2 + 1)} + \frac{s+2}{8((s+2)^2 + 1)} + \frac{1}{8((s+2)^2 + 1)}.
\end{aligned}
$$

By taking the inverse Laplace transform on both sides we obtain

$$y(t) = \frac{\sin t}{8} - \frac{\cos t}{8} + \frac{e^{-2t}}{8}(\cos(t) + \sin(t)), \ t \geq 0,$$

is the solution to the given problem.   □

Example 8.4.4. Solve the boundary value problem:

$$
\begin{cases}
y'' - y = 1, \ t \in (0, 1), \\
y'(0) = 1, \ y(1) = 2.
\end{cases}
$$

Solution. Let $y(0) = \alpha$ and $Y(s)$ denote the Laplace transform of $y(t)$. After taking the Laplace transform, the given equation becomes

$$s^2 Y(s) - \alpha s - 1 - Y(s) = \frac{1}{s},$$

or

$$Y(s) = \frac{\alpha s^2 + s + 1}{s(s^2 - 1)}.$$

As in the previous examples, using the partial fractions, we get

$$Y(s) = -\frac{1}{s} + \frac{\alpha + 2}{2(s-1)} + \frac{\alpha}{2(s+1)}.$$

By taking the inverse Laplace transform on both sides, we have

$$y(t) = -1 + \left(\frac{\alpha}{2} + 1\right)e^t + \frac{\alpha}{2}e^{-t}, \ t \geq 0. \tag{8.32}$$

We now find α such that the boundary condition at $x = 1$ is satisfied, i.e., $y(1) = 2$. Thus we obtain

$$y(1) = -1 + \left(\frac{\alpha}{2} + 1\right)e + \frac{\alpha}{2e} = 2,$$

or

$$\alpha = \frac{2e(3 - e)}{e^2 + 1}, \ t \geq 0.$$

Substituting α in (8.32), we get

$$y(t) = -1 + \left(\frac{3e+1}{e^2+1}\right)e^t, + \left(\frac{e(3-e)}{e^2+1}\right)e^{-t}, \ t \geq 0,$$

is the solution to the given problem. □

Example 8.4.5. Solve the initial value problem:

$$\begin{cases} x'(t) = y(t), \ t > 0 \\ y'(t) = 2x(t) + y(t), \ t > 0, \\ x(0) = 2, \ y(0) = 1. \end{cases}$$

Solution. From the results proved in Section 5.3, it clear that the general solution to the system of homogeneous ODE with constant coefficients is of exponential order. Hence the Laplace transform of x, x', y, and y' exists. Let $X(s)$ and $Y(s)$ denote the Laplace transforms of $x(t)$ and $y(t)$, respectively. By taking the Laplace transform for the given system, we get

$$\begin{cases} sX(s) - 2 = Y(s), \\ sY(s) - 1 = 2X(s) + Y(s). \end{cases} \tag{8.33}$$

We now solve (8.33) to obtain

$$X(s) = \frac{2s-1}{s^2-s-2}, \ Y(s) = \frac{s+4}{s^2-s-2}.$$

Using the partial fractions we readily get that

$$X(s) = \frac{1}{s-2} + \frac{1}{s+1}, \ Y(s) = \frac{2}{s-2} - \frac{1}{s+1}.$$

By taking the inverse Laplace transform it follows that

$$x(t) = e^{2t} + e^{-t}, \ y(t) = 2e^{2t} - e^{-t}, \ t \geq 0$$

is the solution to the given system. □

Example 8.4.6. Solve the initial value problem:

$$\begin{cases} x'(t) = -y(t) + 1, \ t > 0, \\ y'(t) = x(t) - 3, \ t > 0, \\ x(0) = 4, \ y(0) = 0. \end{cases}$$

Solution. In view of Exercise 5.15, both $x(t)$ and $y(t)$ are functions of exponential order. Let $X(s)$ and $Y(s)$ denote the Laplace transforms of $x(t)$ and

$y(t)$, respectively. As in the previous example, by taking the Laplace transform of the given system to get

$$\begin{cases} sX(s) - 4 = -Y(s) + \dfrac{1}{s}, \\[2mm] sY(s) = X(s) - \dfrac{3}{s}. \end{cases} \tag{8.34}$$

We now solve (8.34) to obtain

$$X(s) = \frac{4s^2 + s + 3}{(s^2 + 1)s}, \quad Y(s) = \frac{s+1}{(s^2 + 1)s},$$

or

$$X(s) = \frac{3}{s} + \frac{s+1}{s^2 + 1}, \quad Y(s) = \frac{1}{s} - \frac{s-1}{s^2 + 1}.$$

By applying the inverse Laplace transform on both sides of the above equation, we find that

$$x(t) = 3 + \cos t + \sin t, \quad y(t) = 1 - \cos t + \sin t, \quad t > 0,$$

is the required solution. □

Exercise 8.1. Find the Laplace transform of each of the following functions:

(*i*) $\sin^4(at)$, $a \neq 0$.

(*ii*) $\cos^5(2t)$.

(*iii*) $e^{-9t} + \sin(7t)\cos(8t)$.

(*iv*) $f(t) = \begin{cases} t, & 0 \leq t < a, \\ a, & t > a. \end{cases}$

(*v*) $f(t) = \begin{cases} \sin t, & 0 \leq t < 2\pi, \\ \cos(3t), & t > 2\pi. \end{cases}$

Exercise 8.2. Find the Laplace transform of each of the following functions:

(*i*) $e^{2t}t^{\frac{3}{2}}$.

(*ii*) $e^{at}\sin^3(2t)$, $a \in \mathbb{R}$.

(*iii*) $e^{-t}\sin(7t)\cosh(8t)$.

(*iv*) $t^3\cosh^3(t)$.

Exercise 8.3. Find the Laplace transform of each of the following functions:

(*i*) $t^2 e^{4t}\sin(at)$, $a \neq 0$.

(*ii*) $te^{at}\cos^3(2t)$, $a \neq 0$.

(*iii*) $\dfrac{e^{-t}\sin^2(7t)}{t}$.

(iv) $\dfrac{e^{-t} - 1}{t}$.

(v) $\dfrac{1 - \cos(2t)}{t^2}$.

Exercise 8.4. Find the Laplace transform of each of the following functions:

(i) $\displaystyle\int_0^t \tau^2 e^{4\tau} \sin(\tau) d\tau$.

(ii) $\displaystyle\int_0^t \tau^4 e^{4\tau} d\tau$.

(iii) $t \displaystyle\int_0^t \tau \cos^2(4\tau) d\tau$.

(iv) $t \displaystyle\int_0^t \tau \sin(\tau) d\tau$.

Exercise 8.5. Find the Laplace transform of each of the following functions:

(i) $\dfrac{1}{t} \displaystyle\int_0^t e^{-\tau} \sin(\tau) d\tau$.

(ii) $t \displaystyle\int_0^t \dfrac{\sin \tau}{\tau} d\tau$.

Exercise 8.6. Let $\alpha \in \mathbb{R}$. Use the periodicity of $\sin(\alpha t)$ and $\cos(\alpha t)$ to find their Laplace transforms.

Exercise 8.7. Find the Laplace transform of each of the following periodic functions:

(i) $f : [0, \infty) \to \mathbb{R}$, $f(t) = t$, $t \in [0, t_0]$ and $f(t + t_0) = f(t)$, $t > 0$.

(ii) $f : [0, \infty) \to \mathbb{R}$. $f(t) = \begin{cases} t, \ 0 \leq t < t_0, \\ 2t_0 - t, \ t_0 \leq t \leq 2t_0. \end{cases}$

with $f(t + 2t_0) = f(t)$. $t > 0$.

(iii) $f : [0, \infty) \to \mathbb{R}$, $f(t) = \begin{cases} t, \ 0 \leq t < t_0, \\ 2t_0 - t, \ t_0 \leq t \leq 3t_0, \\ t - 4t_0, \ 3t_0 \leq t \leq 4t_0, \end{cases}$

with $f(t + 4t_0) = f(t)$, $t > 0$.

Exercise 8.8. Find the Laplace transform of each of the following functions:

(i) $f : [0, \infty) \to \mathbb{R}$, $f(t) = \begin{cases} 0, \ 0 \leq t < t_0, \\ e^{3t} \sin(5t), \ t > t_0. \end{cases}$

(ii) $f : [0, \infty) \to \mathbb{R}$, $f(t) = \begin{cases} 0, \ 0 \leq t < t_0, \\ te^t, \ t > t_0. \end{cases}$

Exercise 8.9. Compute $f * g$ if $f, g : [0, \infty) \to \mathbb{R}$ are given by:

(i) $f(t) = t, \; g(t) = t^2$.

(ii) $f(t) = t^2, \; g(t) = t^2$.

(iii) $f(t) = \sin(nt), \; g(t) = \sin(mt), \; m, n \in \mathbb{N}$.

(iv) $f(t) = e^{mt}, \; g(t) = e^{nt}, \; m, n \in \mathbb{Z}$.

(v) $f(t) = \sin(2t), \; g(t) = t^2$.

Exercise 8.10. Prove the following statements:

(i) $\underbrace{1 * 1 * \cdots * 1}_{n-\text{convolutions}} = \dfrac{t^n}{n!}, \; n \in \mathbb{N}$.

(ii) Suppose $f : [0, \infty) \to \mathbb{R}$ is a piecewise continuous function. Then show that

$$\underbrace{\int_0^t \int_0^{s_{n-1}} \cdots \int_0^{s_2} \int_0^{s_1} f(s_0)\,ds_0 ds_1 \cdots ds_{n-2} ds_{n-1}}_{n-\text{integrals}} = \int_0^t \frac{(t - \sigma)^{n-1}}{(n-1)!} f(\sigma)\,d\sigma.$$

Hint: Prove (i) by induction. For (ii) show that both the left and right hand sides are equal to $\left(\underbrace{1 * 1 * \cdots * 1 * 1}_{(n-1)-\text{times}} \right) * f$.

Exercise 8.11. Find the inverse Laplace transform of each of the following functions:

(i) $\dfrac{1}{(s + 2)^4}$.

(ii) $\dfrac{s}{s^2 + 6s + 20}$.

(iii) $\dfrac{e^{-8s}}{s^2 + 6s + 20}$.

(iv) $\dfrac{e^{-2s}}{(s + 3)^2}$.

Exercise 8.12. Find the inverse Laplace transform of each of the following functions:

(i) $\dfrac{3s}{(s^2 + 2)(s^2 + 6)}$.

(ii) $\dfrac{as + b}{s^2 + ms + n}$, where $a, b, m, n \in \mathbb{R}$.

(iii) $\dfrac{1}{s(s + a)^4}, \; a > 0$.

(iv) $\log\left(\dfrac{s + 2}{s + 9} \right)$.

Exercise 8.13. Solve the following initial value problems:

(*i*) $y'' + 2y' + y = e^t$, $y(0) = 0$, $y'(0) = 0$.

(*ii*) $y'' + 4y' - 21y = e^{2t}$, $y(0) = 1$, $y'(0) = 0$.

(*iii*) $y'' - 3y' + 2y = \sin t$, $y(0) = 0$, $y'(0) = 1$.

Exercise 8.14. Find the solutions to the following boundary value problems:

(*i*) $y'' - 4y = \sin(4t)$, $y(0) = 0$, $y(1) = 1$.

(*ii*) $y'' + 9y = e^t$, $y(0) = 1$, $y(\frac{\pi}{4}) = 0$.

(*iii*) $y'' + 9y = 5$, $y'(0) = 1$, $y(1) = 1$.

Exercise 8.15. Solve the following initial value problems

$$(i) \begin{cases} x' = x + 2y + 1, \\ y' = y, \\ x(0) = 0, \ y(0) = 1. \end{cases} \quad (ii) \begin{cases} x' = -2y, \\ y' = x + y, \\ x(0) = 0, \ y(0) = 1. \end{cases} \quad (iii) \begin{cases} x' = x + 2y, \\ y' = 4x - y, \\ x(0) = 1, \ y(0) = 1. \end{cases}$$

Chapter 9

Numerical Methods

9.1 Introduction

So far we have seen many methods to solve different types of ODEs and studied the qualitative as well as quantitative behavior of solutions to ODEs. However, the methods available to find solutions and to analyze their behavior are limited; whereas there are uncanny ODEs which arise in the modeling of physical/biological/chemical phenomena. Therefore, many times we have to compromise and be content with an 'approximate solution' or 'numerical solution' to the Cauchy problem/ IVP/ boundary value problem under consideration. In fact, at times we would prefer a numerical solution to an analytical solution.

One such situation is when the solution is given in the form of a power series. Consider the case where $\phi(x) = \sum\limits_{n=0}^{\infty} a_n x^n$ is the solution to the given initial value problem which models a physical quantity. Then, in order to find an approximate value of $\phi(M_i)$ for M_i large, $1 \leq i \leq n$, we need to take many terms in the power series, increasing the cost of computation.

On the other hand, if the solution to the given Cauchy problem/IVP/BVP under consideration is given by integrals which cannot be written in closed form (for instance $y(x) = \int_0^x e^{-t^2}\, dt$), then also such solution is of little practical importance. It requires using quadrature formulae to evaluate the integrals at each point where the solution needs to be found.

Therefore in this chapter, we present various methods to find approximate solutions to IVPs. We begin with presenting the first order methods due to Euler and later introduce higher order methods which give better results in less iterations. Toward the end of the chapter, we explain how to extend these methods to find numerical solutions to IVP for systems of ODEs.

9.2   Euler methods

Throughout the chapter, we assume that $f : [a, b] \times \mathbb{R} \to \mathbb{R}$ is an infinite differentiable function with bounded partial derivatives. Consider the initial value problem (IVP)

$$\begin{cases} \dfrac{dy}{dx} = f(x, y(x)), \ x \in (a, b), \\ y(a) = y_0. \end{cases} \tag{9.1}$$

From the hypotheses on f it follows that there exists a unique solution to (9.1). The main step in finding a numerical solution to this IVP is to replace the term $\frac{dy}{dx}$ with another suitable term. We rewrite the ODE in (9.1) as

$$\lim_{h \to 0} \frac{y(x + h) - y(x)}{h} = f(x, y(x)), \ x \in (a, b).$$

Notice that for any computing machine, due to its finite memory and precision, it is impossible to handle the limiting process. For the same reason, $y(x)$ cannot be found at every point in $[a, b]$. Therefore we define a new equation for which the domain is a finite set and there is no limiting process. Of course, this new equation should be 'close' to the given equation (9.1).
Firstly, we set

$$x_0 = a, \ x_k = x_0 + kh, \ 0 \le k \le n,$$

where $h = \frac{b-a}{n}$. Here h is referred to as the *step size*. These x_k's are called the *grid points*. The set of grid points is the domain (which is a finite set) for the new equation that we are going to present here in contrast with the domain of (9.1).
Secondly, we use the Taylor theorem to arrive at

$$y(x_{k+1}) - y(x_k) = (x_{k+1} - x_k)y'(x_k) + O(|x_{k+1} - x_k|^2),$$

where O denotes the standard 'big O'[1]. After rearranging, we get

$$y'(x_k) = \frac{y(x_{k+1}) - y(x_k)}{h} + O(h).$$

Therefore at the grid points, (9.1) becomes

$$\frac{y(x_{k+1}) - y(x_k)}{h} = f\big(x_k, y(x_k)\big) + O(h), \ 0 \le k \le n - 1. \tag{9.2}$$

Here

$$\frac{y(x_{k+1}) - y(x_k)}{h} - f\big(x_k, y(x_k)\big)$$

---

[1] We say that $f(x) = O(g(x))$ as $x \to \alpha$ if $\lim\limits_{x \to \alpha} \dfrac{f(x)}{g(x)}$ exists in $\mathbb{R}$. Throughout the chapter, we take $\alpha = 0$ and we just write $f(x) = O(g(x))$.

is called the local truncation error. In general, the local truncation error is small (large) if h is small (large). The term $\dfrac{y(x_{k+1}) - y(x_k)}{h}$ is called the forward difference approximation of y' at the grid point x_k and is denoted by

$$y'(x_k) \approx \frac{y(x_{k+1}) - y(x_k)}{h}, \quad 0 \leq k \leq n-1. \tag{9.3}$$

Similarly, the backward difference and the central difference approximation of y' at the grid points are given by

$$y'(x_k) \approx \frac{y(x_k) - y(x_{k-1})}{h}, \quad 1 \leq k \leq n, \tag{9.4}$$

$$y'(x_k) \approx \frac{y(x_{k+1}) - y(x_{k-1})}{h}, \quad 1 \leq k \leq n-1, \tag{9.5}$$

respectively. In view of (9.1)–(9.3), we have

$$\frac{y(x_{k+1}) - y(x_k)}{h} \approx f\big(x_k, y(x_k)\big).$$

We now consider the system of equations

$$\begin{cases} \dfrac{Y_{k+1} - Y_k}{h} = f(x_k, Y_k), \ 0 \leq k \leq n-1, \\[2mm] Y_0 = y_0, \end{cases}$$

which is a finite difference equation. By rearranging the terms we obtain

$$\begin{cases} Y_{k+1} = Y_k + hf(x_k, Y_k), \ 0 \leq k \leq n-1, \\[2mm] Y_0 = y_0. \end{cases} \tag{9.6}$$

This is called the *Euler formula* or Euler forward difference approximation to (9.1). The method of finding an approximate solution to (9.1) using (9.6) is called the *Euler method*.

In this method, Y_1 is calculated using the known quantity Y_0. By induction Y_{k+1} is given in terms of already known quantity Y_k. Thus the Euler method is an *explicit method*.

Similarly, using (9.4), we get the Euler backward difference formula:

$$\begin{cases} Y_k = Y_{k-1} + hf(x_k, Y_k), \ 1 \leq k \leq n, \\[2mm] Y_0 = y_0. \end{cases} \tag{9.7}$$

From (9.7), we notice that for each k, one needs to solve a nonlinear equation to find Y_k. Since Y_k is implicitly given by the (potentially) nonlinear system (9.7), this method is an *implicit method*. Of course, if (9.1) is a linear equation then (9.7) will become an explicit method (the details are left to the reader).

Definition 9.2.1. If ϕ is the solution to IVP (9.1) and Y_k is an approximate solution using a numerical method, then the error in the numerical method at the grid point x_k is denoted by e_k and is defined as $e_k = |\phi(x_k) - Y_k|$. The relative error and the percentage error of the method at x_k are denoted by $\tilde{e}_k$ and E_k, respectively, they are given by

$$\tilde{e}_k = \frac{|\phi(x_k) - Y_k|}{|\phi(x_k)|}, \quad E_k = \frac{|\phi(x_k) - Y_k|}{|\phi(x_k)|} \times 100, \quad (9.8)$$

whenever $\phi(x_k) \neq 0$.

The relative error $\tilde{e}_k$ or percentage error E_k is a better measure of accuracy than the error e_k because it takes into consideration the magnitude of the exact solution $\phi(x_k)$. The percentage error plays a central role while comparing various methods. Henceforth, in each of the examples in which we compute the numerical solution, we also compute the percentage error.

Caution: In this chapter, all the calculations are done taking nine significant digits after the decimal point but in general we display only four/five/six digits after the decimal point in Y_k, $\phi(x_k)$ and E_k. Therefore, if the reader computes E_k using the values of Y_k and $\phi(x_k)$ displayed in any of the tables, then it may not agree with the value of corresponding E_k given in the same table.

We now present some examples which use Euler's method, Euler's backward difference formula to find approximate solutions to IVPs.

Example 9.2.2. Find an approximate solution to $xy' = y + y^2$, $x > 1$, $y(1) = -2$, at the grid points in $(1, 2]$ using the Euler method (the Euler forward difference formula) with $h = 0.05$.

Solution. By comparing the given ODE with (9.1), we have $f(x, y) = \dfrac{y + y^2}{x}$. The Euler method using this f is

$$Y_0 = 5, \quad Y_{k+1} = Y_k + h\left(\frac{Y_k + Y_k^2}{x_k}\right), \quad k = 0, 1, \ldots, \quad (9.9)$$

where $x_k = 1 + kh$, $k = 0, 1, \ldots$. On substituting $h = 0.05$ in (9.9) and rearranging the terms, we find that

$$Y_{k+1} = Y_k + \frac{Y_k + Y_k^2}{20 + k}. \quad (9.10)$$

On the other hand, we notice that the given ODE is a Bernoulli equation and using (2.33) the exact solution to the given initial value problem is computed as

$$\phi(x) = \frac{2x}{1 - 2x}, \quad x > 1.$$

It is often convenient to arrange the values at the grid points as shown in

TABLE 9.1: Numerical
solution using the Euler method
with $h = 0.05$

x_k	Y_k	$\phi(x_k)$	E_k
1.05	−1.9000	−1.9091	0.4762
1.10	−1.8186	−1.8333	0.8052
1.15	−1.7509	−1.7692	1.0357
1.20	−1.6937	−1.7143	1.1983
1.25	−1.6448	−1.6667	1.3130
1.30	−1.6024	−1.6250	1.3931
1.35	−1.5652	−1.5882	1.4479
1.40	−1.5325	−1.5556	1.4840
1.45	−1.5033	−1.5263	1.5061
1.50	−1.4772	−1.5000	1.5176
1.55	−1.4537	−1.4762	1.5211
1.60	−1.4325	−1.4545	1.5185
1.65	−1.4131	−1.4348	1.5112
1.70	−1.3954	−1.4167	1.5004
1.75	−1.3792	−1.4000	1.4870
1.80	−1.3642	−1.3846	1.4715
1.85	−1.3504	−1.3704	1.4546
1.90	−1.3376	−1.3571	1.4365
1.95	−1.3258	−1.3448	1.4177
2.00	−1.3147	−1.3333	1.3984

Table 9.1. The first column has the grid points $x_k = 1.05, 1.10, \ldots, 2.00$. The approximate solution at the corresponding grid points using (9.10) is shown in the second column. For example, the first entry -1.9000 in the second column is the approximate value of the solution at the grid point 1.05. The third column represents the solution at the grid points. For instance, the first entry of the third column is $\phi(1.05) = \frac{2 \times 1.05}{1 - 2 \times 1.05} = -1.9091$. Finally, the last column shows the percentage error. Notice that the relative percentage error using this method is low for this problem. In fact, it is around 1.4% at $x_k = 2.00$. □

Example 9.2.3. Find an approximate solution to $y' = 3(x - y)$, $x > 0$, $y(0) = 1$, at the grid points in $(0, 1]$ using the Euler backward difference formula with (*i*) $h = 0.1$, (*ii*) $h = 0.05$.

Solution. The given ODE is a linear equation and hence using (2.31) its solution is given by

$$\phi(x) = x - \frac{1}{3} + \frac{4}{3}e^{-3x}, \ x > 0.$$

By comparing with (9.1), we obtain that $f(x, y) = 3(x - y)$. The Euler backward difference formula (9.7) with this f is

$$Y_0 = 1, \ Y_k = Y_{k-1} + 3h(x_k - Y_k), \ k \in \mathbb{N}, \tag{9.11}$$

where $x_k = kh$.

(i) If $h = 0.1$, then (9.11) reduces to

$$Y_k = \frac{10Y_{k-1} + 0.3k}{13}, \quad k \in \mathbb{N}.$$

As before we present the approximate solution, exact solution, and the relative

TABLE 9.2:   Numerical results using the Euler backward difference formula with $h = 0.1$

x_k	Y_k	$\phi(x_k)$	E_k
0.1	0.7923	0.7544	5.0215
0.2	0.6556	0.5984	9.5595
0.3	0.5736	0.5088	12.7359
0.4	0.5335	0.4683	13.9335
0.5	0.5258	0.4642	13.2706
0.6	0.5429	0.4871	11.4639
0.7	0.5792	0.5299	9.2865
0.8	0.6301	0.5876	7.2317
0.9	0.6924	0.6563	5.5046
1.0	0.7634	0.7330	4.1382

percentage error in Table 9.2. The percentage error in this method oscillates and at $x_k = 1$ it is about 4.1%. We next reduce the step size and see how the percentage error changes.

(ii) If $h = 0.05$, then from (9.11) we obtain

$$Y_0 = 1, \quad Y_k = \frac{20Y_{k-1} + 0.15k}{23}, \quad k \in \mathbb{N}.$$

The numerical results in this case are displayed in Table 9.3. From Table 9.3, it is evident that the percentage error is significantly reduced. Moreover, from Tables 9.2 and 9.3, one can observe that the percentage error at the grid points $0.1, 0.2, \ldots, 1.0$ is roughly halved by reducing the step size from 0.1 to 0.05. □

Example 9.2.4. Find an approximate solution to $y' = -2xy$, $x > 0$, $y(0) = 1$, at the grid points in $(0, 1]$ using the Euler backward difference formula with (i) $h = 0.1$, (ii) $h = 0.05$.

Solution. Using the method of separation of variables, the solution to the given problem is given by

$$\phi(x) = e^{-x^2}, \quad x > 0.$$

In this example, we have $f(x, y) = -2xy$ and the Euler backward difference formula (9.7) gives

$$Y_0 = 1, \quad Y_k = Y_{k-1} - 2hx_kY_k, \quad k \in \mathbb{N}, \tag{9.12}$$

where $x_k = kh$.

TABLE 9.3: Numerical results using the Euler backward difference formula with $h = 0.05$

x_k	Y_k	$\phi(x_k)$	E_k
0.05	0.8761	0.8643	1.3664
0.10	0.7749	0.7544	2.7085
0.15	0.6934	0.6668	3.9766
0.20	0.6290	0.5984	5.1116
0.25	0.5796	0.5465	6.0532
0.30	0.5431	0.5088	6.7505
0.35	0.5179	0.4833	7.1735
0.40	0.5025	0.4683	7.3200
0.45	0.4957	0.4623	7.2164
0.50	0.4962	0.4642	6.9096
0.55	0.5033	0.4727	6.4570
0.60	0.5159	0.4871	5.9152
0.65	0.5334	0.5064	5.3331
0.70	0.5551	0.5299	4.7483
0.75	0.5805	0.5572	4.1865
0.80	0.6092	0.5876	3.6638
0.85	0.6406	0.6208	3.1883
0.90	0.6744	0.6563	2.7630
0.95	0.7104	0.6938	2.3871
1.00	0.7481	0.7330	2.0577

(i) If $h = 0.1$, then from (9.12) we have

$$Y_k = \frac{Y_{k-1}}{1 + 0.02k}, \quad k \in \mathbb{N}.$$

Table 9.4 has the numerical solution and the percentage error when $h = 0.1$. From the fourth column of Table 9.4, we find that the percentage error oscillates and at $x = 1$ it is about 3%.

(ii) If $h = 0.05$, then from (9.12) we obtain

$$Y_0 = 1, \ Y_k = \frac{Y_{k-1}}{1 + 0.005k}, \quad k \in \mathbb{N}.$$

The numerical results in this case are displayed in Table 9.5. As in the previous example, we conclude from Tables 9.4 and 9.5 that the relative percentage error is halved if the step size is reduced from 0.1 to 0.05. In particular, at $x_k = 1$ the percentage error is around 1.57% which is nearly half of the same when $h = 0.1$. $\qquad\square$

In the examples considered so far, the percentage error was small and was further reduced by decreasing the step size. The next example is a contrast where the percentage error is high though the step size is reduced to $h = 0.05$.

Example 9.2.5. Find an approximate solution to $y' = 3y$, $x > 0$, $y(0) = 5$, at

TABLE 9.4:   Numerical
results using the Euler backward
difference formula with $h = 0.1$

x_k	Y_k	$\phi(x_k)$	E_k
0.1	0.9804	0.9900	0.9755
0.2	0.9427	0.9608	1.8844
0.3	0.8893	0.9139	2.6923
0.4	0.8234	0.8521	3.3673
0.5	0.7486	0.7788	3.8791
0.6	0.6684	0.6977	4.1985
0.7	0.5863	0.6126	4.2970
0.8	0.5054	0.5273	4.1457
0.9	0.4283	0.4449	3.7147
1.0	0.3569	0.3679	2.9726

the grid points in $(0, 1]$ using the Euler method (the Euler forward difference formula) with (i) $h = 0.1$, (ii) $h = 0.05$.

Solution. In the given problem we have $f(x, y) = 3y$, $Y_0 = 5$. The Euler method using this f is

$$Y_{k+1} = Y_k + 3hY_k, \ k = 0, 1, \ldots. \tag{9.13}$$

On the other hand, the solution to the given initial value problem is

$$\phi(x) = 5e^{3x}, \ x > 0.$$

(i) Since $h = 0.1$, from (9.13) we get

$$Y_{k+1} = 1.3Y_k = (1.3)^2 Y_{k-1} = \cdots = (1.3)^{k+1} Y_0.$$

The numerical results are presented in Table 9.6 and the relative percentage error using this method is very high for this problem. In fact, it is as high as 31% at $x = 1.0$. Therefore the approximate values of the solution are not satisfactory with this step size.

(ii) We now compute the approximate solution, the exact solution and the relative percentage error with $h = 0.05$ and present in Table 9.7. From Table 9.7, we conclude that with reduced step size also the approximate solution is not close to the exact solution.   □

From the examples that we have seen so far, we observe that the accuracy of the approximate solution can be improved by taking smaller values of step size with high computational overhead. In Example 9.2.5, we need to reduce h to very small quantity to get a reasonable approximation. Decreasing the step size beyond certain stage can also lead to a type of error, namely round-off error[2]. We now explain the impact of round-off error due to very small h in detail.

---

[2]Roughly speaking, the error in the arithmetic computation due to finite precision of computing machines is called the round-off error.

TABLE 9.5: Numerical results using the Euler backward difference formula with $h = 0.05$

x_k	Y_k	$\phi(x_k)$	E_k
0.05	0.9950	0.9975	0.2484
0.10	0.9852	0.9900	0.4926
0.15	0.9706	0.9778	0.7300
0.20	0.9516	0.9608	0.9583
0.25	0.9284	0.9394	1.1752
0.30	0.9013	0.9139	1.3785
0.35	0.8709	0.8847	1.5658
0.40	0.8374	0.8521	1.7350
0.45	0.8013	0.8167	1.8840
0.50	0.7631	0.7788	2.0105
0.55	0.7234	0.7390	2.1124
0.60	0.6824	0.6977	2.1877
0.65	0.6408	0.6554	2.2341
0.70	0.5988	0.6126	2.2496
0.75	0.5571	0.5698	2.2321
0.80	0.5158	0.5273	2.1794
0.85	0.4754	0.4855	2.0893
0.90	0.4361	0.4449	1.9598
0.95	0.3983	0.4056	1.7885
1.00	0.3621	0.3679	1.5732

Suppose the computer used for our computations has 15 digits of precision. Let $h = 10^{-17}$ in Example 9.2.5. Then from (9.13) and the limited precision of the computer we obtain that

$$Y_1 = (1 + 3 \times 10^{-17})5 = 5.$$

Similarly, we get that $Y_k = 1$, $k \in \mathbb{N}$. Hence the percentage error at $x_{10^{17}} = 1$ is

$$E_{10^{17}} = \frac{|100.4277 - 1|}{100.4277} \times 100 = 99\% \text{ (approximately)}.$$

This numerical disaster is due to the round-off error. Thus decreasing the step size, so that certain 'important' terms in the computation of the approximate solution are neglected, can lead to increase in errors which in turn causes adverse effects. For a detailed analysis of round-off errors refer to [4]. Though reduction of step size decreases the local truncation error, it can create two types of problems. They are (i) increase of computation cost/time due to increase in the number of arithmetic calculations (indirectly increases the round-off errors), (ii) direct increase in the round-off errors as explained before. In all the examples and exercises that we consider henceforth, we do not take h too small and assume that the round-off errors are negligible. To summarize the discussion, if h is large then the local truncation error is large,

TABLE 9.6:   Numerical results
using the Euler method with $h = 0.1$

x_k	Y_k	$\phi(x_k)$	E_k
0.1	6.5000	6.7493	3.6936
0.2	8.4500	9.1106	7.2508
0.3	10.9850	12.2980	10.6766
0.4	14.2805	16.6006	13.9759
0.5	18.5647	22.4084	17.1533
0.6	24.1340	30.2482	20.2134
0.7	31.3743	40.8308	23.1604
0.8	40.7865	55.1159	25.9986
0.9	53.0225	74.3987	28.7319
1.0	68.9292	100.4277	31.3643

and if h is too small then the round-off error is large. In order to minimize both the local truncation error as well as the round-off error we need to use higher order methods, in particular, multi-step methods.

We now introduce two-step Euler method (or improved Euler method) for (9.1). For, we first write the backward Euler formula (9.7) as

$$\begin{cases} Y_{k+1} = Y_k + hf(x_{k+1}, Y_{k+1}), \ 0 \le k \le n-1, \\ Y_0 = y_0. \end{cases} \tag{9.14}$$

On taking the average of the Euler forward and backward difference methods (9.6) and (9.14), we obtain

$$\begin{cases} Y_{k+1} = Y_k + \dfrac{h}{2}\left[f(x_{k+1}, Y_{k+1}) + f(x_k, Y_k)\right], \ 0 \le k \le n-1, \\ Y_0 = y_0. \end{cases} \tag{9.15}$$

We notice that the numerical method prescribed in (9.15) is also an implicit method. In order to avoid solving nonlinear equations in (9.15), the following technique due to Heun is used where we replace the unknown on the right hand side with a suitable simpler approximation. First, we compute Y_1 using an explicit method like the Euler method. This value is called as the *predictor*. Then we substitute the value of Y_1 on the right hand side of (9.15) to find the *corrector* Y_1. We proceed in the same way to find Y_2, Y_3 and so on. To be more precise, the improved Euler method (the Heun method) is given by

$$\begin{cases} Z_{k+1} = Y_k + hf(x_k, Y_k), \ 0 \le k \le n-1, \\ Y_{k+1} = Y_k + \dfrac{h}{2}\left[f(x_{k+1}, Z_{k+1}) + f(x_k, Y_k)\right], \ 0 \le k \le n-1, \\ Y_0 = y_0. \end{cases} \tag{9.16}$$

TABLE 9.7: Numerical results
using the Euler method with $h = 0.05$

x_k	Y_k	$\phi(x_k)$	E_k
0.05	5.7500	5.8092	1.0186
0.10	6.6125	6.7493	2.0268
0.15	7.6044	7.8416	3.0247
0.20	8.7450	9.1106	4.0125
0.25	10.0568	10.5850	4.9902
0.30	11.5653	12.2980	5.9580
0.35	13.3001	14.2883	6.9159
0.40	15.2951	16.6006	7.8640
0.45	17.5894	19.2871	8.8025
0.50	20.2278	22.4084	9.7314
0.55	23.2620	26.0349	10.6509
0.60	26.7513	30.2482	11.5610
0.65	30.7639	35.1434	12.4618
0.70	35.3785	40.8308	13.3534
0.75	40.6853	47.4387	14.2360
0.80	46.7881	55.1159	15.1096
0.85	53.8063	64.0355	15.9743
0.90	61.8773	74.3987	16.8301
0.95	71.1589	86.4389	17.6773
1.00	81.8327	100.4277	18.5158

This is an example of *predictor-corrector* method where first the value of Y_{k+1} is predicted (as Z_{k+1}) and later it is corrected. Thus the approximate solution using the improved Euler method or the Heun method is given by system (9.16). We use it to compute the numerical solutions to the equations in the following examples.

Example 9.2.6. Find an approximate solution to $y' = 3y$, $x > 0$, $y(0) = 5$, at the grid points in $(0, 1]$ using the improved Euler method (the Heun method) with (i) $h = 0.1$, (ii) $h = 0.05$.

Solution. Here $f(x, y) = 3y$ and $\phi(x) = 5e^{3x}$, $x > 0$. We assume that (Y_k) is the approximate solution (using the Heun method) at the grid points (hk), $k \in \mathbb{N} \cup \{0\}$. Then Y_k is given by

$$
\begin{cases}
Z_{k+1} = Y_k + 3hY_k, \ 0 \le k \le n - 1, \\[2mm]
Y_{k+1} = Y_k + \dfrac{3h}{2}\big[Z_{k+1} + Y_k\big], \ 0 \le k \le n - 1, \\[2mm]
Y_0 = 5.
\end{cases}
\qquad (9.17)
$$

In order to compute Y_1 from Y_0, one needs to compute Z_1. It can be done by taking $k = 0$ in the first equation of (9.17). Using Y_0, Z_1, and $k = 0$ in the second equation of (9.17), we compute the numerical solution Y_1 at the grid point $x_1 = h$. Next, we use this Y_1 to compute Z_2. Then we use Y_1 and Z_2 to compute the approximate solution Y_2 at the grid point $x_2 = 2h$. We proceed in this way to compute the rest of the required terms in the sequence (Y_k).

(i) When $h = 0.1$, system (9.17) becomes

$$\begin{cases} Z_{k+1} = 1.3Y_k, \ 0 \le k \le 9, \\ Y_{k+1} = 1.15Y_k + 0.15Z_{k+1}, \ 0 \le k \le 9, \\ Y_0 = 5. \end{cases} \qquad (9.18)$$

The numerical results (without Z_k) are shown in Table 9.8. From Table 9.8,

TABLE 9.8:   Numerical results using the Heun method with $h = 0.1$

x_k	Y_k	$\phi(x_k)$	E_k
0.1	6.7250	6.7493	0.3599
0.2	9.0451	9.1106	0.7186
0.3	12.1657	12.2980	1.0760
0.4	16.3629	16.6006	1.4320
0.5	22.0080	22.4084	1.7868
0.6	29.6008	30.2482	2.1404
0.7	39.8131	40.8308	2.4926
0.8	53.5486	55.1159	2.8436
0.9	72.0229	74.3987	3.1933
1.0	96.8708	100.4277	3.5417

it is clear that this method gives better results than the Euler method. In particular, the approximate value at the grid point $x_{10} = 1$ is 96.8708 and the percentage error E_{10} using this method is about 3.5%. This is a substantial improvement over the error obtained using the Euler method with the same step size (recall from Table 9.6 in Example 9.2.5 that $E_{10} = 31.36\%$). At other grid points also we find the same phenomena.

(ii) We first put $h = 0.05$ in (9.17) to get the system of linear equations. Next, by following the same procedure explained before, one gets the numerical solution which is given in Table 9.9. As expected, the numerical results in Table 9.9 are indeed far better than those obtained using the Euler method with $h = 0.05$ (see Table 9.7 in Example 9.2.5). In the Heun method with $h = 0.05$, the percentage error is less than 1% at all the grid points. In fact, the approximate solution at $x = 1$ using this method is 99.4225 which is very close to the exact solution. Moreover, the relative percentage error at this grid point is about 1% which is satisfactory. □

Example 9.2.7. Find an approximate solution to $y' = -2xy$, $x > 0$, $y(0) = 1$, at $x = 1$ using the improved Euler method with (i) $h = 0.1$, (ii) $h = 0.05$.

TABLE 9.9:   Numerical results
using the Heun method with $h = 0.05$

x_k	Y_k	$\phi(x_k)$	E_k
0.05	5.8063	5.8092	0.0503
0.10	6.7425	6.7493	0.1005
0.15	7.8297	7.8416	0.1508
0.20	9.0923	9.1106	0.2010
0.25	10.5584	10.5850	0.2512
0.30	12.2610	12.2980	0.3013
0.35	14.2380	14.2883	0.3515
0.40	16.5339	16.6006	0.4016
0.45	19.2000	19.2871	0.4517
0.50	22.2960	22.4084	0.5017
0.55	25.8912	26.0349	0.5518
0.60	30.0662	30.2482	0.6018
0.65	34.9144	35.1434	0.6518
0.70	40.5443	40.8308	0.7017
0.75	47.0821	47.4387	0.7516
0.80	54.6741	55.1159	0.8016
0.85	63.4903	64.0355	0.8514
0.90	73.7281	74.3987	0.9013
0.95	85.6168	86.4389	0.9511
1.00	99.4225	100.4277	1.0009

Solution. In the given problem $f(x, y) = -2xy$. The solution to the IVP given in the exercise is $\phi(x) = e^{-x^2}$, $x > 0$. The numerical solution Y_k using the Heun method is obtained by solving

$$
\begin{cases}
Z_{k+1} = Y_k - 2hx_kY_k, \ 0 \le k \le n-1, \\
Y_{k+1} = Y_k - hx_k(Z_{k+1} + Y_k), \ 0 \le k \le n-1. \qquad (9.19) \\
Y_0 = 1.
\end{cases}
$$

(i) The numerical results when $h = 0.1$ are presented in Table 9.10. From the last column of this table, we notice that the values of E_k are small compared to the same computed using the Euler backward formula in Example 9.2.4. In fact, the values of E_k in Table 9.10 are around one tenth of the same presented in Example 9.2.4 with the same step size (see Table 9.4). Moreover, the relative percentage errors at the grid points $0.1, 0.2, \ldots$ are one fifth of the same computed in Example 9.2.4 with $h = 0.05$ (see Table 9.5).

(ii) The numerical results when $h = 0.05$ are presented in Table 9.11. From Table 9.11, it is easy to observe that the approximate solution matches with the solution to the IVP in the given problem up to three decimal places at every grid point. Furthermore, the percentage error is also very small. It is at most one third of the same using this method with $h = 0.1$ (see Table 9.10).

□

TABLE 9.10:   Numerical results
using the Heun method with $h = 0.1$

x_k	Y_k	$\phi(x_k)$	E_k
0.1	0.99000	0.99005	0.0050
0.2	0.96070	0.96079	0.0097
0.3	0.91381	0.91393	0.0128
0.4	0.85204	0.85214	0.0122
0.5	0.77876	0.77880	0.0046
0.6	0.69777	0.69768	0.0139
0.7	0.61292	0.61263	0.0486
0.8	0.52785	0.52729	0.1058
0.9	0.44572	0.44486	0.1930
1.0	0.36905	0.36788	0.3191

TABLE 9.11:   Numerical results
using the Heun method with $h = 0.05$

x_k	Y_k	$\phi(x_k)$	E_k
0.05	0.9975	0.99750	0.0003
0.10	0.9900	0.99005	0.0006
0.15	0.9777	0.97775	0.0009
0.20	0.9608	0.96079	0.0011
0.25	0.9394	0.93941	0.0013
0.30	0.9139	0.91393	0.0012
0.35	0.8847	0.88471	0.0010
0.40	0.8521	0.85214	0.0004
0.45	0.8167	0.81669	0.0006
0.50	0.7788	0.77880	0.0021
0.55	0.7390	0.73897	0.0043
0.60	0.6977	0.69768	0.0073
0.65	0.6555	0.65540	0.0112
0.70	0.6127	0.61263	0.0163
0.75	0.5699	0.56978	0.0227
0.80	0.5275	0.52729	0.0305
0.85	0.4857	0.48554	0.0402
0.90	0.4451	0.44486	0.0518
0.95	0.4058	0.40555	0.0656
1.00	0.3682	0.36788	0.0818

Example 9.2.8. Find an approximate solution to $y' = 3(x - y)$, $x > 0$, $y(0) = 1$, at the grid points in $(0,1]$ using the improved Euler method with $h = 0.2$.

Solution. In this example we have $f(x,y) = 3(x - y)$. Recall from Example 9.2.3 that the exact solution to the given initial value problem is

$$\phi(x) = x - \tfrac{1}{3} + \tfrac{4}{3}e^{-3x}, \ x > 0.$$

As before, we first write the system which gives an approximate solution using the Heun method as

$$\begin{cases} Z_{k+1} = Y_k + 3h(x_k - Y_k), \ 0 \le k \le n-1, \\ Y_{k+1} = Y_k + \dfrac{3h}{2}(x_k + x_{k+1} - Z_{k+1} - Y_k), \ 0 \le k \le n-1, \\ Y_0 = 1. \end{cases} \qquad (9.20)$$

The numerical results are shown in Table 9.12. An approximate solution to

TABLE 9.12: Numerical results
using the Heun method with $h = 0.2$

x_k	Y_k	$\phi(x_k)$	E_k
0.2	0.6400	0.5984	6.9491
0.4	0.5152	0.4683	10.0246
0.6	0.5268	0.4871	8.1613
0.8	0.6176	0.5876	5.0933
1.0	0.7542	0.7330	2.8827

the same IVP is found in Example 9.2.3. From Tables 9.2 and 9.12, it is evident
that the numerical results are better in this method with $h = 0.2$ compared
to the Euler backward difference method though we took smaller step size
($h = 0.1$) in that method. $\qquad\Box$

9.3 The Runge–Kutta Method

From the examples in Section 9.2, it is clear that the Heun method gives
better approximation of solutions to IVPs compared to the Euler forward and
backward difference methods for a given step size. However, in many practical
situations, we need to take the step size h to be very small to get the desired
accuracy while using the Heun method. As there are some disadvantages of
taking very small step sizes (see Section 9.2), we need to use higher order
methods. Now-a-days plenty of higher order methods are available and the
simpler ones among them are due to Runge–Kutta. A fourth order Runge–
Kutta formula to find an approximate solution to (9.1) is given by

$$\begin{cases} Y_{k+1} = Y_k + \dfrac{h}{6}(m_k^{(1)} + 2m_k^{(2)} + 2m_k^{(3)} + m_k^{(4)}), \ 0 \le k \le n-1, \\ Y_0 = y_0, \end{cases}$$

where

$$\begin{cases} m_k^{(1)} = f(x_k, Y_k), \quad m_k^{(2)} = f\left(x_k + \dfrac{h}{2}, Y_k + \dfrac{hm_k^{(1)}}{2}\right), \\ m_k^{(3)} = f\left(x_k + \dfrac{h}{2}, Y_k + \dfrac{hm_k^{(2)}}{2}\right), \quad m_k^{(4)} = f(x_k + h, Y_k + hm_k^{(3)}). \end{cases}$$

Notice that at each grid point x_k, we need to compute four quantities $m_k^{(1)}$, $m_k^{(2)}$, $m_k^{(3)}$, and $m_k^{(4)}$. Now let us compute approximate solutions to the IVPs given in examples in Section 9.2.

Example 9.3.1. Find an approximate solution to $y' = 3y$, $x > 0$, $y(0) = 5$, at the grid points in $(0,1]$ using the Runge–Kutta method with $h = 0.5, 0.2$, and 0.1.

Solution. Recall from Examples 9.2.5 and 9.2.6 that for this problem when $h = 0.1$, the Euler method gave unsatisfactory results and the Heun method gave approximate solution with percentage error being 3.4% at $x = 1$. The system given by the Runge–Kutta method for this problem is

$$
\begin{cases}
Y_0 = 5, \ Y_{k+1} = Y_k + \dfrac{h}{6}(m_k^{(1)} + 2m_k^{(2)} + 2m_k^{(3)} + m_k^{(4)}), \ k \in \mathbb{N}, \\[2mm]
m_k^{(1)} = 3Y_k, \ m_k^{(2)} = 3(Y_k + \dfrac{hm_k^{(1)}}{2}), \\[2mm]
m_k^{(3)} = 3(Y_k + \dfrac{hm_k^{(2)}}{2}), \ m_k^{(4)} = 3(Y_k + hm_k^{(3)}).
\end{cases}
$$

(i) When $h = 0.5$, we have only two grid points in $(0,1]$, namely $x_1 = 0.5$ and $x_2 = 1$. The numerical results can be found in Table 9.13. From Table 9.13

TABLE 9.13:   Numerical results using the Runge–Kutta method with $h = 0.5$

x_k	Y_k	$\phi(x_k)$	E_k
0.5	21.9922	22.4084	1.8576
1.0	96.7313	100.4277	3.6807

and Table 9.8 in Example 9.2.6, it is evident that the accuracy obtained using the Heun method with $h = 0.1$ can be achieved by the Runge–Kutta method with $h = 0.5$ at the grid points $x_1 = 0.5$, $x_2 = 1$.

(ii) Approximate solution when $h = 0.2$ is presented in Table 9.14. From Table 9.14, we observe that the resulting approximate value at $x = 1$ is 100.2298. The percentage error for this method is about 0.2% which is a substantial improvement over all the methods seen so far for this IVP.

(iii) Table 9.15 displays the results using a step size $h = 0.1$. The percentage error decreased to 0.016%. From Tables 9.14 and 9.15, it follows that the relative percentage error with $h = 0.1$ is roughly one tenth of that found at the grid points $0.2, 0.4, 0.6, 0.8$, and 1.0 with $h = 0.2$.   □

In this example, it appears that the Runge–Kutta method is capable of producing accurate results without very small step size. As a result, the computational cost and the round-off errors are reduced. The next couple of examples confirm this.

TABLE 9.14: Numerical
results using the Runge–Kutta
method with $h = 0.2$

x_k	Y_k	$\phi(x_k)$	E_k
0.2	9.1070	9.1106	0.0394
0.4	16.5875	16.6006	0.0789
0.6	30.2125	30.2482	0.1183
0.8	55.0290	55.1159	0.1577
1.0	100.2298	100.4277	0.1971

TABLE 9.15: Numerical
results using the Runge–Kutta
method with $h = 0.1$

x_k	Y_k	$\phi(x_k)$	E_k
0.1	6.7492	6.7493	0.0016
0.2	9.1103	9.1106	0.0032
0.3	12.2974	12.2980	0.0047
0.4	16.5995	16.6006	0.0063
0.5	22.4067	22.4084	0.0079
0.6	30.2454	30.2482	0.0095
0.7	40.8263	40.8308	0.0110
0.8	55.1089	55.1159	0.0126
0.9	74.3881	74.3987	0.0142
1.0	100.4118	100.4277	0.0158

Example 9.3.2. Find an approximate solution to $y' = -2xy$. $x > 0$. $y(0) = 1$, at $x = 1$ using the Runge–Kutta method with (i) $h = 0.5$, (ii) $h = 0.2$.

Solution. In the given problem $f(x, y) = -2xy$. The approximate solution (Y_k) at the grid points is given by

$$
\begin{cases}
Y_0 = 1, \ Y_{k+1} = Y_k + \dfrac{h}{6}(m_k^{(1)} + 2m_k^{(2)} + 2m_k^{(3)} + m_k^{(4)}), k \in \mathbb{N}, \\
m_k^{(1)} = -2x_k Y_k, \ m_k^{(2)} = -2(x_k + \dfrac{h}{2})(Y_k + \dfrac{hm_k^{(1)}}{2}), \\
m_k^{(3)} = -2(x_k + \dfrac{h}{2})(Y_k + \dfrac{hm_k^{(2)}}{2}), \ m_k^{(4)} = -2(x_k + h)(Y_k + hm_k^{(3)}).
\end{cases}
$$

(i) When $h = 0.5$ there are only two grid points in $(0, 1]$ and the approximate solution at those points is given in Table 9.16. The numerical results are satisfactory with the percentage error being small (approximately 0.04%). From Tables 9.10 and 9.16, it is clear that the error in the approximate solution

TABLE 9.16: Numerical
solution using the Runge–Kutta
method with $h = 0.5$

x_k	Y_k	$\phi(x_k)$	E_k
0.5	0.778645	0.778800	0.0199
1.0	0.368031	0.367879	0.0414

using this method is about one tenth of the same when we use the Heun
method with $h = 0.1$.

(*ii*) The approximate solution when $h = 0.2$ is given in Table 9.17. One can

TABLE 9.17: Numerical
solution using the Runge–Kutta
method with $h = 0.5$

x_k	Y_k	$\phi(x_k)$	E_k
0.2	0.960789	0.960789	0.0000
0.4	0.852142	0.852143	0.0001
0.6	0.697675	0.697676	0.0001
0.8	0.527297	0.527292	0.0010
1.0	0.367903	0.367879	0.0066

observe that this method gives a very good approximation (Y_k and $\phi(x_k)$ agree
up to three digits after the decimal point for $k = 1, 2, 3, 4, 5$). In particular,
for the grid points 0.2, 0.4, 0.6, and 0.8 they agree up to five digits after the
decimal point.   □

   In the next example, we compute the approximate solution of a problem
using all the methods that are discussed so far.

Example 9.3.3. Find an approximate solution to $xy' = y + y^2$, $x > 1$,
$y(1) = -2$, at $x = 2$ with $h = 0.2$ using (*i*) the Euler method (the Euler
forward difference formula), (*ii*) the Euler backward difference formula, (*iii*)
the Heun method, and (*iv*) the Runge–Kutta method.

Solution. Recall that the solution for this problem is $\phi(x) = \dfrac{2x}{1 - 2x}$, $x > 1$.
The numerical results using various methods are provided in Tables 9.18 to
9.21. From Tables 9.18 to 9.20, it is clear that the Heun method is better
than the Euler methods (both forward and backward). In fact, in the Heun
method both Y_5 and $\phi(2)$ agree up to two digits after the decimal point. On
the other hand, when we use the Runge–Kutta method, the approximate value
and the exact value of the solution at $x = 2$ agree up to four decimal places.
Therefore the relative percentage error is the least (as low as 0.001%) for the
Runge–Kutta method. This demonstrates the superiority of the higher order
methods over Euler forward/backward difference methods.   □

TABLE 9.18:   Numerical
results using the Euler forward
difference method

x_k	Y_k	$\phi(x_k)$	E_k
1.20	-1.6000	-1.7143	6.6667
1.40	-1.4400	-1.5556	7.4286
1.60	-1.3495	-1.4545	7.2229
1.80	-1.2905	-1.3846	6.7949
2.00	-1.2489	-1.3333	6.3346

TABLE 9.19:   Numerical
results using the Euler backward
difference method

x_k	Y_k	$\phi(x_k)$	E_k
1.2	-1.7720	-1.7143	3.3668
1.4	-1.6264	-1.5556	4.5573
1.6	-1.5261	-1.4545	4.9187
1.8	-1.4530	-1.3846	4.9362
2.0	-1.3974	-1.3333	4.8069

In fact, all the methods that are discussed so far can be easily extended to
systems of ODEs. To find an approximate solution to any higher order IVP,
we first write the given higher order ODE as a system of first order ODEs
and use any numerical method to find an approximate solution to the first
order system. Conversion of a higher order IVP to a first order IVP is already
explained in Chapter 5, Section 5.1. We now focus on the methods to find
approximate solutions to the systems of ODEs. In order to do that, consider
the system for $t > 0$,

$$\begin{cases} x'(t) = f(t, x(t), y(t)), \\ y'(t) = g(t, x(t), y(t)), \\ x(a) = x_0, \ y(a) = y_0. \end{cases} \tag{9.21}$$

Let the grid points be $t_0, t_1, \ldots$, where $t_k = a + kh$, $k \in \mathbb{N} \cup \{0\}$. Then the
Euler method to find an approximate solution to (9.21) is

$$\begin{cases} X_{k+1} = X_k + hf(t_k, X_k, Y_k), \ k \in \mathbb{N} \cup \{0\}, \\ Y_{k+1} = Y_k + hg(t_k, X_k, Y_k), \ k \in \mathbb{N} \cup \{0\}, \\ X_0 = x_0, \ Y_0 = y_0. \end{cases} \tag{9.22}$$

TABLE 9.20: Numerical
results using the Heun method

x_k	Y_k	$\phi(x_k)$	E_k
1.2	-1.7200	-1.7143	0.3333
1.4	-1.5613	-1.5556	0.3675
1.6	-1.4595	-1.4545	0.3433
1.8	-1.3889	-1.3846	0.3098
2.0	-1.3370	-1.3333	0.2783

TABLE 9.21: Numerical results
using the Runge–Kutta method

x_k	Y_k	$\phi(x_k)$	E_k
1.2	-1.714245	-1.714286	0.0024
1.4	-1.555523	-1.555556	0.0021
1.6	-1.454520	-1.454545	0.0018
1.8	-1.384594	-1.384615	0.0015
2.0	-1.333316	-1.333333	0.0013

Similarly, we can write the Runge–Kutta method for system (9.21) as

$$
\begin{cases}
X_{k+1} = X_k + \dfrac{h}{6}(\mu_k^{(1)} + 2\mu_k^{(2)} + 2\mu_k^{(3)} + \mu_k^{(4)}),\ 0 \le k \le n-1, \\[2mm]
Y_{k+1} = Y_k + \dfrac{h}{6}(m_k^{(1)} + 2m_k^{(2)} + 2m_k^{(3)} + m_k^{(4)}),\ 0 \le k \le n-1, \\[2mm]
X_0 = x_0,\ Y_0 = y_0,
\end{cases} \tag{9.23}
$$

where

$$
\begin{cases}
\mu_k^{(1)} = f(t_k, X_k, Y_k), \\[2mm]
m_k^{(1)} = g(t_k, X_k, Y_k), \\[2mm]
\mu_k^{(2)} = f\left(t_k + \dfrac{h}{2}, X_k + \dfrac{h\mu_k^{(1)}}{2}, Y_k + \dfrac{hm_k^{(1)}}{2}\right), \\[2mm]
m_k^{(2)} = g\left(t_k + \dfrac{h}{2}, X_k + \dfrac{h\mu_k^{(1)}}{2}, Y_k + \dfrac{hm_k^{(1)}}{2}\right), \\[2mm]
\mu_k^{(3)} = f\left(t_k + \dfrac{h}{2}, X_k + \dfrac{h\mu_k^{(2)}}{2}, Y_k + \dfrac{hm_k^{(2)}}{2}\right), \\[2mm]
m_k^{(3)} = g\left(t_k + \dfrac{h}{2}, X_k + \dfrac{h\mu_k^{(2)}}{2}, Y_k + \dfrac{hm_k^{(2)}}{2}\right), \\[2mm]
\mu_k^{(4)} = f(t_k + h, X_k + h\mu_k^{(3)}, Y_k + hm_k^{(3)}), \\[2mm]
m_k^{(4)} = g(t_k + h, X_k + h\mu_k^{(3)}, Y_k + hm_k^{(3)}).
\end{cases} \tag{9.24}
$$

Example 9.3.4. Find an approximate solution to $y''(t) + y(t) = \sin t$, $y(0) = 0$, $y'(0) = 1$ with $h = 0.2$ at $x = 1$ using the Euler method, and the Runge–Kutta method.

Solution. Using the method of undetermined coefficients (see Section 3.1.2.3), the exact solution to the given system is given by

$$y(t) = \frac{3}{2}\sin(t) - \frac{t}{2}\cos(t), \ t \in \mathbb{R}.$$

In order to apply the Euler method or Runge–Kutta method, we first write the given initial value problem as a system of first order ODEs. For, on setting $x(t) = y'(t)$, the given problem becomes

$$\begin{cases} x'(t) = -y(t) + \sin t, \ t > 0, \\ y'(t) = x(t), \\ y(0) = 0, \ x(0) = 1. \end{cases} \tag{9.25}$$

Let (X_k, Y_k) denote the approximation of solution (x, y) to (9.25) at the grid points t_k, $k \in \mathbb{N}$. Let E_y and E_x denote the percentage error in the approximation of y and x, respectively. On comparing with the standard form (9.21), we obtain

$$f(t, x, y) = -y + \sin t, \ g(t, x, y) = x.$$

An approximate solution to (9.25) using Euler method (9.22) with $h = 0.2$ is computed using the formula

$$\begin{cases} X_{k+1} = X_k - 0.2(Y_k - \sin t_k), \ 0 \le k \le 4, \\ Y_{k+1} = Y_k + 0.2X_k, \ 0 \le k \le 4, \\ X_0 = 1, \ Y_0 = 0. \end{cases}$$

On substituting f, g, and h in (9.23)–(9.24), one can compute the approximate solution using the Runge–Kutta method. Tables 9.22 and 9.23 display the numerical results obtained using the Euler method and the Runge–Kutta method, respectively.

TABLE 9.22: Numerical results using the Euler method

t_n	X_n	$y'(t_n)$	Y_n	$y(t_n)$	E_x	E_y
0.2	1.0000	0.9999	0.2000	0.2000	0.0066	0.0013
0.4	0.9997	0.9989	0.4000	0.3999	0.0790	0.0212
0.6	0.9976	0.9947	0.5999	0.5994	0.2904	0.0974
0.8	0.9906	0.9836	0.7995	0.7974	0.7022	0.2657
1.0	0.9741	0.9610	0.9976	0.9921	1.3627	0.5571

TABLE 9.23:   Numerical results using the Runge–Kutta method

t_n	X_n	$y'(t_n)$	Y_n	$y(t_n)$	$10^3 \times E_x$	$10^3 \times E_y$
0.2	0.99993354	0.99993351	0.19999778	0.19999734	0.0033	0.2202
0.4	0.99894480	0.99894466	0.39991611	0.39991531	0.0033	0.1983
0.6	0.99472870	0.99472836	0.59936398	0.59936302	0.0349	0.1595
0.8	0.98364987	0.98364914	0.79735230	0.79735145	0.0737	0.1069
1.0	0.96103916	0.96103780	0.99205576	0.99205532	0.1418	0.0445

From these tables, it is evident that the Runge–Kutta method gives far better results over the Euler method for the first order systems also. In fact, when we use the Runge–Kutta method, the approximate solution and the exact solution at $x = 1$ agree up to six digits after the decimal point. Subsequently, the percentage error decreases by the factor of 10^3 when we use the same step size.   □

Exercise 9.1. Use the Euler method to find approximate solutions at the given points to the following initial value problems. The step size h is given in each of the following problems separately. Moreover, compute the percentage error in each problem.

(i) If ϕ is a solution to $y' = y(4 - y^2)$, $y(0) = 1$, then find an approximate value of $\phi(2)$ by taking $h = 0.04$.

(ii) If ϕ is a solution to $y' = y + \sin(2x)$, $y(0) = 3$, then find an approximate value of $\phi(1)$ by taking $h = 0.05$.

(iii) If ϕ is a solution to $y' = (4x + y - 1)^2$, $y(0) = 0$, then find an approximate value of $\phi(5)$ by taking $h = 0.25$.

(iv) If ϕ is a solution to $y' = \cos y$, $y(0) = 0$, then find an approximate value of $\phi(3)$ by taking $h = 0.1$.

Exercise 9.2. Solve all the problems given in Exercise 9.1 using the Heun method and compute the percentage errors. Compare the results obtained using the Euler method and the Heun method.

Exercise 9.3. Find approximate solutions to the following IVPs in $[0,2]$ using the backward Euler method with $h = 0.1$:

(i) $y' = xy + y^2$, $y(0) = 1$.   (ii) $y' = y - y^2$, $y(0) = 0.1$.

Exercise 9.4. Plot the approximate solutions to the following IVPs in $[0,20]$ using the Runge–Kutta method with $h = 0.2$ and $x(0) = 0$, $y(0) = 1$ in the phase plane:

(i) $\begin{cases} x' = x + y, \\ y' = x - y + t^2. \end{cases}$

(ii) $\begin{cases} x' = y - x^3, \\ y' = -3x - 5y^5. \end{cases}$

$$(iii) \quad \begin{cases} x' = -y + x(1 - 3x^2 - 3y^2), \\ y' = x + y(1 - 3x^2 - 3y^2). \end{cases}$$

Exercise 9.5. Find an approximate solution to the following IVPs in $[0,5]$ using the Runge–Kutta method with $h = 0.2$:

(i) $y'' + 4y = 0$, $y(0) = 1$, $y'(0) = 0$.

(ii) (The van der Pol equation) $y'' + (1 - y^2)y' + y = 0$, $y(0) = y'(0) = 0$.

(iii) (The Duffing equation) $y'' + y' - y^3 = 0$, $y(0) = y'(0) = 0$.

Exercise 9.6. Generalize the Euler method, the Heun method and the Runge–Kutta method for $n \times n$ system of ODEs. Let ϕ be the solution to $y^{(iv)} - y = x$, $y(0) = 0$, $y'(0) = 0$, $y''(0) = 0$, $y'''(0) = 1$. Then find approximate values of $\phi(3)$ using each of these methods with $h = 0.1$.

Appendix A

A.1 Metric spaces

Definition A.1.1. A nonempty set X is said to be a metric space if there exists a function $d : X \times X \to [0, \infty)$ such that:

(i) $d(x, y) = 0$ if and only if $x = y$,

(ii) (symmetry) $d(x, y) = d(y, x)$, $x, y \in X$,

(iii) (triangle inequality) $d(x, y) \leq d(x, y) + d(y, z)$, $x, y, z \in X$.

This function d is called a metric on X. We denote by (X, d) the metric space X along with the metric d.

Definition A.1.2. A sequence (ξ_n) in the metric space (X, d) is said to be a Cauchy sequence, if for a given $\epsilon > 0$, there exists $n_0 \in \mathbb{N}$ such that $d(\xi_n, \xi_m) < \epsilon$ whenever $m, n \geq n_0$.

Definition A.1.3. A sequence (ξ_n) in the metric space (X, d) is said to be a convergent sequence in X, if there exists $\xi \in X$ and for a given $\epsilon > 0$, there exists $n_0 \in \mathbb{N}$ such that $d(\xi_n, \xi) < \epsilon$ whenever $n \geq n_0$. Moreover this ξ is called a limit of (ξ_n) and we denote this by $(\xi_n) \to \xi$ in X.

In fact, it is well known that in a metric space, if the sequence (ξ_n) converges then it has a unique limit.

Definition A.1.4. A metric space (X, d) is said to be complete if every Cauchy sequence in X is a convergent sequence in X, i.e., if (ξ_n) is a Cauchy sequence in X, then there exists $\xi \in X$ such that $\xi_n \to \xi$ in (X, d).

For example, $(\mathbb{R}, d)$ is a complete metric space where the metric is given by $d(x, y) = |x - y|$, $x, y \in \mathbb{R}$.

Definition A.1.5. Let (X, d) be a metric space. We say that F is a closed subset of X, if (ξ_n) is a sequence in F such that $\xi_n \to \xi$ in X then $\xi \in F$.

Theorem A.1.6. *If (X, d) is any complete metric space and $Y \subseteq X$ is closed set then (Y, d) is a complete metric space.*

Proof. The proof is left to the reader as an easy exercise. $\square$

Definition A.1.7. Let (X, d_X) and (Y, d_Y) be metric spaces. A function $T : X \to Y$ is said to be a contraction mapping if there exists $\alpha \in (0, 1)$ such that

$$d_Y(Tx_1, Tx_2) \leq \alpha d_X(x_1, x_2), x_1, x_2 \in X.$$

For example, consider $f : ([-1, 1], d) \to (\mathbb{R}, d)$, $f(x) = \frac{x}{2}$, where d is given by $d(x_1, x_2) = |x_2 - x_1|$, $x_1, x_2 \in [-1, 1]$. One can easily verify that f is a contraction mapping.

Definition A.1.8. Let (X, d_X), (Y, d_Y) be metric spaces. We say that a function $f : (X, d_X) \to (Y, d_Y)$ is continuous if for every $(x_n) \to x$ in X, we have $\big(f(x_n)\big) \to f(x)$ in Y.

Remark A.1.9. From the definitions, it immediately follows that every contraction map is a continuous map.

Definition A.1.10. Let X be a nonempty set and $T : X \to X$ a function. A point $x \in X$ is said to be a fixed point of T if $Tx = x$.

Theorem A.1.11 (Banach fixed point theorem). *Assume that (X, d) is a complete metric space and $T : X \to X$ is a contraction mapping. Then there exists a unique fixed point of T.*

Proof. We begin with the proof of existence of a fixed point. We consider an arbitrary point ξ_0 in X. If ξ_0 is a fixed point of T, then we are done. Therefore we suppose ξ_0 is not a fixed point, i.e., $T(\xi_0) \neq \xi_0$. We show that the sequence $(T^n(\xi_0))$ defined by

$$T^n(\xi_0) = T(T^{n-1}\xi_0)), n > 1,$$

is a Cauchy sequence whose limit is a fixed point.
Since T is a contraction mapping, there exists $\alpha \in (0, 1)$ such that

$$d(T(\xi), T(\eta)) < \alpha d(\xi, \eta), \ \xi \neq \eta, \ \xi, \eta \in X.$$

We first notice from the definition of a contraction mapping that for $n > m$,

$$
\begin{aligned}
d\big(T^n(\xi_0), T^m(\xi_0)\big) &\leq \alpha d\big(T^{n-1}(\xi_0), T^{m-1}(\xi_0)\big) \\
&\leq \alpha^2 d\big(T^{n-2}(\xi_0), T^{m-2}(\xi_0)\big) \\
&\vdots \\
&\leq \alpha^m d(T^{n-m}(\xi_0), \xi_0).
\end{aligned}
\tag{A.1}
$$

For $k \geq 1$, from the triangle inequality we obtain

$$
\begin{aligned}
d(T^k(\xi_0), \xi_0) &\leq d\big(T^k(\xi_0), T^{k-1}(\xi_0)\big) + \cdots + d(T(\xi_0), \xi_0) \\
&\leq (\alpha^{k-1} + \alpha^{k-2} + \cdots + 1)d(T(\xi_0), \xi_0).
\end{aligned}
\tag{A.2}
$$

In view of (A.1)–(A.2), it follows that

$$
\begin{aligned}
d\big(T^n(\xi_0), T^m(\xi_0)\big) &\leq \alpha^m(\alpha^{n-m-1} + \alpha^{n-m-2} + \cdots + 1)d(T(\xi_0), \xi_0) \\
&\leq \frac{\alpha^m}{1-\alpha} d(T(\xi_0), \xi_0), \ n > m. \tag{A.3}
\end{aligned}
$$

We next suppose $\epsilon > 0$ is a given number. Since $\alpha^k \to 0$ as $k \to \infty$ and $d(T(\xi_0), \xi_0) \neq 0$, we choose $n_0 \in \mathbb{N}$ such that

$$
\alpha^{n_0} < \frac{(1-\alpha)\epsilon}{d(T(\xi_0), \xi_0)}.
$$

Therefore for $n, m > n_0$, we have

$$
d\big(T^n(\xi_0), T^m(\xi_0)\big) < \epsilon,
$$

which proves that $\big(T^n(\xi_0)\big)$ is a Cauchy sequence. Since X is complete, there exists $\hat{x} \in X$ such that

$$
T^n(\xi_0) \to \hat{x}, \ n \to \infty. \tag{A.4}
$$

From (A.4), owing to the continuity of T, we have

$$
T^{n+1}(\xi_0) \to T(\hat{x}), \ n \to \infty. \tag{A.5}
$$

Finally, we observe that the sequence in (A.5) is a subsequence of the sequence in (A.4), and hence we get

$$
T(\hat{x}) = \hat{x}.
$$

This proves existence of a fixed point.

For uniqueness, on contrary we assume that x_1 and x_2 are distinct fixed points of T. Then

$$
d(x_1, x_2) = d\big(T(x_1), T(x_2)\big) \leq \alpha d(x_1, x_2).
$$

which is a contradiction to $\alpha < 1$. Hence there exists a unique fixed point to T. This completes the proof. $\qquad\square$

Appendix B

B.1   Another proof of the Cauchy–Lipschitz theorem

We want to prove the Cauchy–Lipschitz theorem using the contraction mapping theorem. In fact, we present a proof due to R. Cacciopoli (1930) which gives both the existence as well as the uniqueness of a solution to

$$\begin{cases} y'(x) = f(x, y(x)), \ x \in J, \\ y(x_0) = y_0, \end{cases} \tag{B.1}$$

in a single stroke using the Banach fixed point theorem. The main idea of the proof is to identify appropriate T and X which satisfy the hypotheses of Theorem A.1.11 so that the fixed point of T is the solution to Cauchy problem (B.1). Before we state the Cauchy-Lipschitz theorem we advise the reader to recall Theorem 2.2.10 and the notation therein.

Theorem B.1.1 (Cauchy-Lipschitz). *We assume that f is continuous on*

$$R := \{(x, y) : |x - x_0| \le a, \ |y - y_0| \le b\}, \ a, b > 0$$

and $M := \sup_{(x,y) \in R} |f(x, y)|$. We next assume that f satisfies the Lipschitz condition, i.e., there exists $L > 0$ such that

$$|f(x, y_1) - f(x, y_2)| \le L|y_1 - y_2|, \ (x, y_1), \ (x, y_2) \in R. \tag{B.2}$$

Then Cauchy problem (B.1) has a unique solution on $[x_0 - \epsilon, x_0 + \epsilon]$, where

$$\epsilon < \min\left\{a, \frac{b}{M}, \frac{1}{L}\right\}. \tag{B.3}$$

Proof. In order to prove the existence and uniqueness result to (B.1), in view of Lemma 2.2.4, it is enough to prove the same for (2.36). For, we first notice that we have the same function ϕ (the required solution) on both sides of (2.36). This motivates us to identify that the solutions to (2.36) are precisely the fixed points of the operator

$$T[u] = y_0 + \int_{x_0}^{x} f(t, u(t))dt. \tag{B.4}$$

We now find the domain of this operator such that T satisfies the hypotheses of the contraction mapping theorem, i.e., Theorem A.1.11. To that end, we first fix $\epsilon > 0$ satisfying (B.3), and set $J_1 = [x_0 - \epsilon, x_0 + \epsilon]$. We next prove that

$$\tilde{C} := \left\{ u \in C(J_1) : \sup_{x \in J_1} |u(x) - y_0| \le M\epsilon \right\}$$

is a closed subset of $\big(C(J_1), d_\infty\big)$.

Let (u_n) be a sequence in $\tilde{C}$ such that $u_n \to u$ in $(C(J_1), d_\infty)$. We need to show that $u \in \tilde{C}$. For, we suppose $\delta > 0$ is arbitrary. Then there exists $n_0 \in \mathbb{N}$ such that $d_\infty(u_n, u) < \delta$, $n \ge n_0$. Now for $n \ge n_0$, we consider

$$
\begin{aligned}
d_\infty(u, y_0) &\le d_\infty(u, u_n) + d_\infty(u_n, y_0) \\
&\le \delta + M\epsilon.
\end{aligned}
$$

Since $\delta > 0$ is arbitrary, we obtain that $d_\infty(u, y_0) \le M\epsilon$, or $u \in \tilde{C}$. Hence $\tilde{C}$ is a closed subset of $\big(C(J_1), d_\infty\big)$. Thus in view of Theorems 2.2.10 and A.1.6, it follows that $(\tilde{C}, d_\infty)$ is a complete metric space.

On the other hand, using (B.3), we prove that the following hold:

(a) If $u \in \tilde{C}$, then $(t, u(t)) \in R$, $t \in J_1$.

(b) If $u \in \tilde{C}$, then $T[u] \in \tilde{C}$.

Proof of (a): It is straightforward from the definition.

Proof of (b): In view of (a) and continuity of f, it follows that $T[u]$ is continuous. We fix $x \in [x_0, x_0 + \epsilon]$ and consider

$$
\begin{aligned}
|T[u](x) - y_0| &= \left| \int_{x_0}^{x} f(t, u(t)) dt \right| \\
&\le \int_{x_0}^{x} |f(t, u(t))| dt \\
&\le \int_{x_0}^{x} M dt \le M\epsilon. \qquad \text{(B.5)}
\end{aligned}
$$

It is easy to show that (B.5) holds for $x \in [x_0 - \epsilon, x_0]$ also. Hence $d_\infty(T[u], y_0) \le M\epsilon$ and this establishes $T[u] \in \tilde{C}$. This establishes (b).

From (a) and (b), we conclude that $T : \tilde{C} \to \tilde{C}$ is well defined. To complete the proof it is enough to show that T is a contraction mapping. For $x \in [x_0, x_0 + \epsilon]$,

$u_1, u_2 \in \tilde{C}$, we consider

$$
\begin{aligned}
\left| T[u_1](x) - T[u_2](x) \right| &= \left| \int_{x_0}^{x} \left[f(t, u_1(t)) - f(t, u_2(t)) \right] dt \right| \\
&\leq \int_{x_0}^{x} \left| f(t, u_1(t)) - f(t, u_2(t)) \right| dt \\
&\leq L \int_{x_0}^{x} \left| u_1(t) - u_2(t) \right| dt \\
&\leq L d_\infty(u_1, u_2) \int_{x_0}^{x} dt \\
&\leq L \epsilon d_\infty(u_1, u_2). \quad\quad\quad\quad\quad\quad (\text{B.6})
\end{aligned}
$$

Using the same argument, we can prove (B.6) for $x \in [x_0 - \epsilon, x_0]$ as well. Therefore we get

$$
d_\infty(T[u_1], T[u_2]) \leq L \epsilon d_\infty(u_1, u_2), \quad u_1, u_2 \in \tilde{C},
$$

which establishes that T is a contraction. Hence T has a unique fixed point in $\tilde{C}$, i.e., there exists a unique $\phi \in \tilde{C}$ such that

$$
\phi(t) = y_0 + \int_{x_0}^{x} f(t, \phi(t)) dt, \ t \in J_1.
$$

This completes the proof. $\qquad\qquad\qquad\qquad\qquad\qquad\qquad\qquad\qquad\qquad\square$

Remark B.1.2. From the proof of the Banach fixed point theorem, we notice that the required fixed point $\hat{x}$ of T is the limit of $(T^n(x_0))$, where $x_0 \in X$ is arbitrary. Hence $(T^n(x_0))$ can be viewed as a sequence of approximations to $\hat{x}$. In the context of ODEs, we have identified the solution to (B.1) as the fixed point of the operator T given by (B.4). Therefore the sequence of iterations $(T^n[u_0])$ where u_0 is any function in $\tilde{C}$ approximates the solution to Cauchy problem in $C(J_1, d_\infty)$, $J_1 = [x_0 - \epsilon, x_0 + \epsilon]$. When u_0 is the constant function given by the initial data y_0, then $(T^n[u_0])$ is precisely the Picard sequence of successive approximations to the solution for Cauchy problem (B.1).

Appendix C

C.1 Some useful results from calculus

Definition C.1.1. Let $\Omega \subseteq \mathbb{R}^2$ be an open set and $F : \Omega \to \mathbb{R}$ be a function. We say that F is a C^1-function if both $\frac{\partial F}{\partial x}$ and $\frac{\partial F}{\partial y}$ are continuous. Similarly, if all the second order partial derivatives, namely $\frac{\partial^2 F}{\partial x^2}$, $\frac{\partial^2 F}{\partial y \partial x}$, $\frac{\partial^2 F}{\partial x \partial y}$, and $\frac{\partial^2 F}{\partial y^2}$ are continuous, then we say that F is a C^2-function.

If F is a C^2-function then the second order mixed derivatives are the same, i.e.,

$$\frac{\partial^2 F}{\partial y \partial x} = \frac{\partial^2 F}{\partial x \partial y}.$$

Theorem C.1.2 (Implicit function theorem). *Let $\Omega \subseteq \mathbb{R}^2$ be an open set and $H : \Omega \to \mathbb{R}$, $(x,y) \mapsto H(x,y)$ a C^1-function. Assume that $\frac{\partial H}{\partial y}(x_0, y_0) \neq 0$. Then there exist two open sets V_{x_0} and $W_{(x_0,y_0)}$ in $\mathbb{R}$ and $\mathbb{R}^2$, respectively, and a function $h : V_{x_0} \to \mathbb{R}$ such that*

(i) $x_0 \in V_{x_0}$, $(x_0, y_0) \in W_{(x_0,y_0)}$,

(ii) the function h is a C^1-function,

(iii) $h(x_0) = y_0$.

(iv) $H(x, h(x)) = H(x_0, y_0)$. $x \in V_{x_0}$,

(v) If $H(x,y) = H(x_0, y_0)$ for some $(x,y) \in W_{(x_0,y_0)}$, then $y = h(x)$.

In fact, we say that h is implicitly defined by the relation $H(x,y) = H(x_0, y_0)$.

Next, we present a technical result which gives a formula for the derivative of the determinant of a matrix whose entries are functions of single variable.

Theorem C.1.3. *Let J be an open interval in $\mathbb{R}$ and $X : J \to M_n(\mathbb{R})$ be differentiable. Denote the i-th row of $X(t)$ by $\mathbf{x}_i(t)$, i.e., $\mathbf{x}_i(t) = \begin{pmatrix} x_{i1}(t) & x_{i2}(t) & \cdots & x_{in}(t) \end{pmatrix}$. Assume that $\det \begin{pmatrix} \mathbf{x}_1(t), \ldots, \mathbf{x}_n(t) \end{pmatrix}$ denotes the determinant of the matrix $X(t)$. Then we have*

$$\frac{d}{dt} \det \begin{pmatrix} \mathbf{x}_1(t), \ldots, \mathbf{x}_n(t) \end{pmatrix} = \sum_{i=1}^{n} \det \begin{pmatrix} \mathbf{x}_1(t), \ldots, \frac{d\mathbf{x}_i(t)}{dt}, \ldots, \mathbf{x}_n(t) \end{pmatrix}. \quad \text{(C.1)}$$

Proof. We first fix $t \in J$ and $h \in \mathbb{R}$ be such that $t + h \in J$. We introduce the functions

$$p_n(t, h) := \det\big(\mathbf{x}_1(t+h), \ldots, \mathbf{x}_n(t+h)\big),$$

$$p_0(t, h) := \det\big(\mathbf{x}_1(t), \ldots, \mathbf{x}_n(t)\big),$$

$$p_k(t, h) := \det\big(\mathbf{x}_1(t+h), \ldots, \mathbf{x}_k(t+h), \mathbf{x}_{k+1}(t), \ldots, \mathbf{x}_n(t)\big), \; 1 \le k \le n-1.$$

From the definition of the derivative we obtain that

$$\frac{d}{dt} \det\big(\mathbf{x}_1(t), \ldots, \mathbf{x}_n(t)\big) = \lim_{h \to 0} \frac{1}{h}\big(p_n(t,h) - p_0(t,h)\big)$$

$$= \sum_{k=1}^{n} \lim_{h \to 0} \frac{1}{h}\big(p_k(t,h) - p_{k-1}(t,h)\big). \quad \text{(C.2)}$$

We know that the determinant is an n-linear function. Moreover, it is continuous with respect to each of its arguments. Thus for $1 \le k \le n$, we have

$$\lim_{h \to 0} \frac{1}{h}\big(p_k(t,h) - p_{k-1}(t,h)\big) = \lim_{h \to 0} \det\big(\mathbf{x}_1(t+h), \ldots, \mathbf{x}_{k-1}(t+h),$$

$$\frac{\mathbf{x}_k(t+h) - \mathbf{x}_k(t)}{h}, \ldots, \mathbf{x}_n(t)\big)$$

$$= \det\Big(\lim_{h \to 0}\mathbf{x}_1(t+h), \ldots, \lim_{h \to 0}\mathbf{x}_{k-1}(t+h),$$

$$\lim_{h \to 0} \frac{\mathbf{x}_k(t+h) - \mathbf{x}_k(t)}{h}, \ldots, \mathbf{x}_n(t)\Big)$$

$$= \det\Big(\mathbf{x}_1(t), \ldots, \frac{d}{dt}\mathbf{x}_k(t), \ldots, \mathbf{x}_n(t)\Big). \quad \text{(C.3)}$$

The required result (C.1) follows directly from (C.2)–(C.3).   □

Bibliography

[1] T. Amaranath. *An elementary course in partial differential equations.* Jones & Bartlett Learning, Sudbury, second edition, 2008.

[2] V.I. Arnold. *Ordinary differential equations.* MIT Press, Cambridge, 1973.

[3] W.E. Boyce and R.C. DiPrima. *Elementary differential equations and boundary value problems.* John Wiley & Sons, New York, seventh edition, 2001.

[4] R.L. Burden and J.D. Faires. *Numerical Analysis.* Cengage Learning, Boston, ninth edition, 2011.

[5] M.L. Cartwright and J.E. Littlewood. On the nonlinear differential equations of the second order. i. the equation $\ddot{y} + k(1 - y^2)\dot{y} + y = b\lambda k \cos(\lambda t + a)$; k large. *J. Lond. Math. Soc.*, 20:180–189, 1945.

[6] E. Coddington and R. Carlson. *Linear ordinary differential equations.* SIAM, Philadelphia, 1997.

[7] E. Coddington and N. Levinson. *Theory of ordinary differential equations.* McGraw–Hill, New York, 1955.

[8] E.A. Coddington. *Introduction to ordinary differential equations.* Prentice–Hall of India, New Delhi, 1974.

[9] R. Grimshaw. *Nonlinear ordinary differential equations.* Blackwell Scientific Publications, Boston, 1990.

[10] P. Hartman. *Ordinary differential equations*, volume 38 of *Classics in Applied Mathematics.* SIAM, Philadelphia, second edition, 2002.

[11] M.W. Hirsch, S. Smale, and R.L. Devaney. *Differential equations, dynamical systems, and an introduction to chaos.* Elsevier Academic Press, Amsterdam, 2004.

[12] S.B. Hsu. *Ordinary differential equations with applications.* World Scientific, New Jersey, second edition, 2013.

[13] D.W. Jordan and P. Smith. *Nonlinear ordinary differential equations: an introduction for scientists and engineers.* Oxford University Press, Oxford, fourth edition, 2007.

[14] A.C. King, J. Billingham, and S.R. Otto. *Differential equations, Linear, Nonlinear, Ordinary, Partial.* Cambridge University Press, Cambridge, 2003.

[15] D.L. Kreider, R.G. Kuller, D.R. Ostberg, and F.W. Perkins. *An introduction to linear analysis.* Addison–Wesley, Massachusetts, 1966.

[16] E. Kreyszig. *Introductory functional analysis with applications.* Wiley & Sons, New York, 1978.

[17] Ajit Kumar and S. Kumaresan. *A basic course in real analysis.* CRC Press, Hoboken, 2014.

[18] S. Lang. *Undergraduate analysis.* Springer–Verlag, New York, second edition, 1997.

[19] M. Lerch. Sur un point de la théorie des fonctions génératrices d'abel. *Acta Mathematica*, 27:339–351, 1903.

[20] E. Lorenz. Deterministic nonperiodic flow. *Journal of the Atmospheric Sciences*, 20:448–464, 1963.

[21] B.D. MacCluer. *Elementary functional analysis.* Springer, New York, 2008.

[22] T.R. Malthus. *An Essay on the Principles of Population.* J. Johnson, London, 1798.

[23] J.D. Murray. *Mathematical Biology.* Springer-Verlag, Berlin, 1993.

[24] T. Myint-U. *Ordinary differential equations.* North-Holland, New York, 1978.

[25] T. Myint-U and L. Debnath. *Linear partial differential equations for scientists and engineers.* Birkhäuser, Berlin, 2009.

[26] R.K. Nagle, E.B. Saff, and A.D. Snider. *Fundamentals of differential equations and boundary value problems.* Addison–Wesley, Reading, Mass., seventh edition, 2000.

[27] A.K. Nandakumaran, P.S. Datti, and R.K. George. *Ordinary differential equations: Principles and applications.* Cambridge-IISc Series, Cambridge Universty Press, New Delhi, 2017.

[28] L. Perko. *Differential equations and dynamical systems.* Springer, New York, 2001.

[29] B. Perthame. *Transport equations in biology*. Frontiers in Mathematics, Birkhäuser Verlag, Basel, 2007.

[30] J.F. Ritt. *Integration in finite terms: Liouville's theory of elementary methods*. Columbia Universty Press, New York, 1948.

[31] W. Rudin. *Principles of mathematical analysis*. McGraw–Hill, New York, 1976.

[32] G.F. Simmons. *Differential equations with applications and historical notes*. CRC Press, Boca Raton, third edition, 2017.

[33] I. Stakgold. *Boundary value problems of mathematical physics*, volume 1. Macmillan, New York, 1972.

[34] G. Strang. *Introduction to linear algebra*. Wellesley-Cambridge Press, Wellesley, fifth edition, 2016.

[35] I.I. Vrabie. *Differential equations: an introduction to basic concepts, results and applications*. World Scientific, River Edge, 2004.

[36] G.N. Watson. *A treatise on the theory of Bessel functions*. Cambridge University Press, London, second edition, 1944.

[37] H. Widom. *Lectures on integral equations*. Van Nostrand-Reinhold, New York, 1969.

Index

Airy equation, 102
Allee model, 50
analytic function, 224
autonomous
    scalar equation, 44
    system, 169

Banach fixed point theorem, 308
Bendixson's negative criterion, 212
Bernoulli equation, 19, 54, 286
Bessel
    function of order α, 244
    function of the first kind, 244
Bessel equation, 243
boundary conditions:
    Dirichlet, 125
    periodic, 109
    homogeneous, 111
    non-homogeneous, 104, 105, 107, 119
    separated, 109
boundary value problem, 103

Cauchy
    problem, 20
    sequence, 23
Cauchy–Lipschitz, 23, 138
Cauchy–Schwarz inequality, 137, 174
center, 182
closed
    curve, 208
    path, 208
comparison test, 141, 227, 239
comparison theorem, 38, 39
comparison theorem, Sturm, 92
complete orthonormal basis, 128
contraction mapping, 308

convolution, 266–268
critical point, 171
    stable, 172
    unstable, 172
    asymptotically stable, 172

derivative of the determinant, 84, 159, 315
discrete, 49
Duffing equation, 5
Dulac negative criterion, 221

eigenfunction, 120
eigenvalue, 120
error, 286
    local truncation, 285
    percentage, 286
    relative, 286
Euler
    backward difference, 285
    equation, 75
    forward difference, 285
    method, 285
    method, improved, 292
evolution matrix, 165
exact equations, 13
existence
    global, 32, 57, 138
existence and uniqueness, 23
explicit method, 285
exponential
    order, function of, 253
    shift rule, 73
exponential of
    a diagonal matrix, 145
    a Jordan matrix, 146
    a matrix in Jordan form, 148